AF597679

Phthalocyanines

Properties and Applications

Volume 3

Phthalocyanines
Properties and Applications

Volume 3

Edited by

C. C. Leznoff and A. B. P. Lever

C. C. Leznoff
Department of Chemistry
York University
4700 Keele Street
Downsview, Ontario
Canada M3J1P3

A. B. P. Lever
Consulting & Editing Services, Ltd.
455 Sentinel Road #1010
Downsview, Ontario
Canada M3J1V5

Library of Congress Cataloging-in-Publication Data

CIP pending.

© 1993 VCH Publishers, Inc.

This work is subject to copyright.

All rights reserved, whether the whole or part of the material is concerned, specifically those of translation, reprinting, re-use of illustrations, broadcasting, reproduction by photocopying machine or similar means, and storage in data banks.

Registered names, trademarks, etc., used in this book, even when not specifically marked as such, are not to be considered unprotected by law.

Printed in the United States of America

ISBN 1-56081-638-4 VCH Publishers
ISBN 3-527-89638-4 VCH Verlagsgesellschaft

Printing History:
10 9 8 7 6 5 4 3 2 1

Published jointly by

VCH Publishers, Inc.
220 East 23rd Street
New York, New York 10010-4606

VCH Verlagsgesellschaft mbH
P.O. Box 10 11 61
D-6490 Weinheim
Federal Republic of Germany

VCH Publishers (UK) Ltd.
8 Wellington Court
Cambridge CB1 1HZ
United Kingdom

Contents

Contributors

A.B.P. Lever[a], Elena R. Milaeva[b], and Gabor Speier[c], [a]Department of Chemistry, York University, North York, Ontario, Canada; [b]Department of Organic Chemistry, Moscow State University, Moscow, USSR; [c]Department of Organic Chemistry, University of Veszprem, 8201 Veszprem, Hungary.

M.M. Nicholson, 2209 California Boulevard, San Marino, California 91108, U.S.A.

Branimir Simic-Glavaski, Electrical Engineering and Applied Physics Department, Case Western Reserve University, Cleveland, Ohio 44106, U.S.A.

W.E. Smith and B.N. Rospendowski, Department of Pure and Applied Chemistry, University of Strathclyde, Glasgow G1 1XL, Scotland, U.K.

Martin J. Stillman, Department of Chemistry, University of Western Ontario, London, Ontario N6A 5B7, Canada.

Series Preface

Since their synthesis early this century, phthalocyanines have established themselves as blue and green dyestuffs par excellence. They are an important industrial commodity (output 45,000 tons in 1987) used primarily in inks (especially ballpoint pens), coloring for plastics and metal surfaces, and dyestuffs for jeans and other clothing. More recently their use as the photoconducting agent in photocopying machines heralds a resurgence of interest in these species. In the coming decade, their commercial utility is expected to have significant ramifications. Thus future potential uses of metal phthalocyanines, currently under study, include (I) sensing elements in chemical sensors, (II) electrochromic display devices, (III) photodynamic reagents for cancer therapy and other medical applications, (IV) applications to optical computer read/write discs, and related information storage systems, (V) catalysts for control of sulfur effluents, (VI) electrocatalysis for fuel cell applications, (VII) photovoltaic cell elements for energy generation, (VIII) laser dyes, (IX) new red-sensitive photocopying applications, (X) liquid crystal color display applications, and (XI) molecular metals and conducting polymers.

In recent years there has been a growth in the number of laboratories exploring the fundamental academic aspects of phthalocyanine chemistry. Interest has been focused, inter alia, on the synthesis of new types of soluble and unsymmetrical phthalocyanines, on the development of new approaches to the synthesis of polynuclear, bridged, and polymeric species, on their electronic structure and redox properties, and their electro- and photocatalytic reactivity.

It is timely therefore to consolidate these academic and applied aspects of phthalocyanine chemistry into a series of monographs that will present our current knowledge of this fascinating area of chemistry. In this series of monographs, we plan to bring together for the first time detailed critical coverage of the whole field of phthalocyanine chemistry in a sequence of chapters that should prove stimulating for industrial, medical, and academic researchers.

C. C. Leznoff
A. B. P. Lever

Preface

The third in this series of volumes continues our coverage of the fundamental and applied chemistry of the phthalocyanine derivatives. The redox character of phthalocyanine species is fundamental to many of their potential industrial applications. Lever, Milaeva, and Speier begin with a detailed, comprehensive discussion of the solution electrochemical behavior of these macrocyclic derivatives, the first time this topic has been reviewed. Nicholson then follows with an account of her detailed explorations into the electrochromic behavior of the rare earth diphthalocyanines showing how the redox properties of these species might be exploited in future display devices. The applied aspect of these species is taken further by Simic-Glavasky in his examination of the possible application of the Raman spectroscopy of these species in information storage molecular electronic devices.

The remaining two chapters present fundamental spectroscopic aspects, with Smith and Rospendowski conducting a general discussion of the Raman spectra of phthalocyanine species, and Stillman continuing his earlier electronic spectroscopic review from Volume 1 with a review of the electronic spectra of phthalocyanine anion and cation radical spectra. These topics have not been previously reviewed.

Together, volumes 1–3 cover a broad spectrum of the physical and chemical aspects of the phthalocyanines and some of their relatives, providing a firm, up-to-date basis for future exploration of these very important species.

C. C. Leznoff
A. B. P. Lever
September 1992

1

The Redox Chemistry of Metallophthalocyanines in Solution

A. B. P. Lever, Elena R. Milaeva, and Gabor Speier

A. Introduction

Metallophthalocyanines (MPcs) have a number of characteristic properties that contribute in a major way to their extraordinary versatility. These include their intense color, redox activity, high thermal stability, and non-toxicity. Ultimately, this research will lead to a place in the emerging field of molecular electronics [1]. Actual and potential applications include photoconductive surfaces, optical information storage media, electrochromic devices, analytical (sensor) devices for industrial, environmental and medical applications, molecular metals, batteries, and electrocatalytic and photocatalytic processes including solar energy conversion and production of chemical materials.

Most of the applications rely critically upon the redox properties of MPc species. The Pc unit is an 18π electron aromatic system that, in its common oxidation state carries two negative charges. This will be designated Pc(-2) [2]. This unit is capable of oxidation or reduction [3-5]; thus oxidation by one or two-electrons yields Pc(-1) and Pc(0), while reduction by one to four electrons forms Pc(-3), Pc(-4), Pc(-5) and Pc(-6). The central metal ion may be incapable of a redox process in the usual electrochemical regime [most main group species and certain transition metal species such as Ni(II), Cu(II), etc.] or may be a transition element that undergoes oxidation or reduction at potentials comparable to the Pc ring processes.

Many MPc systems bind one or two axial ligands; such coordination can have a major effect upon the observed redox activity [6-12]. Such species are designated here as LMPc or L_2MPc, where the placing of L ahead of M implies L binding axially to the central metal M.

Most unsubstituted MPc species have only very limited solubility in virtually all solvents, thereby limiting solution phase redox measurements. However, MgPc is rather more soluble as are many transition metal phthalocyanines that dissolve in donor solvents through an axial interaction between the metal center and the donor solvent. This last statement applies especially to those central metal ions that strongly prefer six-coordination rather than four coordination. Thus, for example, iron(II) and cobalt(II) phthalocyanines are soluble in a wide range of donor solvents, while copper(II) phthalocyanine is very much less soluble.

Ring substitution has proved to be a very effective procedure for rendering these substituted MPc species very soluble in a range of solvents, to an extent that, of course, depends upon the substituents used. Even with such species, additional solubility is conferred by axially coordinating central ions. This has led to systems that are extremely soluble in many organic solvents, for example, the tetraneopentoxyphthalocyanines [13] or in water, for example, the tetrasulfonated phthalocyanines [14] [see Table 11, above list of references, for the abbreviations used in this chapter].

It is our intent, in this chapter, to organize and systematize the redox properties of a large number of MPc species. To facilitate this organization, data will be reported versus the standard calomel electrode (SCE). However, actual studies have utilized a wide range of reference electrodes, including internal references such as the ferricenium/ferrocene couple. Table 1 lists these various electrodes and indicates the correction made in this chapter to adjust potentials versus SCE. Such adjustments are subject to some error depending upon the quality of the reference electrode, internal resistance problems etc.

Many phthalocyanines aggregate to a greater or lesser extent, both in water and organic phase. Such aggregation is influenced by pH, ionic strength, temperature, the amount of electrolyte in solution, etc. [15-18]. Thus care must be taken in distinguishing redox potentials arising from mononuclear MPc species, and from aggregated species. Aggregation is also influenced by the net charge on the MPc unit, being more important with positively charged species than negatively charged ones. Six coordinate MPc species generally do not aggregate because they are kept apart by the axially bound ligands.

The redox properties may be influenced by different axial ligands attached to the metal center. This can be a very important factor in transition metal MPc chemistry, since many transition metal ions will prefer six-coordination and will bind a donor solvent if no other ligands are competing. Thus the redox chemistry in donor, potentially axially binding solvents, can be very different from that of the same MPc species in a nondonor solvent such as dichlorobenzene.

Even although the metal ion in main group phthalocyanines does not possess partially filled d orbitals, the coordination number for these ions can be more than four; indeed a four coordinate planar environment is very unusual for a main group ion. Thus these main group species may also bind additional ligands or coordinating (donor) solvents, forming five- or six-coordinate species in solution. Such binding can influence the observed redox potentials.

Commonly used nondonor solvents that can dissolve MPc species to a sufficient extent, include o-dichlorobenzene (DCB) and dichloroethane (DCE), and in some early studies, nitrobenzene (NO_2Ph), chloronaphthalene (ClNap), and 1-methylnaphthalene (MeNp); this last solvent was generally used at 150^oC. Dichloromethane (CH_2Cl_2) can also be used for the more organic soluble substituted MPc systems such as tetra-t-butylPc (TBuPc) or tetraneopentoxyPc (TNPc). More commonly used donor and potentially coordinating solvents include pyridine (Py), dimethylformamide (DMF), dimethylacetamide (DMA), dimethylsulfoxide (DMSO) and benzonitrile (PhCN), all of which tend to be rather good solvents for MPc species even unsubstituted ones, if they, in fact, coordinate to the central ion. Thus, for example, pyridine is a good solvent for Co(II)Pc, with which it coordinates, while it is a poor solvent for Ni(II)Pc with which it will not coordinate. With any of the solvents listed, extreme care should be taken to ensure that the solvent is dry, if high quality and wide potential range data are sought. In the context of unsubstituted MPc chemistry, the word good conveys that the species is sufficiently soluble for spectroscopic and electrochemical study, but this may mean a solubility of the order of 1×10^{-4} M !

Supporting electrolyte anions can also play an important role if they have donor characteristics. Thus the perchlorate and hexafluorophosphate ions are usually regarded as nondonor species, although this may not always be absolutely true. Halide ions, on the other hand, especially the commonly employed chloride ion may influence the redox properties dramatically if they axially coordinate.

Although early electrochemical measurements were carried out polarographically using dropping or hanging mercury electrodes, most modern electrochemistry employs solid electrodes, usually platinum or graphite, and cyclic voltammetry (CV) or differential pulse voltammetry (DPV). These last two techniques provide a vary rapid assessment of the electrochemical properties of the MPc species. Their analysis can usually rapidly establish whether the electron transfer is electrochemically reversible, or not, and whether there are coupled chemical reactions involved. In general, ring reduction processes are often electrochemically reversible, while ring oxidation processes, especially that associated with Pc(0)/Pc(-1), are often irreversible.

It is not the purpose of this chapter to pursue in detail the electrochemical characteristics of the various redox processes except where the Authors concerned have explored such avenues. The reader is referred to an electrochemical text, for example, Bard and Faulkner [19], for the background to such analysis.

Lanthanide diphthalocyanines will not be dealt with in detail in this chapter, since they are considered in Chapter 5 in this volume [20]. Further, the electrocatalytic properties of metallophthalocyanines towards reactive species such as oxygen, hydrazine, sulfides and mercaptans, that involve a significant body of research will be covered in a later volume in this series.

One could pursue the comparative electrochemistry of the phthalocyanines with their cousins the porphyrins. Such a comparison would have greatly increased the length of this contribution. The interested reader is referred to several important reviews of porphyrin electrochemistry [21-23]. Suffice it to say that there are quite close similarities in the gross behavior of both series of complexes. In general, the lower basicity of the phthalocyanines relative to the porphyrins results in the greater stabilization of the lower oxidation states in the former.

In other words, the higher oxidation states of central transition metal ions are more readily accessible in the porphyrin series than in the phthalocyanine series, or, where formed, higher oxidation state metallophthalocyanines are stronger oxidizing agents than their porphyrin analogues.

Table 1 Reference Electrode Corrections to the Standard Calomel Electrode (SCE).

Electrode used	Correction
NHE	-0.24 V
AgCl/Cl	-0.045 V
Fc^+/Fe	+0.425 V (CH_3CN)[a]
.	+0.49 V (DCB,DCE) [47]
.	+0.40 V (DMF)
.	+0.50 V (Py, DMSO) [97]
.	+0.45 V (CH_2Cl_2) [75,69]
.	+0.47 V (DMA) [46]
$AgClO_4/Ag$ (0.01 M $HClO_4$)	+0.47 V (CH_2Cl_2) [75]

[a] Our laboratory experience is that the ferrocenium/ferrocene correction is best set at 0.425 V versus SCE in acetonitrile.

B. The Electronic Structure of Metallo-phthalocyanine Species

The metallophthalocyanines belong to the point group D_{4h}. The electronic structure of MPc was described by Gouterman and co-workers [24-28] and discussed in depth by Stillman and Nyokong in volume 1 of this series [29]. The HOMO level is $1a_{1u}(\pi)$, the next low lying filled orbital is $1a_{2u}(\pi)$. The LUMO orbital is $1e_g(\pi^*)$ and the next is $1b_{1u}(\pi^*)$ (Figure 1). Transitions from the two upper filled π orbitals to $1e_g(\pi^*)$ yield the so-called Q (near 600-750 nm) and Soret (or B) π--->π^* (near 300-450 nm) transitions. These both involve an ^1E_u excited state, but they are not significantly mixed (unlike the situation in the porphyrin series) because the $1a_{1u}$ and $1a_{2u}$ orbitals are fairly well separated in energy. For a complete analysis see the earlier discussion by Stillman and Nyokong [29]. In main group Pcs the redox activity is directly associated, in oxidation, by the successive removal of the electrons from the HOMO, $1a_{1u}$, while up to four additional electrons are readily added to $1e_g$ (LUMO) (reduction), terminating in the Pc(-6) species. The Pc(-3), Pc(-4), Pc(-5) and Pc(-6) ring reduced species have the ground state electron configurations $(a_{1u})^2e_g$, S= 1/2; $(a_{1u})^2(e_g)^2$ S = 0, $(a_{1u})^2(e_g)^3$, S = 1/2; and $(a_{1u})^2(e_g)^4$, S = 0, respectively. The Pc(-3) and Pc(-5) ions with an uneven number of electrons show paramagnetism [32] as free radical anions of the phthalocyanine ligand at g factors near that of the free-electron.

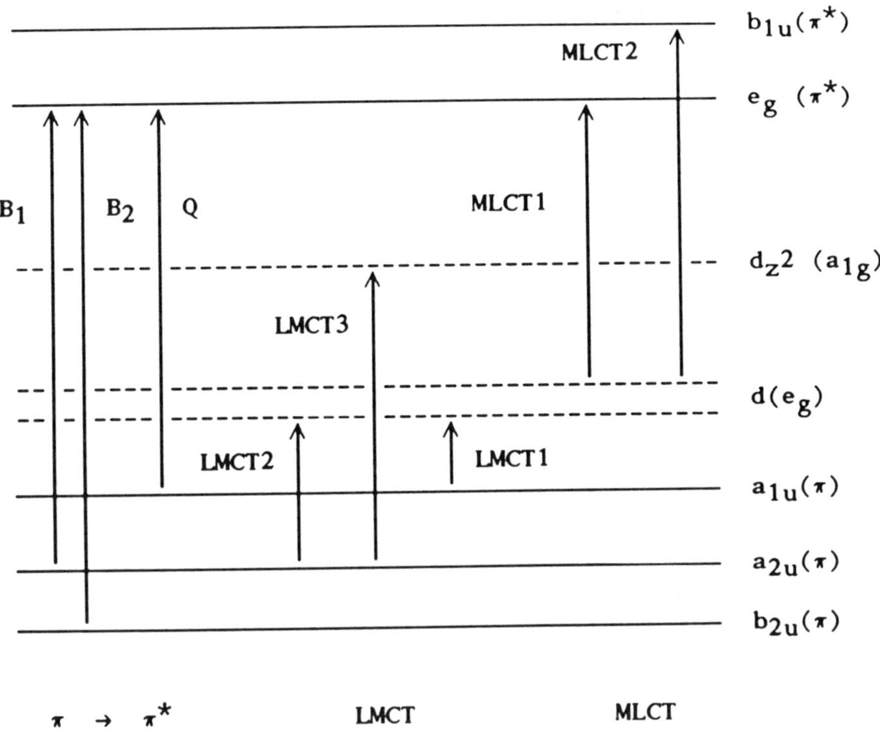

Figure 1 The conventional scheme of the energy levels in MPc and the various transitions (Q, Soret, LMCT, MLCT bands) [7,26-28,30,31].

The ground state for the Pc(-4) species, as an open shell, could be expected to be a triplet, but experimental investigation has shown it to be a singlet [32]. With diamagnetic central ions such as Mg(II), Ni(II), the Pc(-4) species shows no ESR spectrum, but with paramagnetic species such as Co(II) [32] and Cu(II) [32,33] the ESR spectrum characteristic of the central ion itself is observed. If the Pc(-4) species were in fact an S = 1 fragment, then coupling, for example, to the Cu(II) center would yield either a S = 3/2 species (that would not give a typical Cu(II) type spectrum) or, via coupling, an S = 1/2 species, but centered on the Pc ring, giving rise to a free radical spectrum. Thus the Pc(-4) species is likely S = 0. However, some anomalies in the spacing of the successive reduction processes, for some metal ions, may suggest that the Pc(-4) is not always S = 0, but can, with some central ions, be S = 1. The first two reduction processes are generally separated by about 0.4 - 0.5 V (see, for example, Figure 2). With electropositive central ions such as

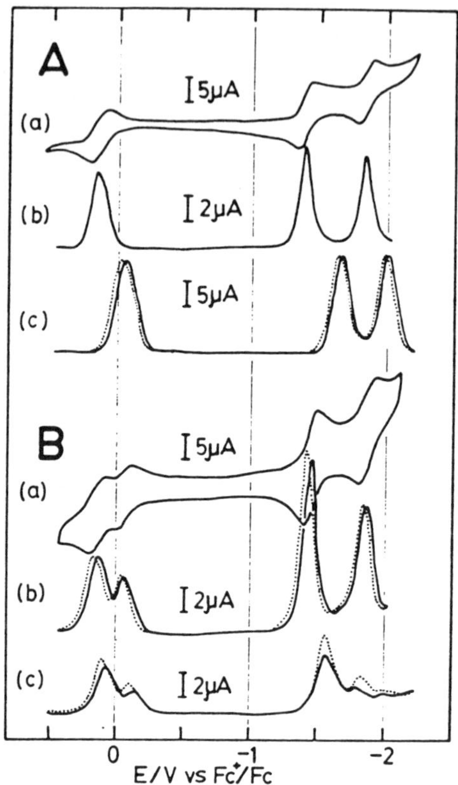

Figure 2 Cyclic and or differential pulse voltammetry of A) Zn[TNPc(-2)] and B) Ant$[ZnTrNPc(-2)]_2$ in DMF (curves a and b) and in DCB (curves c). Scan rates 100 mV/s for cyclic voltammetry, and 5 mV/s for differential pulse voltammetry (DPV). In the DPV curves, the solid and dotted lines indicate cathodic and anodic scans respectively. Reproduced with permission from Ref. [126].

hydrogen or magnesium, the third reduction occurs some 0.8 V more negative than the second. However, with more polarizing central metal ions, the separation is smaller, ≈ 0.5 - 0.6 V for Zn(II) (but see Zn(II)Pc discussion to follow) and as low as 0.4 V for Al(III). It increases to about 0.8 V again for Ni(II) and Cu(II). The fourth reduction is usually observed some 0.4 - 0.5 V negative of the third process.

The redox properties of the transition metal phthalocyanines differ from those of main group MPc due to the fact that metal d levels may be positioned between the HOMO (π) and LUMO (π^*) orbitals of the phthalocyanine ligand. This has the spectroscopic consequence that one or more metal to ligand (MLCT), or ligand to metal (LMCT) charge-transfer transitions may be observed in the visible or the near-infrared region [7].

The redox consequence is that oxidation or reduction of the metal may occur at potentials similar to those of ring oxidation or reduction. It is important also to recognize that such internal metal redox processes greatly influence the potentials for ring redox. Thus, for example, the ring oxidation of a species such as Co(II)Pc(-2) (to $[Co(II)Pc(-1)]^{+}$) will lie at a significantly less positive potential than the oxidation of $[Co(III)Pc(-2)]^{+}$ to $[Co(III)Pc(-1)]^{2+}$ (see the following) due to polarization of the phthalocyanine ring by the oxidized metal center.

The oxidized or reduced phthalocyanine species with an uneven number of electrons in the ligand and a diamagnetic metal center exhibit ESR signals due to paramagnetism and therefore can be examined by ESR spectroscopy. Such species generally show a narrow signal at $g \approx 2.0$ near the free-electron g value, characteristic of organic free radicals. For example, the Pc(-1) species has a configuration $(1a_{1u})^{1}$. Hyperfine structure is not usually observed on these signals because of the extensive delocalization of the spin density in the phthalocyanine ring and relaxation due to aggregation.

In characterizing Pc anion or cation radicals by ESR spectroscopy, care should be taken in evaluating the presence of a free radical ESR signal because many diamagnetic phthalocyanine compounds show a weak ESR signal at $g \simeq 2.0$ in neutral (not oxidized or reduced) form in the solid state and sometimes in solution. The origin of the ESR signal in these phthalocyanines was discussed by several authors [34-40] in terms of impurities, broken π bonds or defect structures. The dependence of the number of radical species in diamagnetic MPc and the intensity of the ESR signal in the presence of molecular oxygen leads to the belief [41] that the appearance of paramagnetism is a charge-transfer interaction between the phthalocyanine molecule and dioxygen. Thus the ESR signals for MPc and free, unmetallated phthalocyanine are attributed to partial oxidation of the phthalocyanine ring [42].

C. Main Group Phthalocyanine Electrochemistry

For main group phthalocyanines the first ring oxidation is separated from the first ring reduction by approximately 1.5 V (Table 2) which corresponds approximately to the magnitude of energy difference between the HOMO and LUMO [6,43] (except for metal ions positioned out of the macrocycle plane, for example, Pb^{2+}, Hg^{2+}, Cd^{2+}). However, the individual potentials for the first ring reduction and first ring oxidation do vary remarkably and indeed are functions of the polarizing power of the central metal ion, expressed as charge/radius (ze/r). In general the more polarizing the central metal ions, the

Table 2 Main Group and Transition Group Phthalocyanines Exhibiting Redox Processes occurring at the Ring Alone, the Metal Center remaining Invariant. (versus SCE)

Processes	I	II	III	IV	V	VI	Solv.	Ref.
Main Group MPc Species								
$Pc^{2-}(PrA^{-})_2$			-1.24	-1.55	1.95	-2.19	DMF	4
H_2Pc		0.64	-0.72				DMF	104
H_2Pc			-0.66	-1.06	-1.93	-2.23	DMF	4
H_2[TNPc]			-0.93	-1.31			DCB	47
H_2[TNPc]	1.38	0.9irr	-0.90	-1.20			DCB	48
H_2[TNPc]	1.26	0.77	-0.86	-1.21			DCB	49
H_2[TBuPc]		0.81					PhCN	50
		0.71					$PhNO_2$	51
			-0.7				DMSO	52, 50
	0.94	0.62	-0.82	-1.19			CH_2Cl_2	53
			-0.72	-1.13			CH_2Cl_2	52[b]
			-0.61	-1.10			MeN/150°	52
H_2[TAPc]			-0.93	-1.34			DMF	137
H_2[TsPc]		0.90[c]	-0.53	-0.97	-1.81		DMSO	5, 54
H_2[OCNPc]			-0.1	-0.45	-0.88	-1.50	DMF	33,55
TDBA[LiPc]$^-$			0.17	-1.37			DMF	56
Li_2Pc			0.24	0.11[d]			Acetone	57
Li_2Pc	1.0	0.17	-1.53				THF/ClNap	58
NaPc			-1.06				DMF	104
Mg(II)Pc	1.26	0.65	-0.92	-1.26			DMF	43
			-0.91	-1.39	-2.14	-2.58	DMF	4

Processes	I	II	III	IV	V	VI	Solv.	Ref.
Mg(II)Pc		0.59	-0.99				DMF	56
Mg(II)Pc	1.26	0.61	-0.96				DMF	59
Mg(II)Pc		0.70	-0.95				ClNap	60
Mg(II)Pc	1.11	0.65	-0.90	-1.39			DMA	46
Mg(II)Pc			-0.97	-1.44			MeNp/150°	52
TDBA[FMg(II)Pc]$^{-}$		0.50	-1.10				DMF	56
$(Im)_2$Mg(II)Pc	1.10	0.64	-0.93	-1.47			DMA	46
$(MeIm)_2$Mg(II)Pc	1.21	0.64	-0.92	-1.39			DMA	46
$(Py)_2$Mg(II)Pc	1.16	0.67	-0.91	-1.45			DMA	46
		0.70	-0.95				DMF	54
$(MePy)_2$Mg(II)Pc	1.13	0.67	-0.82	-1.29			DMA	46
Mg(II)[TBuPc]		0.67					PhCN	50
			-0.84				DMSO	50
			-1.15				CH_2Cl_2	52[b]
			-1.04				MeNp/150°	52
Ba(II)Pc	0.77	0.45	-0.49[d]				DMF	6
			-1.08				DMF	6
Zn(II)Pc		0.67	-0.86	-1.30	-1.85	-2.25	DMF	55
Zn(II)Pc			-0.89	-1.33	-2.06	-2.68	DMF	4
Zn(II)Pc		0.67	-0.90				DMF	59
Zn(II)Pc		0.64	-0.94				DMF	56
$(Py)_2$Zn(II)Pc		0.68	-0.89				Py	61
Zn(II)Pc		0.78	-0.80				DMA	60
Zn(II)Pc		0.68					ClNap	62

Processes	I	II	III	IV	V	VI	Solv.	Ref.
Zn(II)Pc		0.76	-0.78	-1.14			MeNp/150°	52
Zn(II)[TsPc]			-0.83	-1.30			DMF	63
Zn(II)[TsPc]			-0.85qr	-1.0qr			H_2O	64
Zn(II)[TNPc]	1.13	0.47	-1.17	-1.55			DCB	65
Zn(II)[TNPc]		0.45	-1.03	-1.45			DMF	65
Zn(II)[TBuPc]			-0.94	-1.29			CH_2Cl_2	52[b]
Zn(II)[TBuPc]		0.66					PhCN	50
Zn(II)[TBuPc]		0.68	-0.92				DMF	tw
Zn(II)[TBuPc]	1.33i	0.61					DCB	50, 66, 67
Zn(II)[TBuPc]		0.71	-0.88				DMF	54
Zn(II)[TBuPc]		0.34					$PhNO_2$	51
Zn(II)[TBuPc]			-0.96				DMSO	50, 52
Zn(II)[TBuPc]		0.75	-0.83	-1.21			MeNp/150°	52
Zn(II)[OCNPc]			-0.15	-0.50	-1.10	-1.35	DMF	33, 55
Zn(II)[OBuxPc]		0.50	-1.06	-1.53			DMF(Py)	68
Zn(II)[Cl_{16}Pc]			-0.5	-0.8			DMF	63
TDBA[FZn(II)Pc]$^-$		0.51	-1.06				DMF	56
Cd(II)Pc	0.88	0.54	-1.17				DMF	43, 59
Hg(II)Pc		0.25	-1.30				DMF	43
Pb(II)Pc		0.67	-0.82	-1.11			DMSO	104
	0.95	0.65	-0.75	-1.04	-1.92		DMF	9
ClAl(III)Pc			-0.53	-0.98	-1.42	-1.98	DMF	4
		0.91	-0.66				DMF	104
ClAl(III)Pc		0.89	-0.695				DMF	56

Processes	I	II	III	IV	V	VI	Solv.	Ref.
ClAl(III)Pc		0.94	-0.66				DMF	59
ClAl(III)Pc		1.15	-0.50				ClNap	60
$(TDBA)_2[[FAl(III)Pc]_2O]^{2-}$	0.87	0.47	-0.99				DMF	56
$TDBA[F_2Al(III)Pc]^-$		0.69	-0.99				DMF	56
ClGa(III)Pc		0.86	-0.74	-1.14			DMF	6, 43
ClGa(III)[TBuPc]		0.99					PhCN	50
ClIn(III)Pc		0.83	-0.72	-0.95			DMF	43
ClIn(III)[TBuPc]		0.9					PhCN	50
$(HO)_2Ge(IV)[TBuPc]^e$			-0.34	-0.84	-1.67	-2.10	THF^f	138
$((nC_6H_{13})_3SiO)_2Si(IV)Pc^g$		1.00	-0.90	-1.48			CH_2Cl_2	69
$(O\text{-}^tAmyl)_2Si(IV)Pc$			-0.54	-1.14			DMF	43
Transition metals (non-redox active)								
OTi(IV)[TBuPc]		0.85	-0.515	-1.02			DMF	6
OV(IV)[TBuPc]		0.94	-0.58	-1.08			DMF	6
Ni(II)Pc		1.05					ClNap	62
Ni(II)Pc			-0.85	-1.23	-2.01	-2.35	DMF^h	4
Ni(II)Pc			-0.72	-1.20			MeNp/150°	52
Ni(II)[TsPc]		0.95	-0.71	-1.14			DMF	6
Ni(II)[TsPc]			-0.67	-1.13			DMSO	64
Ni(II)[TsPc]				$-1.34^c{}_{irr}$			H_2O	64
Ni(II)[TBuPc]		0.67					$PhNO_2$	51
		1.06	-0.89				DMF	54
			-0.85				DMSO	52

Processes	I	II	III	IV	V	VI	Solv.	Ref.
Ni(II)[TBuPc]			-0.76	-1.24			MeNp/150°	52
			-0.89	-1.31			CH_2Cl_2	52[b]
Ni(II)[TAPc]		0.23[i]	-1.00	-1.45			DMSO	70
Cu(II)Pc		0.98					ClNap	62
Cu(II)Pc			-0.84	-1.18	-2.01	-2.28	DMF[h]	4
Cu(II)Pc			-0.70	-1.14			MeNp/150°	52
Cu(II)[DsPc]			-0.79	-1.18			DMSO	64
Cu(II)[DsPc]				-1.10[c]	-1.25ads		H_2O	64
Cu(II)[TsPc]		0.87[c]	-0.745	-1.14			DMF	6
Cu(II)[TsPc]		0.84	-0.84				DMF	56
Cu(II)[TNPc]			-0.88	-1.26			DCE	47
Cu(II)[TBuPc]		0.64					$PhNO_2$	51
			-0.91	-1.34			DMF	tw
		0.87					DCB	66
			-0.86	-1.26			CH_2Cl_2	52[b]
			-0.81				DMSO	52
			-0.74	-1.18			MeNp/150°	52
			-0.89				DMF	6
Cu(II)[OCNPc]			-0.2	-0.63	-1.08	-1.25	DMF	33, 55
Cu(II)[TcPc]		1.26					ACN	71
$Me_2O(5)[Cu(II)TrNPc]_2$		0.82		-1.04	-1.31		DCB	47
Pd(II)[TBuPc]		0.72					$PhNO_2$	51
Pt(II)Pc			-0.67	-1.05			MeNp/150°	52
Pt(II)[TBuPc]		0.73					$PhNO_2$	51

Processes	I	II	III	IV	V	VI	Solv.	Ref.
Py(CO)Ru(II)Pc		0.87					CH_2Cl_2	12
Py_2Ru(II)Pc	1.36	0.73					CH_2Cl_2	12
$(4\text{-MePy})_2$Ru(II)Pc	1.36	0.71					CH_2Cl_2	12
(4-MePy)(CO)Ru(II)Pc		0.84					CH_2Cl_2	12
$(4\text{-BuPy})_2$Ru(II)Pc	1.28	0.66					CH_2Cl_2	12
$(CH_3CN)_2$Ru(II)Pc		0.68					CH_2Cl_2	12
$(DMF)_2$Ru(II)Pc		0.76					CH_2Cl_2	12
(DMF)(CO)Ru(II)Pc		0.86					CH_2Cl_2	12
$(DMSO)_2$Ru(II)Pc		0.85					CH_2Cl_2	12

[a] Where data have been reported by two laboratories using the same solvent and report numbers agreeing within 20 mV, only the average is cited in the table. The processes are: I MPc(0)/MPc(-1); II MPc(-1)/MPc(-2); III MPc(-2)/MPc(-3); IV MPc(-3)/MPc(-4); V MPc(-4)/MPc(-5); VI MPc(-5)/MPc(-6). Charge excluded for clarity. [b] Data reported against Fc^+/Fc internal reference. We have corrected using +0.45 V versus SCE; the authors report 0.55 V but this seems to provide, for the most part, poorer agreement with related data. [c] 2-electron process. [d] Assignment very uncertain since the potential is too close to the next more positive wave. [e] Presumed soluble product from "defragmentation" of a μ-oxogermanium(IV) phthalocyanine polymer. [f] The potential at -2.10 V is uncertain due to chemical instability; note that the potentials recorded for this species were internally calibrated against Fc^+/Fc assuming the latter to lie at 0.307 V vs SSCE. [g] See data for oligomeric silicon Pcs in Table 3. [h] Pr_4NOH supporting electrolyte. [i] authors suggest Ni(III)/Ni(II).
See Tables 4 and 5 for further data for zinc phthalocyanines. Axial ligand abbreviations are those commonly used and are not listed. All phthalocyanine, and solvent, abbreviations can be found in Table 11.

easier it is to reduce the ring, and the more difficult to oxidize the ring. A linear plot of these quantities, the first reduction and oxidation potentials versus ze/r, has been obtained for many MPc species (M = Zn^{2+}, Mg^{2+}, In^{3+}, Ga^{3+}, Al^{3+}, Si^{4+}) in DMF [43]

The linear relationships are:

$$E^{o}_{ox} = 1170 - 11.7\ (r/ze)$$

$$E^{o}_{red} = -385 - 12\ (r/ze) \qquad (1)$$

for the first oxidation and reduction potentials versus SCE (E^{o} in mV, r in pm using ionic radii from the listing of Shannon and Prewitt [44]). Cd(II), Hg(II), and Pb(II)Pc lie well off these lines as a consequence of being too large to fit well inside the Pc ring. It is certainly true that there will be a similar relationship for transition metal central ions, with the higher oxidation state ions causing more ready ring reduction provided that central metal ion reduction does not occur first (at less negative potentials); however, no explicit linear correlations concerning this prediction have been reported.

Axial ligation has been studied for some main group MPcs, especially magnesium, [8, 45, 46] (Table 2) however, the effects are rather small for the main groups (in contrast to the transition groups where axial ligation can effect large redox potential changes.) Although the axial ligation can be monitored spectroscopically [8, 45], the electrochemical response in the main groups is small.

The low solubility of most main group Pcs has limited the availability of acceptable quality electrochemical data. A few more soluble ring substituted species are available and are cited in Table 2. In general, however, aside from Mg and ZnPc, main group MPc solution electrochemistry has been rather neglected.

i. Metal-Free Phthalocyanines

A range of metal-free Pcs has been explored and all oxidation states between $[H_2Pc(0)]^{2+}$ and $[H_2Pc(-5)]^{3-}$ have been reported. The first oxidation and reduction potentials for the variously substituted species fall within fairly narrow ranges (Table 2) with separation of about 1.4 V. However, H_2[TNPc] in the aromatic solvent DCB, curiously, is a little more difficult to reduce and a little more difficult to oxidize than average, with separation of some 1.6 V. The irreversibility of the first oxidation noted in [48] arises from aggregation effects. The Pc(-2) anion is present in the propylammonium salt (Table 2) (and also in the lithium salt, see the following) and is more difficult to reduce because of the net negative charge [4].

On the other hand, the octacyano substituted H_2[OCNPc] is much easier to reduce. Its electrochemistry, however, is somewhat anomalous. The first three reduction potentials are separated by about 0.3 V, a rather small potential gap, especially between the second and third steps; however, the fourth reduction step is displaced some 0.62 V more negative than the third rather than the more usual 0.3 - 0.4 V.

ii. The Alkali Metal Phthalocyanines

The low solubility and sensitivity to water has precluded many studies of the electrochemistry of these species. It would be advantageous to synthesize organic soluble ring substituted Pc derivatives of these metals; except for Li_2[TBuPc] [72], such are currently unknown.

The alkali metal complexes provide the interesting example of the dilithium salt [56-58, 73], Li_2Pc(-2) which would be expected to contain the highly oxidizable $[Pc(-2)]^{2-}$ anion with electrostatic forces binding the Li^+ ions. Indeed oxidation occurs at only $\approx$ 0.15 V versus SCE, giving the free radical one-electron oxidation product LiPc(-1), which is quite a stable species and whose crystal structure has been recorded [73]. A poorly defined wave corresponding to the second oxidation process to $[LiPc(0)]^+$ has also been reported [58], as has reduction to the first anion radical, $[Pc(-3)]^{3-}$, occurring at a significantly more negative potential, as a consequence of excess negative charge [58].

iii. The Alkaline Earth Metal Phthalocyanines

The heavier metals, Ca, Ba, Sr have hardly been explored and again would bear study as organic soluble species. Magnesium Pc however, has been studied extensively.

MgPc is a nice example of a well-behaved metallophthalocyanine in that the central metal is not redox active and all redox processes therefore occur on the phthalocyanine ring. In all, six redox processes have been observed, spanning $[MgPc(0)]^{2+}$ to $[MgPc(-6)]^{4-}$.

The effect of imidazoles, pyridines, and cyanide as ligands in the six-coordinate L_2Mg(II)Pc(-2) species (L = Py, 4MePy, Im, MeIm, CN^-) on the spectroscopic and redox properties has been studied [46] using electronic absorption, MCD spectroscopy, and electrochemistry in solutions. The axial ligands cause a shift in all absorption bands compared with that of the position of the parent Mg(II)Pc. The band center energies are red-shifted in a sequence that follows a decrease in the σ-donor and π-acceptor strengths of the ligand: H_2O > CN^- > Me-py > py > Me-im > im; however, the overall shifts are very small, of the order of a few hundred wavenumbers. The unligated species will be axially solvated by solvent donor molecules, and may be coordinated by trace water in nondonor solvents.

However, axial ligands actually have no significant effect upon the ring redox processes exhibited by L_2Mg(II)Pc (Table 2). This is not peculiar to the hard acid magnesium since, as will be discussed, axial ligation has very little effect upon ring localized redox processes even with transition metal MPc species.

The $[FMg(II)Pc(-2)]^-$ anion has also been studied [56] and has been shown to be rather more easily oxidized, and somewhat more difficult to reduce than other L_2Mg(II)Pc systems (Table 2). This suggests significant reduction of the net electropositve nature of the hard Mg(II) ion by strong binding to fluoride.

Data for Ba(II)Pc are reported in Table 2. On the basis of the polarizing power of the central ion [43], this species should be easier to oxidize than Mg(II)Pc, and it is, but it is more difficult to reduce than Mg(II)Pc. The oxidation potential at 0.45 V is approximately correct as predicted by Eq.(1). Two reduction potentials are reported at -0.49 and -1.08 V [6]. The first is far too positive to be associated with formation of $[Ba(II)Pc(-3)]^-$ but the second is consistent therewith. The process at -0.49 V must arise from an impurity.

iv. Aluminum, Gallium, Indium, and Thallium Phthalocyanines

In line with their greater polarizing power, XAl(III)Pc(-2) species [4, 56, 59, 60] are easier to reduce than Mg(II)Pc(-2) but more difficult to oxidize (Table 2). In parallel with the $[FMg(II)Pc(-2)]^-$ analogue, the $[F_2Al(III)Pc(-2)]^-$ [56] is relatively easier to oxidize and more difficult to reduce. A binuclear complex, $[FAl(III)Pc(-2)]_2O$ shows two oxidation processes consistent with formation of a mixed-valence binuclear species (see the following) [56].

XGa(III)Pc(-2) and XIn(III)Pc(-2) [6, 43, 50] behave analogously to XAl(III)Pc(-2). There is no evidence for reduction of the metal to Ga(I) or In(I)Pc(-2) species, although canonical contributions of XIn(I)Pc(-2) to XIn(III)Pc(-4) could possibly be relevant. Contributions from Tl(I) could be more important in XTl(III)Pc(-2) reduction processes, but data are unavailable.

v. Germanium, Silicon and Tin Phthalocyanines

Electrochemical data have been reported [138] (Table 2) for the decomposition product of μ-oxobis(tetra-t-butylphthalocyanatogermanium(IV)), likely $(HO)_2$Ge(IV)[TBuPc], in THF.

There have been many papers describing the electrochemistry and electronic structure of monomeric and oligomeric silicon phthalocyanines [43, 69, 74-79]. Data are reported in Table 3 (and shown in Figure 3) appertaining to

$(nC_6H_{13})_3SiO$ derivatives [69], while data for t-$BuMe_2SiO$ systems [75] are very closely similar and are not reported here. Data for the same $(nC_6H_{13})_3SiO$ derivatives are also included in [77] reported versus a ClO_4^- anodized silver reference electrode. If these are corrected to SCE by normalizing to the data for the monomeric species in [69], then the results for the dimer, trimer and tetramer differ substantially and by a non constant error from those in [69]. However, the individual potential energy differences between successive redox processes agree in both reports, for all species. This lack of agreement can only be explained by assuming that one of the reference systems was drifting from one experiment to the next. Since SCE electrodes are usually drift free, while silver wire electrodes are known to have drift problems, we suppose that the data in [69] are correct, especially as they agree with data for the closely related species in [75].

Although an early report suggested that the first oxidation and reduction waves of the dimer corresponded each with two-electron processes [76], it is now generally agreed that all these processes (Figure 3) are one-electron in nature and therefore that mixed-valence species are generated (see Section E, viii.).

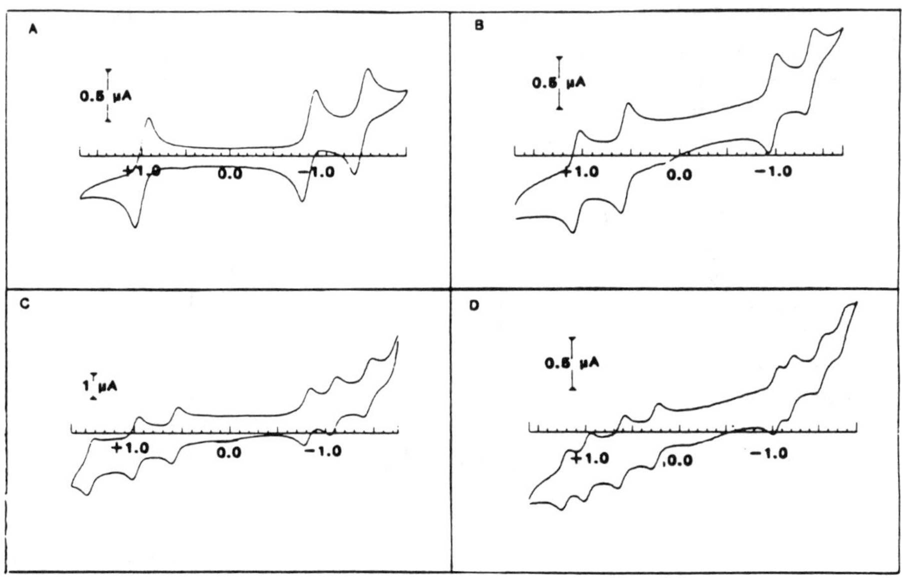

Figure 3 Cyclic voltammetry of oligomeric silicon phthalocyanines, $(n\text{-}C_6H_{13})_3SiO(SiPcO)_nSi(n\text{-}C_6H_{13})_3$ (n = 1-4) in 0.1 M TBA (BF_4) in dichloromethane. All potentials versus Ag reference electrode. Scan rate 100 mV/s. A) Monomer, B) Dimer, C) Trimer and D) Tetramer. Reproduced with permission from Ref. [69].

It becomes progressively easier both to reduce and to oxidize these species as one proceeds from monomer to dimer to trimer to tetramer (possibly not true for the first reduction of the tetramer), although this effect is more marked for oxidation than reduction. The separation between the first oxidation and reduction waves necessarily decreases in this sequence.

This separation is a measure of the LUMO-HOMO gap, that is evidently decreasing. This is to be expected if both the π and π^* levels of adjacent SiPc rings overlap.

The phthalocyanine rings interact through spatial overlap of the π-orbitals [75, 78, 80], rather than through bonds, as for example in a homologous series of conjugated aromatics. The steady decrease in the first oxidation potential from monomer to tetramer shows a progressive stabilization of the (mono)cation whose charge is probably delocalized to some extent over the oligomeric system. The much smaller effect for the first reduction potential shows that, in contrast, the (mono)anion is not significantly stabilized [75].

Table 3 Silicon Phthalocyanine Oligomers Showing Mixed-Valence Behavior (In Dichloromethane). Species $(nC_6H_{13})_3SiO(SiPcO)_nSi(n\text{-}C_6H_{13})_{3a}$ [69]

	n = 1	n = 2	n = 3	n = 4
4th Oxidation				1.38
3rd Oxidation			1.47	1.15
2nd Oxidation		1.20	1.00	0.79
1st Oxidation	1.00	0.71	0.59	0.43
1st Reduction	-0.90	-0.81	-0.78	-0.84
2nd Reduction	-1.48	-1.21	-1.06	-0.98
3rd Reduction			-1.41	-1.30
4th Reduction				-1.54

[a] Diffusion coefficient values are also included in this report [69].

However, the mean oxidation potential (midway between the 2, 3 and 4 oxidation waves of the dimer, trimer and tetramer, respectively) does not differ significantly from that of the monomer oxidation. This mean potential refers to the free energy for forming the dication of the dimer, the trication of the trimer, and the tetra-cation of the tetramer. Its invariance and similarity to the oxidation potential of the monomer shows that once all the rings have been oxidized by one-electron, there is no net stabilization of the oligomer.

The mean of the first reduction, however, defined in a similar fashion, becomes progressively more negative as oligomerization proceeds (see Figure 3 and also Figure 3 in [75]). Thus the change in free energy for placing a negative charge on each ring of the monomer, dimer etc. is consistent with increasing interelectronic repulsion destabilizing the anion radicals so formed.

Data for t-amyloxy silicon phthalocyanine, dissolved in DMF, are also reported in [43] (Table 3). This species appears to be about 300 mV easier to reduce than the $(n\text{-}C_6H_{13})_3SiO$ species as a consequence of the change in axial ligand and/or solvent. This datum fits the polarization plot [43], while a value of about -0.9 V for the first reduction process would be a poor fit to the regression analysis.

vi. Phosphorus, Arsenic, Antimony, and Bismuth Phthalocyanines

While species with these central ions are known, little is known about them. Electrochemical studies could prove very interesting.

vii. Zinc Phthalocyanines

Zn(II)Pc species have been the object of intense electrochemical study [33, 48-52, 54-56, 59-68, 81] (Figures 2, 4). As with magnesium, the central ion is well behaved and redox inactive so that all oxidation states from $[Zn(II)Pc(0)]^{2+}$ to $[Zn(II)Pc(-6)]^{4-}$ are fairly readily observable. Indeed the actual redox potentials are very closely similar to those of Mg(II)Pc.

The question of the separation of the second and third reduction process was raised previously. Two data sets [4, 55] for Zn(II)Pc in DMF cover the entire reduction region to $[Zn(II)Pc(-6)]^{4-}$ but unfortunately the numbers for the third and fourth reduction processes differ quite dramatically in the two reports (Table 2). In terms of the relative separations between successive reduction processes, the data in [55] seem more reliable.

Zn(II)[TBuPc] has been studied by many workers (see Table 2). Data are fairly consistent, although the datum for oxidation in $PhNO_2$ seems too low [51].

Similar to the fluoroaluminum system, $[FZn(II)Pc(-2)]^-$ is easier to oxidize and more difficult to reduce than most other Zn(II)Pc species. Two other species bear special comment. Data for a series of tetrasubstituted zinc phthalocyanines are shown in Table 4, while similar data for octasubstituted Zinc species are shown in Table 5.

The octacyano species, Zn(II)[OCNPc] [33, 55] is very much easier to reduce due to the electron attracting cyano substituents, with reduction processes shifted some 0.8 V more positive than most other Zn(II)Pc species. This octacyano species is strongly aggregated in DMF solution, and the aggregation-disaggregation equilibrium is slow relative to the electrochemical time scale. Thus redox processes due to both aggregated and unaggregated Zn(II)[OCNPc] can simultaneously be observed [55] (Figure 4, processes 1a,1b), with the aggregated species being reduced first (process 1a). Two different anion species, $[Zn(II)[OCNPc(-3)]]^-$ can be observed by reduction at -0.1 and -0.3 V showing that the monoanion species also exists as a mixture of aggregated and unaggregated species. The Pc(-3) and more reduced species are however, disaggregated, probably because of their larger negative charges.

The perchloro species, $Zn(II)[Cl_{16}Pc(-2)]$ [63] also has its reduction processes shifted dramatically positive, by the electron-withdrawing effect of the chloro substituents, but only by about 0.5 V, rather less than for the octacyano derivative. The low solubility of the perchloro species precludes a very detailed study.

Table 4 Oxidation Potentials of some Ring Tetra-Substituted MPc Species in Dichlorobenzene, except where indicated.

R	$R_4Pc(-1)/R_4Pc(-2)$ Zn(II)	$R_4Pc(-1)]/R_4Pc(-2)$ Ref.	Co(II)	Ref.
H			0.86(ACN)	71
			0.845	83
MeO	0.625	82	0.725	83
			0.69(ACN)	71
t-Bu	0.685	82	0.745	83
PhO	0.755	82	0.8[illegible]	83
Ph	0.755	82	0.835	83
PhS	0.785	82	0.865	83
NH_2			0.68(ACN)	71
CO_2H			1.20(ACN)	71
NO_2			1.26(ACN)	71
Neopentoxy	0.47	65	0.52	49

viii. Cadmium, Mercury, and Lead Phthalocyanines

Another situation arises with phthalocyanines containing cadmium, mercury and lead as the central ion. These compounds show anomalous redox behavior [6, 9, 43]. The ionic radii of these elements are too large to lie within the plane of the phthalocyanine core, and as indicated above, their redox potentials do not adhere to the predictions in Eqs.(1). Since the metal sits outside of the phthalocyanine ring, the complex shows a tendency to demetallation in redox reactions. This was observed by Kadish and co-workers [9] during the electrochemical reduction of Pb(II)Pc in DMF. There are three successive reversible one-electron reduction steps observable on the cyclic voltammetric time scale. However, if the potential is held just negative of the first reduction (to $[Pb(II)Pc(-3)]^-$) then demetallation occurs over a period of minutes, lead metal is deposited onto the electrode and metal-free phthalocyanine (or its anion radical) is generated in solution.

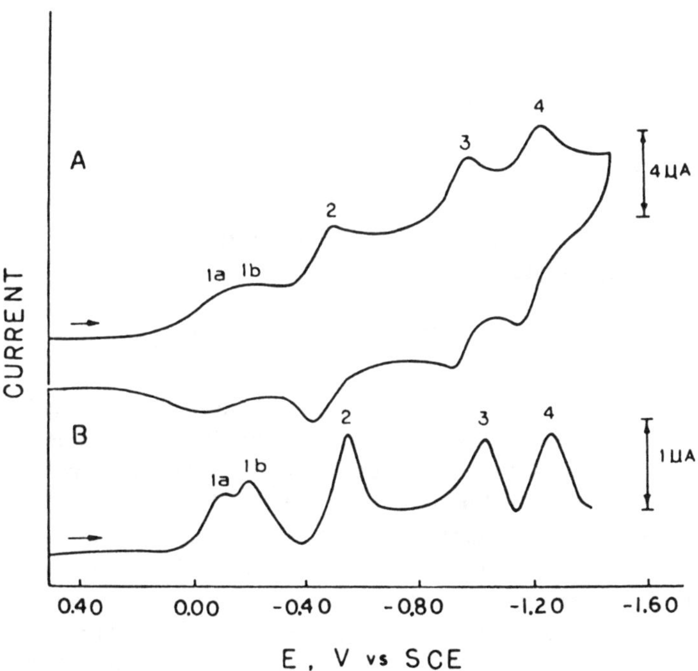

Figure 4 Voltammetry of Zinc Octacyanophthalocyanine (5×10^{-4} M) [33, 55], at a Hg electrode in DMF/(0.1 M TBAP). A) Cyclic voltammetry, 0.2 V/s; and B) Differential pulse voltammetry at 10 mV/s. Reproduced with permission from Ref. [55].

This behavior is reminiscent of that of Ag(II)TNPc (see the following) that is also demetallated when reduced to Ag(I)[TNPc(-2)], for the same reason. Pb(II)Pc shows two reversible oxidations to the mono- and dication radical species [9].

Table 5 Oxidation Potentials of some Ring Octa-Substituted MPc Species in Acetonitrile.

R	$Zn(II)[R_8Pc(-1)]^+$ / $Zn(II)[R_8Pc(-2)]$	Ref.	$Co(II)[R_8Pc(-1)]^+$ / $Co(II)[R_8Pc(-2)]$	Ref.
MeO			0.73	71
Me			0.79	71
CO_2H			1.61	71
CN			1.76	71
BuO	0.50(DMF)	68		
H	0.67(DMF)	33	0.86	71

D. Redox Inactive Transition Metal Phthalocyanines

i) Titanium and Vanadium Phthalocyanines

OTi(IV)Pc [86] is very insoluble in most solvents. Its solution electrochemistry is unknown, but its surface electrochemistry, while not discussed here, has been explored in some detail [87]. It is, however, possible to form organic solvent soluble OM(IV)[TBuPc] species [6]. The central ion is fairly strongly polarizing and somewhat easier to reduce than, say, Ni(II) or Cu(II)Pc (Table 2). However, they are not especially difficult to oxidize. The separation between first oxidation and first reduction, and the separation between the two reduction processes are typical for ring redox processes. Had metal reduction occurred, then successive reduction processes are expected to be separated further (for example, see Co(II), Fe(II) to follow). Moreover, reduction would likely have generated species such as $[Ti(III)Pc(-2)]^+$ in which the oxide ligand had been reduced off. The electrochemistry would then most probably have been irreversible. Thus, although spectroelectrochemical data were not obtained it would appear that there is no evidence for reduction of the central titanium or vanadium ions, even although these metal ions are normally readily reducible.

No solution data exist for Zr, Nb, Hf, or Ta phthalocyanine species. There is a clear need for further development of organic soluble MPc species of the left-hand period transition elements and elucidation of their electrochemical and photophysical properties. The extensive patent literature for OV(IV)Pc as a potential optical recording agent suggests that such further study would be profitable.

ii. Nickel, Palladium, Platinum and Copper Phthalocyanines

Ni(II)Pc(-2) behaves rather like Zn(II)Pc(-2); it is slightly more difficult to oxidize, while reduction to the monoanion radical, Pc(-3), occurs at about the same potential [4-6, 51, 52, 54, 62, 70]. No reduction at the central nickel ion is expected although it is theoretically possible.

The aqueous solution chemistry has been explored with Ni(II)[TsPc] which shows, in water, a two-electron first reduction process [64]. The species is strongly aggregated in water.

Ni(II)[TAPc] [70] is more difficult to reduce than most other Ni(II)Pc species, showing a rather more electron-rich tetraamino-phthalocyanine ligand. An oxidation wave is observed only some 1.2 V positive of the first reduction, rather than the more usual ≈ 1.6 V. The authors had proposed that this was a Ni(III)/Ni(II) couple; however, ring oxidation is more probable [135]. Oxidation of the corresponding Co(II)[TAPc] occurs at only some 0.69 V positive of the first reduction process and the assignment of oxidation to $[(DMSO)_2Co(III)TAPc(-2)]^+$ is almost certainly valid.

Brief details of the reduction of Pd(II)Pc and Pt(II)Pc dissolved in methylnaphthalene at 150° C are reported in Table 2. Little is known of these systems in solution, although surface state electrochemical data for Pt(II)Pc are available [84].

Copper phthalocyanines show well behaved redox processes [4,6, 47, 51, 52, 56, 62, 64, 66] centered on the ligand, from $[Cu(II)Pc(-1)]^+$ to $[Cu(II)Pc(-5)]^{3-}$ at potentials very similar to those of NiPc. The octacyano species, Cu(II)[OCNPc], as with other such species, shows couples shifted some 0.6 - 0.8 V positive of those of other CuPc species [33, 55].

E. Redox Active Transition Metal Phthalocyanines

i. General Introduction

If the transition metal ion concerned has no accessible d orbital levels lying within the $1a_{1u}$ (HOMO) - $1e_g$ (LUMO) gap of a phthalocyanine species, then its redox chemistry will appear very much like that of a main group species. The nickel, palladium, platinum group behave in this fashion, with the M(II) central ion being unchanged as the MPc unit is either oxidized or reduced. Copper(II) also appears invariant in the MPc framework, with reduction occurring at the ring rather than at the copper ion. Silver however, behaves differently, as will be discussed. However, some species vary their electrochemistry according to their environment. For example, the oxidation of Co(II)Pc(-2) can lead to $[Co(II)Pc(-1)]^+$ or $[Co(III)Pc(-2)]^+$ depending upon whether there are any available suitable coordinating species that would stabilize the Co(II) center; these variations are explored in the following.

When a cyclic voltammogram, such as that shown in Figure 2, for Zn(II)[TNPc] is obtained, then, assuming the rest potential [85] is known, the couples are easily assigned to successive ring reductions and ring oxidations of the bulk species. With the corresponding voltammograms for Co(II)[TNPc] (Figure 5) containing a metal center that itself may undergo a redox process, the assignment of the couples is no longer straightforward. From knowledge of the rest potential, it is certainly easy to distinguish net reductions of the bulk, from net oxidations, but the voltammogram does not readily convey information about the site of the redox process, metal center, or ring. To solve this dilemma, the usual procedure is to carry out controlled potential reductions some 200 mV negative of each reduction couple, and some 200 mV positive of each oxidation couple, in order to generate solutions containing bulk quantities of the various reduced and oxidized species.

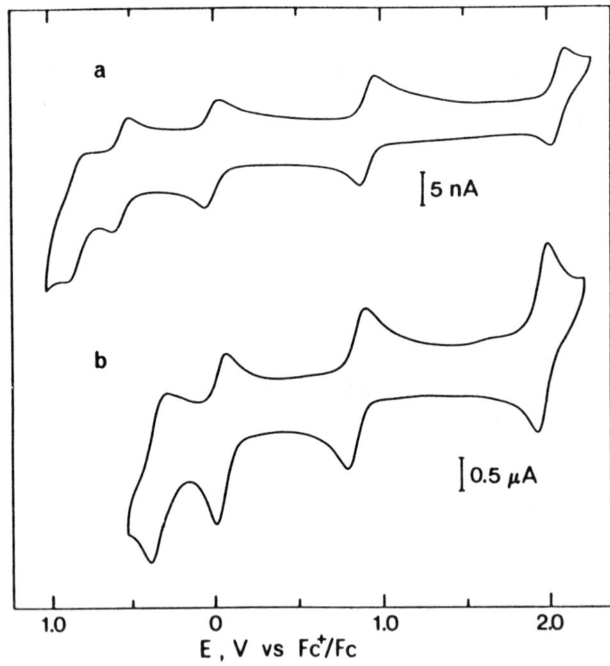

Figure 5 Cyclic voltammetry for Co(II)[TNPc(-2)] (1×10^{-4} M), a) in DCB solution, and b) in DMF solution. Scan rate 50 mV/s, [TBAP] = 0.3 M. Reproduced with permission from ref. [49].

Table 6 Transition Metal Electrochemistry - Systems Displaying M(III), M(II) and M(I) (versus SCE)[a].

Species	Ring Oxidn Process	Metal centered Process M(III)/M(II)	M(II)/M(I)	Ring Reduction. Processes			Condn.	Ref.
Cr(III)[TsPc]	0.51	-0.40		-1.00			DMF	5,6,54
$(Py)_2Cr(II)Pc$	0.95	0.52					Py	61,92a
$(Py)_2Cr(II)Pc$		0.63		-0.90			Py/Cl	92a
$Na[(CN)_2Cr(III)Pc]$	0.90	-0.56		-1.23			Acetone	88
(HO)Cr(III)[TBuPc]	0.70	-0.87		-1.36			DMF	92a
(HO)OMo(V)[TsPc]		0.51(VI/V)	-0.28(V/IV)				DMF	92b
Mn(II)Pc		-0.08	-0.755				DMSO	93,95
Mn(II)Pc		-0.085	-0.70				DMSO/Br	93
Mn(II)Pc		-0.125	-0.765				DMSO/Cl	93
Mn(II)Pc		0.005	-0.785				Py	93,95
Mn(II)Pc		-0.035	-0.71				Py/Br	93
Mn(II)Pc		-0.105	-0.80				Py/Cl	93
Mn(II)Pc		-0.23					1M Py/CH_2Cl_2	
Mn(II)Pc		-0.14	-0.69				DMF	6,60,93
Mn(II)Pc	0.87	-0.14	-0.69	-1.46			DMF	93,95
Mn(II)Pc				-1.50	-2.20	-2.75	DMF	4
Mn(II)Pc		-0.115	-0.78				DMF/Br	93
Mn(II)Pc		-0.155	-0.80				DMF/Cl	93
Mn(II)Pc		-0.13	-0.80				DMA/Br	93
Mn(II)Pc		-0.14	-0.80				DMA/Cl	93
Mn(II)Pc			-0.71	-1.16			MeNp/150°	52

Species	Ring Oxidn Process[a]	Metal centered Process M(III)/M(II)	M(II)/M(I)	Ring Reduction. Processes			Condn.	Ref.
Mn(II)[TsPc]			-0.72	-1.36			DMF	5,6,54
Mn(II)[TBuPc]	0.89	-0.13	-0.69				DMF	54
Mn(II)[TBuPc]		-0.115	-0.89				DMF	6
Na[$(CN)_2$Mn(III)Pc]	0.89	-0.14	-0.91				Acetone	88
Fe(II)Pc		0.37	-0.55	-1.17			DMA	6,60,93
Fe(II)Pc		0.38	-0.55	-1.17			DMA	95
Fe(II)Pc				-1.56	-2.05	-2.25	DMF	4
Fe(II)Pc			-0.70	-1.13			MeNp/150°	52
Fe(II)Pc		0.17	-0.86	-1.19			DMA/Br	97
Fe(II)Pc		-0.15	-0.64	-1.13			DMA/Cl	97
Fe(II)Pc		-0.21					Py/OH	106
Fe(II)Pc		0.46	-0.71				DMSO	95,97
Fe(II)Pc		0.39	-0.77				DMSO/Br	97
Fe(II)Pc		0.35	-0.69				DMSO/Cl	97
Fe(II)Pc			-0.74	-1.15			DMSO/ClO_4	98
Fe(II)Pc			-0.74	-1.14			DMSO/246Coll	98
Fe(II)Pc			-0.80	-1.16			DMSO/24Lut	98
Fe(II)Pc			-0.82	-1.15			DMSO/26Lut	98
Fe(II)Pc			-0.87	-1.20			DMSO/Me_2Im	98
Fe(II)Pc			-0.93	-1.21			DMSO/3Pic	98
Fe(II)Pc			-0.94	-1.20			DMSO/MeIm	98
Fe(II)Pc			-0.95	-1.22			DMSO/Py	98

Fe(II)Pc			-0.99	-1.23		DMSO/4Pic 98	
Fe(II)Pc			-1.17	-1.17[b]		DMSO/NMeIm 98	
Fe(II)Pc			-1.21	-1.21[b]		DMSO/Im 98	
Fe(II)Pc		0.66	-1.07			Py	97
Fe(II)Pc	1.10	0.69	-1.085	-1.39	-1.93	Py	107
Fe(II)Pc		0.19				ClNap	62
Fe(II)Pc		0.74				ACN	71
$[(CN)_2Fe(III)Pc]^-$	0.93	0.14	-0.99	-1.52		Acetone	88
$[(CN)_2Fe(III)Pc]^-$	0.74	0.03				CH_2Cl_2	100
$[(CN)_2Fe(III)Pc]^-$	1.16	0.36	-0.24??			ACN	105
Fe(II)[OMPc]		0.56	-0.92	-1.45qr		Py	137
Fe(II)[TBuPc]			-0.85	-1.16		MeNp/150°	52
Fe(II)[TBuPc]			-0.95	-1.30		CH_2Cl_2	52
Fe(II)[TBuPc]			-1.12			DMSO	52
Fe(II)[TcPc]	1.22					ACN	71
Fe(II)[Cl_{16}Pc]		0.73		-1.11	-1.73	DMF	63
Fe(II)[TsPc]	0.89	0.41	-0.74	-1.08	-1.15	DMF	5,6,54
Ag(II)[TNPc]	1.72i,1.19	0.71i	-0.79i	-1.47		DCB	48
$K_2[(CN)_2Ru(II)Pc]$	0.93	0.34		-0.97		Acetone	88
$K[(CN)_2Rh(III)Pc]$	0.90			-0.90	-1.45	Acetone	88
(MeOH)ClRh(III)Pc	0.87	-0.72				ACN	108
ClRh(III)Pc	0.97	-0.69	-1.47			DCB	109

Solvent conditions include "non-coordinating" supporting electrolyte anion, anion except where another anion is specifically cited. ? signifies experimental data are likely correct but their interpretation is doubtful. ?? signifies the data themselves are doubtful. [a] Assignments made by this author do not always agree with those made by the original authors. [b] Two-electron combined couple. For Cobalt species, see Tables 4,5,7, and 8.

Electronic spectroscopy (including magnetic circular dichroism, MCD) and electron spin resonance (ESR) spectroscopy of these solutions can then usually determine the site of redox. If reduced or oxidized species can be isolated, then other characterization procedures such as FTIR, magnetic susceptibility measurements, and x-ray photoelectron spectroscopy might be employed. Quite frequently the combination of electronic spectra and ESR is sufficient to determine the site of redox especially when studying a metal ion for which there already exists an extensive database.

It is not the purpose of this chapter to delineate in any detail how the distinction between metal centered or ring redox processes can be made. The reader is referred to relevant chapters in these volumes where such information is available.

ii. Chromium, Molybdenum, and Tungsten Phthalocyanines

Although the chemistry of chromium phthalocyanines has been fairly extensively studied [6, 54, 61, 88-90], their electrochemistry is, as yet, poorly characterized (Table 6). Both Cr(II)Pc and XCr(III)Pc species are known, with the former being air sensitive. Thus the first reduction process, for XCr(III)Pc, is almost certainly XCr(III)Pc/Cr(II)Pc. The potential of this first reduction process varies over quite a considerable range (-0.87 - (+0.52) V) (Table 6). The most positive potential appears in pyridine wherein $(Py)_2$Cr(II)Pc is known to be formed as a stable, air-insensitive species [90]. The most negative III/II potential refers to the reduction of HOCr(III)[TBuPc(-2)] [6, 91] in DMF wherein the species $(DMF)_2$Cr(II)[TBuPc(-2)] is likely formed. The DMF group is likely to stabilize Cr(III) to a much greater degree than pyridine consistent with the electrochemical observation. We had previously assigned [6] the potential at -0.87 V to Cr(II)Pc(-2)/Cr(II)Pc(-3) but this is unlikely to be correct since one would then have to assign the oxidation at +0.7 to the Cr(III)/Cr(II) couple, an unreasonably high value. There is no systematic study of the effect of axial ligation on the potential of the Cr(III)/Cr(II) couple, but the potentials for $(Py)_2$Cr(II)Pc and $(DMF)_2$Cr(II)[TBuPc(-2)] (from the hydroxychromium(III) species in DMF) being so disparate (+0.52 and -0.87 V) would make such a study worthwhile.

Some previously unpublished data [92a] for $(Py)_2$Cr(II)Pc dissolved in pyridine in the presence of chloride ion are reported in Table 6. Unfortunately spectroelectrochemical data were not collected at the time. The presence of chloride ion is expected to favor Cr(III). A couple appears at -0.9 V in the $(Py)_2$Cr(II)Pc/Py/Cl^- system and may involve the Cr(III)/Cr(II) couple for a chloride bound species.

Thus the Cr(III)/Cr(II) couple appears to vary over an exceptionally wide range as a function of axial ligand: $(Py)_2$ 0.52; $(CN)_2$ -0.56; $(DMF)_2$ -0.87; Py,Cl ? -0.90 V. Clearly, further work coupled to spectroelectrochemical data is required.

Further reduction (of Cr(II)Pc(-2)) probably yields a $[Cr(II)Pc(-3)]^-$ species, while oxidation (of XCr(III)Pc(-2)) almost certainly yields $[XCr(III)Pc(-1)]^+$ radical cation species, but confirmatory evidence is lacking (Table 6).

Ferraudi has recently published some solution data for O(HO)Mo(V)[TsPc] in DMF [92b], with the apparent identification of the Mo(VI)/Mo(V) and Mo(V)/Mo(IV) redox couples (Table 6). No solution electrochemical data appear to exist for any tungsten phthalocyanines.

iii. Manganese, Technetium, and Rhenium Phthalocyanines

Extensive solution data exist for manganese phthalocyanines, but no data for either technetium or rhenium.

The electrochemistry of Mn(II)Pc has been studied in some depth [6, 52, 60, 93-95]. It shows very little variation with coordinating axial ligand (donor solvent or supporting electrolyte anion). The Mn(III)/Mn(II) oxidation couple lies in the narrow range -0.23 - (+0.005) V for all systems studied. There is a slight stabilization of Mn(II) in the sequence Py > DMSO > DMA = DMF [93]. Coordination by strongly coordinating anions favors Mn(III), in the sequence Cl^- > Br^- > ClO_4^- but the variation is much smaller than for iron phthalocyanine (see the following). There is a rather flat, but linear, correlation with the donor number (Gutmann) of the solvent, with a slope the same as that to be discussed below for Fe(III)Pc/Fe(II)Pc [95].

Two equilibria should be considered:

$$[(Sol)_2Mn(III)Pc(-2)]^+ + e^- \Longleftrightarrow (Sol)_2Mn(II)Pc(-2) \quad (2)$$
$$X^- + [(Sol)_2Mn(III)Pc(-2)]^+ \Longleftrightarrow X(Sol)Mn(III)Pc(-2) + Sol \quad (3)$$

Where Sol is solvent and X^- is a counteranion, and where six-coordinate Mn(III) is inferred from a solution magnetic susceptibility study [93]. Equilibrium (3) shifts the potential to more negative values with increasing stabilization of the X(Sol)Mn(III)Pc(-2) species. Since $[(Sol)_2Mn(III)Pc(-2)]^+$ is easier to reduce than X(Sol)Mn(III)Pc(-2), the equilibrium will shift so as to produce this latter species on the electrode at the reduction couple potential. With high-speed voltammetry one may expect that it should be possible also to see the reduction couple for X(Sol)Mn(III)Pc(-2) if scanning is faster than the rate for re-equilibration to $[(Sol)_2Mn(III)Pc(-2)]^+$. Such an additional couple is seen with X^- = OH^- [94], where couples corresponding to the reduction of both Mn(III) species can be observed. Their total reduction current is constant with respect to scan rate, but the relative ratio of the two currents depends upon scan rate, favoring the X(Sol)Mn(III)Pc(-2) species at higher scan rates. The reduction product, $[HOMn(II)Pc(-2)]^-$ loses OH^-, very rapidly such that its reoxidation is not observed at the highest scan rates studied [94].

A further oxidation couple was only seen in DMF (at +0.87 V) but the product was not stable on spectroelectrochemical oxidation. The couple likely corresponds to $[XMn(III)Pc(-1)]^+$/XMn(III)Pc(-2) rather than oxidation to Mn(IV), but spectroelectrochemical evidence is desirable.

The first reduction potential also occurs within a very narrow range (-0.69 - (-0.80) V) being essentially independent of solvent or counteranion. This argues for reduction to the anion radical, viz Mn(II)Pc(-2)/$[Mn(II)Pc(-3)]^-$ rather than to the d^6 $[Mn(I)Pc(-2)]^-$, which, by analogy with d^6 Fe(III)Pc, is expected to show marked solvent and anion dependence. The electronic spectrum of this reduced species is consistent with formation of the anion radical [93].

Smith, Pilbrow and co-workers [96] have studied the chemical reduction products of Mn(II)[TsPc(-2)], and on the basis of ESR spectroscopy, assign the two successive reduction processes to formation of $Mn(I)[TsPc(-2)]^-$ and $Mn(0)[TsPc(-2)]^{2-}$ (where any charges on the sulfonyl groups are ignored).

While the solvent independence of the first reduction process argues for anion radical formation, the separation between the first and second reduction processes (≈ 0.6 V) is rather large to be ascribed to successive phthalocyanine ring reduction processes that are normally separated by about 0.4 V (see preceding discussion) [4]. The observed separation of 0.77 V (in DMF) is more consistent with the separation of the second and third ring reduced species, but that assumption fails to provide a logical assignment for the -0.7 V couple. If this latter couple is assigned to Mn(II)Pc(-2)/$[Mn(I)Pc(-2)]^-$, then the separation of 0.77 V to the next reduction process, to form $[Mn(I)Pc(-3)]^{2-}$ or $[Mn(0)Pc(-2)]^{2-}$, is not unreasonable, although it is substantially less than the corresponding separation in CoPc chemistry (about 1.1 V, see the following). Note that Clack, Hush and Woolsey [4] quote two further reduction processes, in DMF, whose potentials and separations are consistent with the sequential reduction of the phthalocyanine ring.

In methylnaphthalene at 150° C, Mn(II)Pc shows a normal pair of reduction processes [52] separated by 0.45 V. Possibly in noncoordinating solvents, reduction to $[Mn(II)Pc(-3)]^-$ takes place, while in coordinating solvents, $[S_2Mn(I)Pc(-2)]^-$ is formed, even although little solvent dependence is observed. A complete understanding of this system remains elusive (data are so-assigned in Table 6).

Manganese phthalocyanine can be oxidized to form a bridging oxo species, PcMn(III)-O-Mn(III)Pc whose electrochemistry has been explored in depth [94]. Figure 6 illustrates its intriguing electrochemistry. Detailed analysis of the current of the several couples, as a function of scan rate, reveals that PcMn(III)-O-Mn(III)Pc undergoes a two-electron reduction, at -0.85 V, to yield

$[PcMn(II)\text{-}O\text{-}Mn(II)Pc]^{2-}$, which is unstable with respect to bridge cleavage generating two mononuclear Mn(II)Pc(-2) fragments. These, however, exist on an electrode polarized (at -0.85 V) at a potential negative of the reduction process to form $[Mn(II)Pc(-3)]^{-}$; thus two further electrons are taken up to form two molecules of this anion radical species. Thus PcMn(III)-O-Mn(III)Pc undergoes a four-electron irreversible reduction at this potential. A careful analysis of the high-and low-scan rate data provided evidence for this 2+2 reduction mechanism in distinction to possible alternatives such as a concerted four electron reduction step, or 1+3 combinations [94].

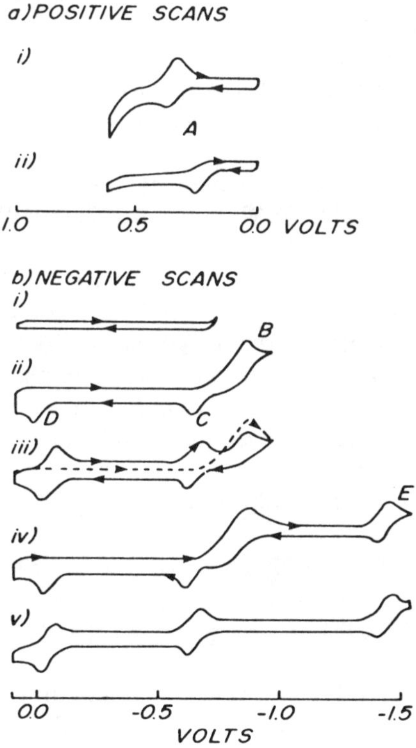

Figure 6 Cyclic Voltammetry of μ-oxobis(phthalocyaninemanganese(III)) [94]. a) oxidation in i) pyridine, and ii) DMF. b) Reduction voltammograms in pyridine/TEAP, i) initial scan +0.1 to -0.8 V; ii) initial scan +0.1 to -1.1 V; iii) initial (dotted) and second (solid) scans +0.1 to -1.1 V; iv) initial scan +0.1 to -1.6 V; v) continuous scan +0.1 to -1.6 V. Scan rates are 0.1 V/s except for v) which is 10 V/s. Reproduced with permission from ref. [49].

In the initial negative going scan (from +0.1 V) (Figure 6, b,iii) only this four-electron reduction process is seen. Successive scans however, reveal couples due to the mononuclear MnPc species formed at the electrode surface (see Scheme I in [94]).

Note that formation of the oxo bridge stabilizes Mn(III)Pc, relative to the mononuclear species, by ≈ 0.75 V.

A one-electron oxidation was observed in pyridine at about +0.35 V, presumably to form $[Pc(-2)Mn(IV)\text{-}O\text{-}Mn(III)Pc(-2)]^+$ but this complex was insufficiently stable to prove its identity spectroscopically [94].

iv. Iron, Ruthenium, and Osmium Phthalocyanines

Iron(II) phthalocyanine is fairly soluble in a wide range of donor solvents (and some nondonor solvents), and this solubility, in distinction to species such as OTiPc, OVPc etc., has led to many studies of its electrochemical properties in solution [4, 6, 10, 54, 60-62, 93, 97-102].

Iron(II) phthalocyanine commonly displays four reversible couples in the range +1.00 - (-2.00) V (Table 6). Oxidation to $[Fe(III)Pc(-2)]^+$ occurs in the range -0.15 - (+0.69) V highly dependent upon the solvent and counteranion. Iron(II) phthalocyanine binds donor solvents to form six-coordinate $(Sol)_2Fe(II)Pc(-2)$ species, while the Fe(III)Pc oxidation product has been proven by electronic and electron spin resonance spectroscopy [97].

Thus, analysis of the scan rate dependence of the anodic and cathodic currents (for the oxidation couple) [93, 103] demonstrates a reversible electron transfer followed by a chemical reaction upon oxidation:

$$(Sol)_2Fe(II)Pc(-2) <===> [(Sol)_2Fe(III)Pc(-2)]^+ + e^-$$

$$[(Sol)_2Fe(III)Pc(-2)]^+ + X^- <===> X(Sol)Fe(III)Pc(-2) \quad (4)$$

$$[X(Sol)Fe(III)Pc(-2)] + e^- <===> [X(Sol)Fe(II)Pc(-2)]^- \quad (5)$$

In this case strong binding of the counteranion in $[X(Sol)Fe(III)Pc(-2)]^+$ greatly influences the potential at that the Fe(III)/Fe(II) process is observed. Reaction (4) must proceed rapidly to the right, but not to the left, relative to the voltammetric time scale, and thus the stability of X(Sol)Fe(III)Pc(-2) determines the observed Fe(III)/Fe(II) potential, shifting to more negative potentials in the sequence:

$ClO_4^- < Br^- < Cl^- < OH^-$

Pyridine stabilizes Fe(II)Pc, through formation of $(Py)_2Fe(II)Pc(-2)$ to such a degree that the oxidation product, $[(Py)_2Fe(III)Pc(-2)]^+$ is unstable in pyridine and there is a marked loss of cathodic current on the return wave. The Fe(III)/Fe(II) couple shifts positively in the sequence:

DMA = DMF < DMSO < Py

and linearly follows the Gutmann donor number for the solvent [95].

Thus the most negative potential is obtained by dissolving Fe(II)Pc in DMA or DMF in the presence of chloride ion. Indeed this combination leads to an air sensitive solution that oxidizes directly in air to the Fe(III) species [97].

In contrast to the situation with Mn(II)Pc, the first reduction process with Fe(II)Pc also shows quite strong solvent dependence shifting negatively in the sequence DMA < DMSO < Py (Table 6), but little dependence upon anion. Electronic and electron spin resonance data clearly show formation of $[Fe(I)Pc(-2)]^-$ species. Moreover, the latter experiment, in the presence of pyridine as solvent (or with imidazole/DMA or Ph_3P/DMSO), clearly identifies a five coordinate $[LFe(I)Pc(-2)]^-$ species [97]. Analysis of the scan-rate dependence of the current then confirms a reversible electron transfer followed by a chemical reaction, consistent with (and generalizing for solvent, Sol):

$$(Sol)_2Fe(II)Pc(-2) + e^- \longleftrightarrow [(Sol)_2Fe(I)Pc(-2)]^- \quad (6)$$

$$[(Sol)_2Fe(I)Pc(-2)]^- \longleftrightarrow [(Sol)Fe(I)Pc(-2)]^- + Sol \quad (7)$$

With increasing donor strength of solvent, the $(Sol)_2Fe(II)Pc(-2)$ species is stabilized. Since anion binding is not involved, there is little dependence thereon. The sequence of solvent dependence is the reverse of that of the Fe(III)/Fe(II) couple, because it is now the higher, rather than the lower, oxidation state that is being preferentially stabilized.

However, the situation is more complex than this as shown by the variable scan rate data shown in Figure 7. Two pairs of couples associated with the Fe(II)/Fe(I) process can be seen. The data may be explained by consideration of two additional equilibria:

$$[(Sol)Fe(I)Pc(-2)]^- \longleftrightarrow (Sol)Fe(II)Pc(-2) + e^- \quad (8)$$

$$(Sol)Fe(II)Pc(-2) + Sol \longleftrightarrow (Sol)_2Fe(II)Pc(-2) \quad (9)$$

Given that six-coordinate $[(Sol)_2Fe(I)Pc(-2)]^-$ would oxidize at a more negative potential (Eq. 6) than five coordinate $[(Sol)Fe(I)Pc(-2)]^-$ (Eq. 8) (due to increased destabilization of the filled d_{z2} orbital), then the former equilibrium (6), is associated with couple B,B' (Figure 7) and the latter with A,A'. Reduction of six-coordinate Fe(II)Pc (Eq. 6) leads to rapid loss of solvent (Eq. 7). At slow scans, on the reverse scan, there is sufficient time for reaction (7) to proceed to the left to a large degree, hence wave B is more prominent than wave A. At high scan rates, there is insufficient time, permitting reaction (8) to dominate, and therefore wave A becomes more intense than wave B. The observation of wave A' at high scan rates demonstrates that equilibrium (9) does not proceed so rapidly to the right [97].

The next reduction process occurs at the phthalocyanine ring, to form $[(Sol)Fe(I)Pc(-3)]^{2-}$. Since Fe(II) is not involved, this second reduction couple shows little solvent or anion dependence. It varies from the first reduction by some 0.2 (Py) to 0.7 V (DMA). The very small separation for pyridine follows from the strong stabilization of Fe(II)Pc by this solvent. The scan rate/current dependence is consistent with a simple electron transfer without any following chemical reaction, consistent with the assignment. Further reduction, likely to form $[(Sol)Fe(I)Pc(-4)]^{3-}$ occurs at a potential ≈ 0.6 V more negative (Table 6).

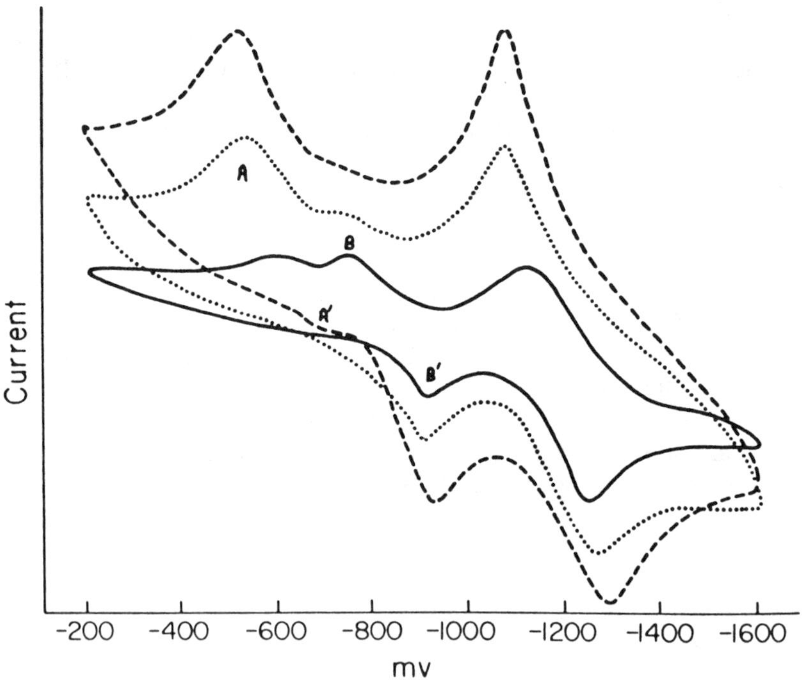

Figure 7 Variable Scan Rate Data for Fe(II)Pc/DMA/TEABr. Scan rates from lower to upper curves are 0.1, 10 and 50 V/s respectively. Wave A does not appear in a voltammogram scanned at 0.01 V/s. Reproduced with permission from Ref. [97].

Kadish, Bottomley, and Cheng [98] have reported a detailed spectroscopic and electrochemical study of Fe(II)Pc in the presence of the bases (L) imidazole, 2-methylimidazole, N-methylimidazole, and a range of mono-di-and trisubstituted pyridines. They report binding constants for these ligands, in DMSO, for L_2Fe(II)Pc(-2) (Log K_1 and Log K_2) and for $[LFe(I)Pc(-2)]^-$, obtained by spectroscopic or electrochemical analysis of ligand-titrated solutions of Fe(II)Pc in DMSO, but the reader should note some dissenting arguments in [130].

Figure 8 shows how couple I, (Eqs. 6,7) shifts toward couple II ($[LFe(I)Pc(-2)]^-/[LFe(I)Pc(-3)]^{2-}$, L = DMSO or imidazole, depending upon concentration of imidazole) with increasing imidazole concentration and merging therewith at concentrations greater than 0.1 M [Im] (or (N-MeIm]). A plot of $E_{1/2}$ versus log [L] gave a slope of -57 mV/pH unit consistent with the loss of one ligand upon reduction. The newly formed process (III) is a two-electron reduction forming $[(Im)Fe(I)Pc(-3)]^{2-}$ directly from $(Im)_2$Fe(II)Pc(-2).

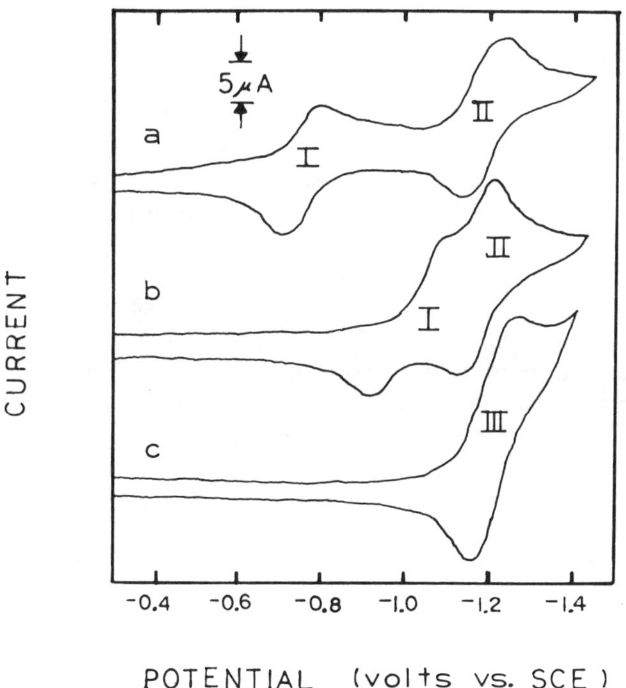

Figure 8 Cyclic voltammetry of iron phthalocyanine (1.18 mM) in Me_2SO/Imidazole/0.1 M TEAP. Scan rate 0.1 V/s; Imidazole concentrations, a) 0, b) 0.01 M, c) 0.95 M. Reproduced with permission from Ref. [98].

Careful analysis of the electrochemical data shows that there are two sequential one-electron processes at the same potential (EE process), with the first (corresponding with Eqs. 6, 7) being rate controlling. The addition of imidazole has no effect upon the potential of couple II.

This observation is in contrast to the titration of 2-methylimidazole, 1,2-dimethylimidazole, and a range of different substituted pyridines in DMSO. At low concentrations (log [L] $<$ 0.05 M) only couple I shifts negatively, and couple II is invariant, but at higher concentrations both couples shift negatively, and with the same slope ($\approx$ -57 mV/pH unit). Thus both the L_2Fe(II)Pc(-2) and $[LFe(I)Pc(-2)]^-$ species lose a ligand L upon reduction finally to form $[(DMSO)Fe(I)Pc(-3)]^{2-}$. With these ligands, couples I and II do not coalesce.

Finally, data for the sterically hindered ligands 2,4-lutidine or 2,4,6-collidine were interpreted in terms of binding of these ligands to Fe(II)Pc(-2) but not to $[Fe(I)Pc(-2)]^-$.

A very early study [62] reported 0.19 V for the Fe(II)Pc oxidation couple in the noncoordinating chloronaphthalene. This datum likely then refers to oxidation of the four coordinate Fe(II)Pc fragment, but whether oxidation occurs at the metal center or phthalocyanine ring was not proven.

While Fe(II)Pc in common with all other unsubstituted metallophthalocyanines, is insoluble in water, it will dissolve therein, in the presence of cyanide ion, forming the somewhat water-soluble $[(CN)_2Fe(II)Pc(-2)]^{2-}$ [18a, 86, 100]. Data for this species (Table 6) have been reported by three groups [100, 88, 105]. It shows an Fe(III)/Fe(II) oxidation wave at 0.14 V (in acetone) (or 0.04 V in CH_2Cl_2), indicating rather strong stabilization of the Fe(III) state by the bound cyanide ligands. The three sets of data, in different solvents, show rather dramatically different potentials that might indicate significant solvatochromism, but the system should be reinvestigated. In acetone [88] the Fe(II)/Fe(I) couple falls at a very typical potential, suggesting that cyanide does not especially favor binding to Fe(II). Curiously, however, the next reduction, forming the $[Fe(I)Pc(-3)]^-$ anion radical species, is found some 0.1 - 0.3 V more negative than all other observations. Ring oxidation is seen at 0.7 - 1.16 V.

Various ring substituted iron phthalocyanines have been studied. The tetrasulfonated species, Fe(II)[TsPc] has potentials very closely similar to those of the unsubstituted species. Its ready solubility permitted a detailed spectroelectrochemical study as a function of pH [101]. The aggregation properties of Co[TsPc] and Fe[TsPc] were studied as a function of oxidation state and the results have been reported in Table 3 of [101]. In summary:

M(I)[TsPc(-2)] species (M = Co,Fe); nonaggregated in acid and base.
Co(II)[TsPc(-2)] species; aggregated in acid and base.
Fe(II)[TsPc(-2)] species; partly aggregated in acid and base.
Co(III)[TsPc(-2)] species; nonaggregated in acid, probably dimeric in base.
Fe(III)[TsPc(-2)] species; aggregated in acid, probably dimeric in base.

These results strictly only apply to the conditions used and, in general, aggregation will diminish with dilution.

The corresponding tetracarboxy species Fe(II)[TcPc] is reported [71] to have an oxidation potential of 1.22 V but this is far too positive to be due to Fe(III)/Fe(II) and must refer to oxidation of the Fe(III)[TcPc] species (Table 6).

The hexadecachlorophthalocyanines have unusual electrochemical properties [63]. The presence of 16 substituting chlorine atoms is expected to greatly stabilize ring reduction and destabilize metal centered oxidation. The former prediction is clearly demonstrated by reduction of Zn(II)[Cl_{16}Pc(-2)] (unequivocally to [Zn(II)[Cl_{16}Pc(-3)]$^-$) occurring some 0.3 V or more positive of the usual reduction potential [63] (Table 2). The observed oxidation potential, for Fe(II)[Cl_{16}Pc(-2)] in DMF, at 0.73 V agrees with the latter prediction, being much more positive than most other Fe(III)/Fe(II) couples. Reduction yields two waves at -1.11 and -1.73 V but the former cannot be assigned to the Fe(II)/Fe(I) couple since it falls considerably negative of this process in all other iron phthalocyanines, rather than positive as expected. Rather these two processes must be assigned to reduction to [Fe(I)[Cl_{16}Pc(-3)]]$^{2-}$ and then to [Fe(I)[Cl_{16}Pc(-4)]]$^{3-}$ occurring at more positive potentials than the corresponding processes in other iron phthalocyanine species (Table 6).

The absence of observation of the Fe(II)/Fe(I) process, and of the corresponding Co(II)/Co(I) process in Co(II)[Cl_{16}Pc(-2)] (see the following), must arise through kinetic sluggishness. The M[F_{16}Pc(-2)] M = Fe(II), Co(II), species behave in a very similar fashion [110].

Ercolani and co-workers [106,111] have explored the electrochemistry of Pc(-2)Fe(III)-O-Fe(III)Pc(-2) and Pc(-2)Fe(III.5)-N-Fe(III.5)Pc(-2). These species are discussed in some detail in another chapter in this volume [112] and so will only be summarized here. The data are reported in Table 5 of ref. [112] together with corresponding tetraphenylporphyrin measurements.

On the voltammetric time scale, Pc(-2)Fe(III)-O-Fe(III)Pc is shown to undergo one-electron oxidation, at 0.47 V in pyridine, to a mixed-valence Fe(III)-O-Fe(IV) species and two successive one-electron reductions (-0.59, -0.95 V) to the mixed-valence Fe(III)-O-Fe(II) and then Fe(II)-O-Fe(II) species with the integrity of the Fe-O-Fe bridge being maintained. Over longer time periods, however, all these reduced or oxidized species cleave and electrochemical waves corresponding to the mononuclear FePc fragments are observed. Given that the Fe(III)/Fe(II) couple for $(Py)_2$Fe(II)Pc is observed at 0.66 V (Table 6), the oxo bridge confers some 1.25 V stability to Fe(III).

The nitrido dimer displays an oxidation couple at 0.0 V in pyridine, forming the Fe(IV)-N-Fe(IV) species, and three successive reductions (-0.83, -1.02, -1.29 V) show bridge integrity. On a longer time scale the first oxidation and reduction processes yield stable species, in contradistinction to the oxo bridged species, but the second and third reduction processes lead to bridge cleavage. Comparing iso-electronic species, the nitrido bridge confers some 1.3 V stability over the corresponding mixed-valence Fe(III)-O-Fe(IV) species.

Ruthenium(II) phthalocyanines have been explored by James and co-workers [12]. Oxidation appears to occur exclusively at the phthalocyanine ring (Table 2). The first half-wave potentials for all complexes studied [12] lie between 0.7-0.9 V (versus SCE) and depend on the nature of the axial ligand. For bispyridine ruthenium phthalocyanines (L_2RuPc(-2)) with substituted pyridines in axial position the value of $E_{1/2}$ decreases in accordance with increasing electron-donor strength of the coordinated ligand in the sequence py > 4Me-py >t-Bu-py (0.77, 0.74, and 0.70 V respectively). When the solvent molecules serve as axial ligands in PcRuL_2, the half-wave potentials show changes depending on the nature of the metal-ligand bond: N-bonded (MeCN) 0.72, O-bonded (DMF) 0.80, and S-bonded (DMSO) 0.89 V. The highest value for the half-wave potential of $(DMSO)_2$RuPc could be consistent with π-electron acceptance by the S-bonded sulfoxide. If one molecule of pyridine in L_2Ru(II)Pc(-2) is replaced by CO the potential is higher, for instance, 0.77 V for $(Py)_2$Ru(II)Pc(-2) and 0.91 V for (Py)(CO)Ru(II)Pc(-2), due to the acceptor effect of the CO ligand.

Data for osmium phthalocyanines are unavailable.

v. Cobalt, Rhodium, and Iridium Phthalocyanines

In common with Fe(II)Pc, Co(II)Pc is soluble in a wide range of donor and nondonor solvents and its solution electrochemistry has been studied extensively [6, 10, 11, 49, 63, 66, 67, 70, 71, 88, 95, 101, 107, 113-116].

Such electrochemistry can conveniently be split into two sections, that referring to donor solvents, and that referring to nondonor solvents. A series of reversible couples are observed and may be summarized:

Donor Solvents	Non-donor solvents
I	I
Co(III)Pc(0)/Co(III)Pc(-1)	Co(III)Pc(0)/Co(III)Pc(-1)
II	II'
Co(III)Pc(-1)/Co(III)Pc(-2)	Co(III)Pc(-1)/Co(II)Pc(-1)
III	III'
Co(III)Pc(-2)/Co(II)Pc(-2)	Co(II)Pc(-1)/Co(II)Pc(-2)
IV	IV
Co(II)Pc(-2)/Co(I)Pc(-2)	Co(II)Pc(-2)/Co(I)Pc(-2)
V	V
Co(I)Pc(-2)/Co(I)Pc(-3)	Co(I)Pc(-2)/Co(I)Pc(-3)
VI	VI
Co(I)Pc(-3)/Co(I)Pc(-4)	Co(I)Pc(-3)/Co(I)Pc(-4)

(11)

(net charges omitted for clarity), where the principle difference lies in whether, for Co(II)Pc, the metal or the ring is oxidized first. Donor solvents (or coordinating counteranions or other ligands in nondonor solvents) strongly favor Co(III)Pc by coordinating along the axis to form a six-coordinate L_2Co(III)Pc species. If such donor molecules are absent, then oxidation to Co(III) is inhibited and ring oxidation occurs first. Table 7 summarizes data for CoPc species in nondonor solvents, while Table 8 collects data for CoPc species in the presence of donors (solvents or added ligands) (see also Figure 5) [49].

Cobalt(III) (in common with Fe(II)) has a much stronger propensity to form six-coordinate (low spin, t_{2g}^6) species than does Fe(III), or any other first row transition metal M(III) species; thus these observations are characteristic of CoPc. Note that the simple replacement of solvent DCB with solvent DMF shifts the Co(III)/Co(II) potential by some 600 mV although the comparison is not strictly valid since the oxidation occurs within Pc(-1) in DCB and Pc(-2) in DMF.

A range of six-coordinate $[X_2$Co(III)Pc(-2)$]^-$ anions has been studied, with variously substituted phthalocyanines, by both the Hanack [88] and Orihashi [71] groups (Table 8); where the same compound has been studied by both groups, there are some rather large discrepancies in oxidation potential due, perhaps, to rather different conditions of measurement.

The products from couples (II) through (V) (11) have all been proved by electronic and/or electron spin resonance spectroscopy and their identity is assured. Couples I and VI are assigned by inference but are likely to be correctly identified.

When Co(II)[TBuPc(-2)] is oxidized at 0.64 V in DCB solution the purple color of the radical cation, $[$Co(II)[TBuPc(-1)]$]^+$ is formed, but when pyridine is added the purple color changes immediately to green producing the characteristic spectroscopic features of Co(III) phthalocyanine, viz $[(Py)_2$Co(III)[TBuPc(-2)]$]^+$ [117]. This is an equilibrium process, and with relatively low concentrations of pyridine in DCB, $\approx 10^{-2}$ M, the process is thermally reversible, the cation radical being regenerated at high temperatures. In a similar experiment, the addition of chloride ion also converts the Co(II) cation radical, $[$Co(II)[TBuPc(-1)]$]^+$, to a chloro cobalt(III) species, probably $[Cl_2$Co(III)[TBuPc(-2)]$]^-$ [67].

More extreme chemistry occurs if hydroxide ion is added, as will be discussed below. Thus the redox couples of the Co(II)[TNPc(-2)] system do depend critically on whether coordinating anions are present. In studying cobalt phthalocyanine electrochemistry care must be taken to exclude extraneous donors (anions or otherwise) except where their presence is explicitly required.

An earlier study [95] explored the effect of varying the donicity of the solvent on the Co(III)/Co(II) potential. In this case pyridine yields the least positive redox potential and DMSO the most. The Co(III)/Co(II) potential is irreversible in DMA so that a datum for this solvent is not available. Nevertheless the shift to more negative potentials from DMSO to Py is the

Table 7 Electrochemical Data for Mononuclear and Polynuclear Cobalt Phthalocyanines (Non-donor Solvents) (Versus SCE).

Species	I	II	III	IV[a]	Ring Reduction[b]	Solv.	Ref.
Co(II)Pc			0.77			ClNap	74
Co(II)Pc		1.12	0.61			DCB	10,66,67
Co(II)Pc				-0.64	-1.34[b]	MeNp/150°	52
Co(II)[TBuPc]				-0.64		MeNp/150°	52
Co(II)[TBuPc]				-0.75		CH_2Cl_2	52
Co(II)[TcPc]	1.20					ACN	71
Co(II)[TNPc(-2)]	1.36	1.08	0.52	-0.42	-1.58	DCB	115,47[c],49
Co(II)[TNPc(-2)]		0.98	0.42	-0.52		DCB/$TBAPF_6$	115
EtMeO(5)[Co(II)TrNPc]$_2$	1.36	(d)	0.54	-0.44	-1.58	DCB	49
Me_2O(5)[Co(II)TrNPc]$_2$		0.55	-0.46	-1.58		DCB	47
Cat(4)[Co(II)TrNPc]$_2$	1.38	(d)	0.52	-0.44	-1.58	DCB	49
C(2)[Co(II)TrNPc]$_2$	1.36	0.82	0.52	-0.45	-1.58	DCB	49
O(1)[Co(II)TrNPc]$_2$	1.40	1.00	0.53	-0.44	-1.58	DCB	49
(-1)[Co(II)TrNPc]$_2$		1.08	0.66	-0.32	-1.20, -1.47[e]	DCB	119
[Co(II)TrNPc]	1.35	0.99	0.54	-0.43	-1.57	DCB	114
Co(II)[TsPc]		0.77	0.19			ClNap	95

[a] Processes: I, Co(III)Pc(0)/Co(III)Pc(-1); II. Co(III)Pc(-1)/Co(II)Pc(-1); III, Co(II)Pc(-1)/Co(II)Pc(-2); IV, Co(II)Pc(-2)/Co(I)Pc(-2). Some waves show some splitting due to the stabilisation of mixed valence species, but these were ill-defined. [b] Reduction processes of Co(I)Pc(-2). [c] The $[Co(II)[TNPc(-1)]^+$/Co(II)[TNPc(-2)] datum reported in [47] appears anomalous. [d] Not resolved. [e] Mixed valence splitting ?. Note that some data collected versus an internal ferrocene couple reference are corrected according to Table 1. Also see Tables 4 and 5.

reverse of the sequence observed for the oxidation of Fe(II)Pc because it is now the higher oxidation state, Co(III), that is being stabilized by the more strongly donor solvents, rather than the lower, as is the case for Fe(II)Pc. A series of substituted pyridines [95] show potentials for both Co(III)/Co(II) and Co(II)/Co(I) shifting to more negative potentials with increasing base strength (pK_a). It is not uncommon for the Co(III)/Co(II) couple to be irreversible because of the large change in spin state, and the likely change in Co-L bond length during this redox process.

Both Fe(II)Pc and Co(II)Pc show M(II)/M(I) potentials that linearly follow the Gutmann donor number with the same sign of the slope since now it is both Co(II) and Fe(II) that are stabilized by the stronger donor solvent [95]. However, nondonor solvents do not fall on the same line [95] as donor solvents, that is, they do not behave simply as very weakly donor solvents. This is a consequence of the fact that in donor solvents the Co(II)Pc will be solvated (five or six-coordinate) while in nondonor solvents it will be unsolvated (four coordinate, different spin state).

A more detailed study of the electrochemistry of Co(II)Pc in DMF reveals more subtle features [49]. A solution of Co(II)[TNPc(-2)] in DMF/ClO_4^- contains several species in equilibrium, viz:

$$\underset{A}{(DMF)Co(II)[TNPc(-2)]} \longleftrightarrow \underset{B}{(DMF)_2Co(II)[TNPc(-2)]} \longleftrightarrow \underset{C}{[DMF(ClO_4)Co(II)[TNPc(-2)]]^-} \qquad (13)$$

which will have three different Co(III)/Co(II) oxidation potentials.

However, if the equilibria are facile only the oxidation of the most easily oxidized species, which should be C, should be observable. It appears however, that both the $[Co(III)[TNPc(-2)]]^+/Co(II)[TNPc(-2)]$ and $[Co(III)[TNPc(-1)]]^{2+}/[Co(III)[TNPc(-2)]]^+$ redox couples are coupled to other equilibria that are relatively slow on the voltammetric time scale and can be probed by variable-scan-rate studies.

The $[Co(III)Pc(-2)]^+/Co(II)Pc(-2)$ process (labelled (IIIa), (IIIc) in Figure 9) has a scan-rate-dependent i_c/i_a ratio approaching unity at higher scan rates and higher perchlorate ion concentrations. This is interpreted in terms of the following processes [49]

$$\underset{C}{[(DMF)ClO_4Co(II)[TNPc(-2)]]^-} \Longleftrightarrow \underset{D}{(DMF)ClO_4Co(III)[TNPc(-2)]} + e^- \qquad (14)$$

$$\underset{\text{D}}{\text{(DMF)ClO}_4\text{Co(III)[TNPc(-2)]}} \longleftrightarrow \underset{\text{E}}{[\text{DMFCo(III)[TNPc(-2)]}]^{+}} + \text{ClO}_4^{-} \qquad (15)$$

Equilibrium (15), which is suppressed in excess perchlorate ion, produces five coordinate species E, which is expected to be much easier to reduce than species D, that is, the Co(III)/Co(II) couple of E will occur at more positive potentials. Thus C is oxidized to D, at (IIIa) (Figure 9). D rearranges to E, at least to a small degree, and E is spontaneously reduced on the positive side of (IIIa) since its reduction potential lies positive of (IIIa). Equilibrium (15) is driven to the right by this reduction process and therefore decreases the intensity of the current at (IIIc). At higher scan rates, there is less time for the rearrangement to occur and more reversible behavior obtains.

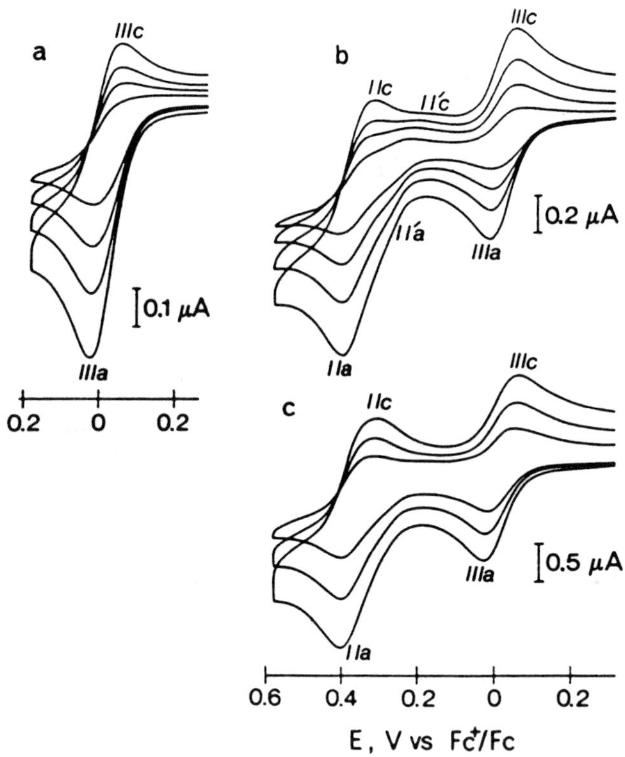

Figure 9 Cyclic Voltammetry of Co(II)[TNPc(-2)] (1×10^{-4} M) in DMF/(0.3 M TBAP), at varying scan rates and switching potentials. a) Co(III)/Co(II) couple at 2, 5, 10 and 20 mV/s; b) Pc(-1)/Pc(-2) and Co(III)/Co(II) couples at 2, 5, 10 and 20 mV/s; c) as (b), at 20, 50 and 100 mV/s. Reproduced with permission from Ref. [49].

Table 8 Electrochemical Data for Mononuclear and Polynuclear Cobalt Phthalocyanines[a] (Donor solvents) (Versus SCE)

Species	I	II	III	IV[a]	Reduct. Proc.[b]	Solvent	Ref.
Co(II)Pc	1.15	0.21	-0.55	-0.91	-1.44	Py	61,107
Co(II)Pc		0.80	-0.20			DMA	60
Co(II)Pc			-0.37	-1.40	-1.80, -2.08, -2.46	DMF	4
Co(II)[TEtPc]	0.94	0.49	-0.32	-1.45		DMF	131[c]
Co(II)[TBuPc]			-0.75			DMSO	52
Co(II)[TcPc]			-0.28	-0.42?		H_2SO_4	116
Co(II)[TcPc]	1.20[i]					ACN	120
Co(II)[TNPc(-2)]	0.78	0.38	-0.45	-1.59		DMF	49
Co(II)[TNPc(-2)]	1.14	0.43	-0.63			DCB/Cl	115
Co(II)[Cl_{16}Pc]		1.09	(d)	-1.21	-1.69	DMF	63
Co(II)[TsPc]		0.04	-0.71			Py	95
Co(II)[TsPc]		0.01	-0.725			4-EtPy	95
Co(II)[TsPc]			-0.85qr	-1.35qr		H_2O	64
Co(II)[TsPc]		0.46	-0.60			DMSO	95
Co(II)[TsPc]		0.43irr	-0.50			DMF	95
Co(II)[TsPc]		0.16	-0.64			3-ClPy	95
Co(II)[TAPc]	0.7,0.5[e]	0.17	-0.52	-1.66		DMSO	70
Co(II)[TAPc]	0.60[e]	0.16	-0.52	-1.74		DMF	132[c]
Co(II)[TAPc]	0.68					ACN	120
Co(II)[TNO_2Pc]	1.26[i]					ACN	120
Co(II)[TMxPc]	0.69[i]					ACN	120
Co(II)[OCNPc]	1.76[i]					ACN	120
Co(II)[OCPc]	1.61[i]					ACN	128

Species	I	II	III	IV[a]	Reduct. Proc.[b]	Solvent	Ref.
Co(II)[OMxPc]	0.74[i]					ACN	120
Co(II)[OMPc]	0.79[i]					ACN	120
Na[(CN)$_2$Co(III)[TAPc]]		0.20[e]	-0.55[f]	-1.15		DMF	132[c]
K[(CN)$_2$Co(III)[TAPc]]	0.55					ACN	71
Na[(CN)$_2$Co(III)[TEtPc]]	0.87		-0.43	-1.48		DMF	131[c]
Na[(CN)$_2$Co(III)Pc]	0.62		-0.94[f]	-1.52		Acetone	88
K[(CN)$_2$Co(III)Pc]	0.88		-0.91[f]	-1.54		Acetone	88,121
K[(CN)$_2$Co(III)Pc]	1.06					ACN	71
Bu_4N[(SCN)$_2$Co(III)Pc]	0.88	-0.34	-1.02	-1.52		Acetone	88
K[(SCN)$_2$Co(III)Pc]	0.94					ACN	71
K[(SCN)$_2$Co(III)Pc]	0.84	-0.32	-0.95	-1.49		Acetone	88
Na[(CN)$_2$Co(III)[OMPc]]	0.92		-0.83[f]	-1.46		Acetone	88
K[(CN)$_2$Co(III)[OMPc]]	0.59					ACN	71
Na[(CN)$_2$Co(III)[OMxPc]]	0.61					ACN	71
Na[(CN)$_2$Co(III)[TBuPc]]	0.77		-0.93[f]	-1.50		Acetone	88
Na[(CN)$_2$Co(III)[NO_2Pc]]	1.30[g]		-0.48[f]	-1.13		Acetone	88
[(CN)Co(III)[TEtPc]]$_n$	0.91		-0.40	-1.49		DMF	131[c]
[(CN)Co(III)[TAPc]]$_n$		0.16[h]	-0.60[f]	-1.82		DMF	132[c]

[a] Processes: I, Co(III)Pc(-1)/Co(III)Pc(-2); II, Co(III)Pc(-2)/Co(II)Pc(-2); III, Co(II)Pc(-2)/Co(I)Pc(-2). Polynuclear species shown in this Table do not display mixed valence behavior. [b] Further reduction processes of Co(I)Pc(-2). [c] Authors of [131,132] correct their ferrocene internal data to SCE using 0.49 V vs SCE. [d] Not observed, probably because of kinetic sluggishness. [e] Irreversible. [f] 2-Electron combined Co(III)-- > Co(II)-- > Co(I). [g] Close to solvent limit. [h] Polymeric species oxidation. [i] May be [Co(II)Pc(-1)]$^+$/Co(II)Pc(-2) with ACN acting as a non-donor solvent. Also see Tables 4,5.

Note that evidence for the kinetic lability of six-coordinate Co(III) macrocycles has been presented [118]. Now, considering the $[Co(III)Pc(-1)]^{2+}/[Co(III)Pc(-2)]^{+}$ process (labeled (II) in Figure 9), two different pairs of couples, II and II', are observed. With increasing scan rate the more positive couple, IIc grows at the expense of the less positive wave IIc', and vice versa. Moreover increasing perchlorate ion concentration favors wave II'. Therefore wave IIc' must be associated with additional bound perchlorate ion and is then reasonably associated with $(ClO_4)_2Co(III)Pc(-1)$

These processes are understood in terms of the following equilibria:

Process II

$$\underset{D}{DMF(ClO_4)Co(III)Pc(-2)} <---> \underset{F}{[DMF(ClO_4)Co(III)Pc(-1)]^{+}} + e^{-} \qquad (16)$$

$$\underset{F}{[DMF(ClO_4)Co(III)Pc(-1)]^{+}} + ClO_4^{-} <---> \underset{H}{(ClO_4)_2Co(III)Pc(-1)} \qquad (17)$$

Process II'

$$\underset{G}{[(ClO_4)_2Co(III)Pc(-2)]^{-}} <---> \underset{H}{(ClO_4)_2Co(III)Pc(-1)} + e^{-} \qquad (18)$$

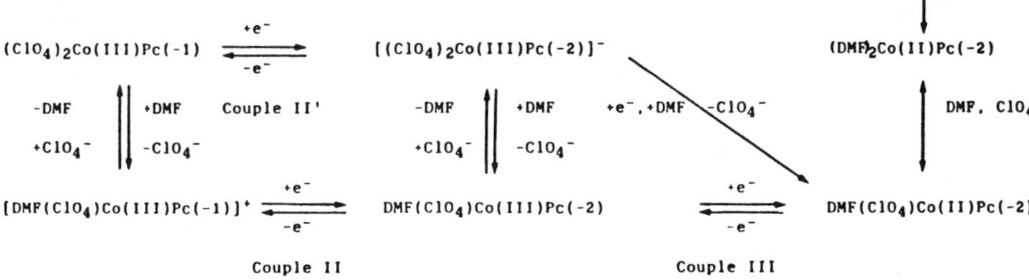

$[Co(III)Pc(0)]^{3+}$ ⇌ (+e⁻ / −e⁻, Couple I) $[Co(III)Pc(-1)]^{2+}$ ⇌ (+e⁻ / −e⁻, Couple II) $[Co(II)Pc(-1)]^{+}$ ⇌ (+e⁻ / −e⁻, Couple III) (a) $Co(II)Pc(-2)$

(a) $Co(II)Pc(-2)$ ⇅ DMF (b) DMF $Co(II)Pc(-2)$ ⇅ DMF $(DMF)_2Co(II)Pc(-2)$ ⇅ DMF, ClO_4^- $DMF(ClO_4)Co(II)Pc(-2)^{-}$

$(ClO_4)_2Co(III)Pc(-1)$ ⇌ (+e⁻ / −e⁻, Couple II') $[(ClO_4)_2Co(III)Pc(-2)]^{-}$ → (+e⁻, +DMF, −ClO_4^-) $DMF(ClO_4)Co(II)Pc(-2)^{-}$

$(ClO_4)_2Co(III)Pc(-1)$ ⇅ (−DMF, +ClO_4^- / +DMF, −ClO_4^-) $[DMF(ClO_4)Co(III)Pc(-1)]^{+}$

$[(ClO_4)_2Co(III)Pc(-2)]^{-}$ ⇅ (−DMF, +ClO_4^- / +DMF, −ClO_4^-) $DMF(ClO_4)Co(III)Pc(-2)$

$[DMF(ClO_4)Co(III)Pc(-1)]^{+}$ ⇌ (+e⁻ / −e⁻, Couple II) $DMF(ClO_4)Co(III)Pc(-2)$ ⇌ (+e⁻ / −e⁻, Couple III) $DMF(ClO_4)Co(II)Pc(-2)^{-}$

Scheme I

At slow scan rates, scanning positively from couple III, we initially observe oxidation D --> F at IIa. This rearranges via (17) to give mainly H which is reduced at II'c. At higher scan rates, there is insufficient time for equilibrium (17) to occur and reduction of species F is observed at IIc. Wave II'a is never obvious because species G never has the opportunity to build up (see Scheme I) [49].

In summary, couples (14), (16) and (18) are observed, in DMF, at +0.38V, +0.78 and +0.70 V, versus SCE, the data having been corrected, using Table 1, from the original experimental data internally referenced to the ferricenium/ferrocene couple.

A spectroelectrochemical study of Co(II)[TsPc(-2)] [101] as a function of pH reveals (Table 7) the varying degrees of aggregation of this species as a function of oxidation state. Unlike Fe(II)TsPc, reduced M(I)TsPc species are nonaggregated, but higher oxidation state Co(III) species are less aggregated than Fe(II) species. This is attributed to the dominant formation of six-coordinate Co(III) species where the axial groups inhibit aggregation.

When hydroxide ion is added to a DMF (or DCB) solution of Co(II)[TNPc(-2)], under nitrogen, the solution is converted, within the time of mixing, into a 1:1 mixture of $[Co(I)[TNPc(-2)]]^-$ and $[(OH)_2Co(III)[TNPc(-2)]]^-$. Thus Co(II)[TNPc(-2)] cannot exist in a DMF (or DCB) solution containing hydroxide ion. Naturally, if air is introduced, there is total conversion to the $[(OH)_2Co(III)[TNPc(-2)]]^-$ species [11]. Some interesting electrochemical observations arise in this system whose voltammetry is shown in Figure 10 (Left).

If a disproportionated solution of Co[TNPc(-2)] in DMF/OH^- is oxidized at a potential ≈ 200 mV positive of couple A (Figure 10 (Left)) then oxidation totally to $[(OH)_2Co(III)[TNPc(-2)]]^-$ occurs. If the solution is polarized ≈ 200 mV negative of couple A, then the solution is totally converted to $[Co(I)[TNPc(-2)]]^-$; in neither case is any intermediate Co(II)[TNPc(-2)] observed.

Couple A has the electrochemical characteristics of a one-electron process having the same current intensity as observed in the absence of hydroxide ion [see Figure 10 (Left) a,b,c] yet clearly a two-electron process occurs when the electrode is polarized at this potential.

To understand this observation, a series of experiments was undertaken. If DMF solutions of Co(II)[TNPc(-2)] are treated with hydroxide ion and left under nitrogen for an hour or so, all the $[(OH)_2Co(III)[TNPc(-2)]]^-$ is reduced to $[Co(I)[TNPc(-2)]]^-$. Such solutions of pure $[Co(I)[TNPc(-2)]]^-$ do not display couple B because Co(II)[TNPc(-2)] is never formed in these solutions.

Initially, prior to addition of hydroxide ion, we may consider two processes, A,B [Figure 10 (Left) a)] that are associated with Co(II)/Co(I) and Co(III)/Co(II) respectively, both showing one-electron reversible behavior.

Voltammetry was explored in concert with the electronic spectroscopy (Figure 10 (Right)), using fresh deaerated solutions containing, necessarily, a

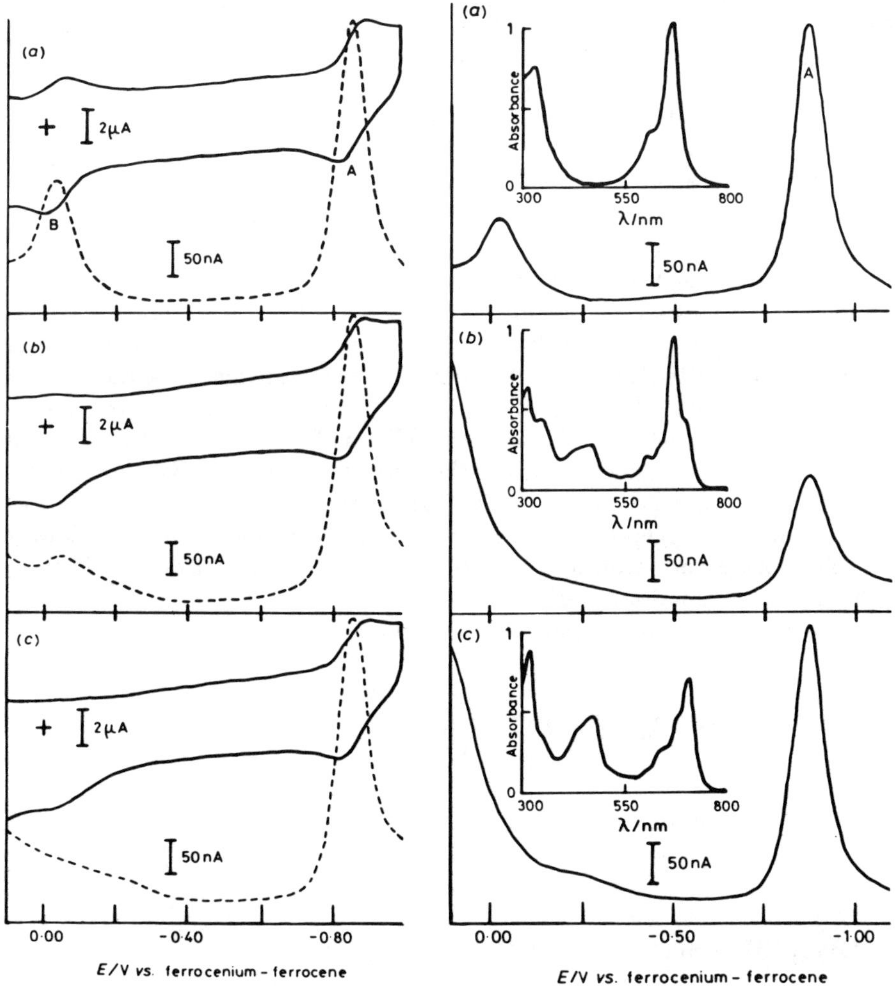

Figure 10 (Left): Cyclic voltammetry (solid line) and differential pulse voltammetry (hatched line) for Co(II)[TNPc(-2)] (9.69×10^{-5} mol dm^{-3}) at a glassy carbon electrode in DMF/LiOH under nitrogen. Concentration of LiOH; a) 0, b) 1.4×10^{-4}, c) 2.8×10^{-4} mol dm^{-3}. B, Right: Electronic spectra and differential pulse voltammetry of Co(II)[TNPc(-2)] (1.01×10^{-4} mol dm^{-3}) under oxygen. The hydroxide concentrations (NBu_4OH-MeOH) are a) 0, b) 5.0×10^{-4} (recorded 30 min after mixing), and c) 5.0×10^{-4} mol dm^{-3} (recorded 2.5 h later). Reproduced with permission from Ref. [11].

mixture of $[(OH)_2Co(III)[TNPc(-2)]]^-$ and $[Co(I)[TNPc(-2)]]^-$. One then did not observe (Figure 10 (Right)) couple B [Co(III)/Co(II)] while the current intensity for couple A [Co(II)/Co(I)] was proportional to the content of $[Co(I)[TNPc(-2)]]^-$. It was possible to conclude that a pure solution of $[(OH)_2Co(III)[TNPc(-2)]]^-$ would not exhibit couple A. Since it is possible to obtain a pure solution of $[(OH)_2Co(III)[TNPc(-2)]]^-$ by oxygenating the $[Co(I)TNPc(-2)]^-$ species in DMF/OH^-, one might be curious as to why such a solution was not directly studied. In fact, couple A lies at too negative a potential to be observed in an oxygenated solution, and if the solution is first deoxygenated, fairly rapid reduction to $[Co(I)[TNPc(-2)]]^-$ takes place. These reactions were interpreted in terms of a model where neither Co(II)[TNPc(-2)] nor $[Co(I)[TNPc(-2)]]^-$ react with hydroxide ion but where there is a very strong stabilization of Co(III)[TNPc(-2)] by binding to hydroxide ion. Thus the potential of couple A should not be affected by hydroxide ion concentration as observed.

The Co(III)/Co(II) couple, B, upon addition of hydroxide ion to the system, became irreversible, losing its cathodic component (Figure 10 (Right) b)) before it essentially disappeared (Figure 10 (Right) c)). This process corresponds to equilibrium (14), species C,D. In the presence of hydroxide ion, species D forms the dihydroxide and is no longer reducible in the region of couple B so the cathodic component disappears.

The following two processes are now relevant:

$$\underset{}{2\,OH^-} + \underset{C}{[(DMF)ClO_4Co(II)[TNPc(-2)]]^-} \overset{E_{1/2}}{<=====>} [(OH)_2Co(III)[TNPc(-2)]]^- + DMF + ClO_4^- + e^- \quad (19)$$

$$[(OH)_2Co(III)[TNPc(-2)]]^- + DMF + ClO_4^- \overset{K}{<--->} 2\,OH^- + \underset{D}{(DMF)ClO_4Co(III)[TNPc(-2)]} \quad (20)$$

and using the electrochemical data to estimate K, the potential for process (19) can be estimated from:

$$E_{1/2}(19) = E_{1/2}(16) - RT/nF[Ln(K) - 2\,Ln(OH)]$$

from which [11]:

$$E_{1/2}(19) - E_{1/2}(16) = -0.059[Log(K) - 2\,Log(OH)] \quad (21)$$

(where it is assumed that the diffusion coefficients of the initial and final redox products are the same, and where the negative sign to the immediate right of the equality differs from [11] because the redox processes have been defined in the opposite sense here).

To explain the observed behavior, the value of $E_{1/2}(19)$ must place this couple negative of couple A. Thus the right-hand side of Eq. (21) must be at least 0.9 V. Given also that even micromolar concentrations of $[OH^-]$ caused a loss of couple B and shifted the Co(III)/Co(II) couple negative of couple A, it was shown that the maximum possible value of K(20) was $\approx 10^{-23}$ [11], a not unreasonable value given the evident strong HO-Co(III) binding.

One more observation needed explanation, namely, why does couple A look like a one-electron process, yet, in reality, two-electrons are consumed ? Consider using a bulk $[Co(I)[TNPc(-2)]]^-$ solution and running a cyclic voltammogram approaching couple A from negative thereof. At couple A, $[Co(I)[TNPc(-2)]]^-$ undergoes a one-electron oxidation to Co(II)[TNPc(-2)]. Hydroxide ion is not bound to the cobalt center; so this species does not directly oxidize to Co(III)Pc, but in a following reaction it does disproportionate to form Co(I) + Co(III). The potential for this process will be the same as for a hydroxide-free environment so long as this disproportionation process is slow on the voltammetric time scale. Indeed disproportionation is slow (minutes) at low OH:CoPc ratios ($\approx$ 5 - 20:1). It is also possible that disproportionation is inhibited at the electrode surface for mechanistic reasons, and does not occur until the Co(II)[TNPc(-2)] diffuses away [11]. Note that $[(CN)_2Co(III)Pc(-2)]^-$ species may also participate in a two-electron reduction to Co(I)Pc [88].

A series of binuclear and tetranuclear cobalt phthalocyanines have been studied [11, 13, 114, 119, 123]. Those whose electrochemical data are reported in Table 7, show no electrochemical evidence for any significant electronic interaction between the two (or more) phthalocyanine units in the molecule exhibiting electrochemical potentials at values closely similar to those of the mononuclear Co[TNPc(-2)] under similar conditions. They do show some spectroscopic evidence for coupling, but it is evident that such coupling is not large enough to effect the electrochemistry. These species also disproportionate in the presence of hydroxide ion, in an intramolecular sense, forming $[Co(I)]_2$ and $[Co(III)]_2$ but not [Co(I)Co(III)] [11]. Another group of binuclear CoPc species, discussed in Section E, viii do exhibit mixed-valence behavior.

Electron withdrawing substituents on the phthalocyanine ring obviously render the ring more difficult to oxidize and easier to reduce, and vice versa for electron-donating substituents (see Section E, ix). Some data are available [4,107] for the further reduction of $[Co(I)Pc(-2)]^-$ as far down as the Pc(-5) species (Tables 7,8), but this region has not been definitively explored.

Some solution data have been reported for Rh(III) phthalocyanines [108, 109]. Oxidation near 0.9 V yields a Rh(III) radical cation species whose photochemistry has been explored [108].

Chlororhodium phthalocyanine, in DMF or DCB, shows an irreversible reduction wave near -0.7 V that has a cathodic but no anodic component [109]. At this potential, there is reduction to monomeric Rh(II)Pc but this rapidly dimerizes at room temperature in solution yielding a dimeric product that is not re-oxidized until about -0.15 V (in DCB/TBAP,[109]). At about - 60° C the dimerization reaction is inhibited and a reversible XRh(III)[TNPc(-2)]/ Rh(II)[TNPc(-2)] couple is observed. Further reduction of the bulk solution exhibits a reversible wave at -1.47 V that probably forms a monomeric $[Rh(I)Pc(-2)]^-$ species although a $[Rh(II)Pc(-3)]^-$ species is also possible. Final details of the solution electrochemistry of the dimeric $[Rh(II)Pc]_2$ species await clarification [109].

No solution data appear available for iridium phthalocyanines.

vi. Silver and Gold Phthalocyanines

Copper phthalocyanine was discussed in Section D, ii, and gold phthalocyanine, while known, has not been studied electrochemically.

Silver tetraneopentoxyphthalocyanine has been the subject of intensive electrochemical study [48]. In common with many other metallophthalocyanines, Ag(II)[TNPc(-2)] is quite strongly aggregated in solution and the aggregation-disaggregation equilibrium is slow on the voltammetry time scale. Conventional CV or DPV yields a broad wave near 0.6 V attributable, on the basis of spectroelectrochemistry, to oxidation to $[Ag(III)[TNPc(-2)]]^+$. A Nernstian analysis of the data shows that the true half-wave potential for oxidation lies at 0.71 V versus SCE, in DCB. Further oxidation yields $[Ag(III)[TNPc(-1)]]^{2+}$ (reversible process) and $[Ag(III)[TNPc(0)]]^{3+}$ (irreversible process). The possibility that a Ag(IV)[TNPc(-2)] species is formed cannot be ruled out. The irreversibility of the third oxidation process may reflect solvent oxidation.

There are two reduction processes which are believed to form $[Ag(I)[TNPc(-2)]]^-$ and, probably, $[Ag(I)[TNPc(-3)]]^{2-}$. However, in parallel with the corresponding chemistry of silver(II) porphyrins [133, 134] the silver(I) species are unstable and hydrolyze to form the metal free species at a rate comparable to the electrochemistry time scale. Thus, upon reduction of Ag(II)[TNPc(-2)] one observes a pair of waves due to the successive reduction of Ag(II)[TNPc(-2)] and a pair of waves due to the successive reduction of the resulting H_2TNPc(-2). The relative intensities of these pairs of waves are affected both by scan rate and temperature, with higher scan rates and lower temperatures favoring observation of the silver reduction couples.

A more detailed consideration of the relative currents of the reduced species suggested that there was an intermediate between the reduction to $[Ag(I)[TNPc(-2)]]^-$ and the formation of reduced H_2[TNPc(-2)]. This was postulated to be a sitting-atop version of the $[Ag(I)[TNPc(-2)]]^-$ species being formed prior to hydrolysis, that is, the large silver(I) sits atop the phthalocyanine ring. Further evidence arises from the fact that even at lower temperatures and higher scan rates, where hydrolysis is largely suppressed, the first reduction couple is irreversible in always showing a larger cathodic than anodic current. A "simple" Ag(II)/Ag(I) couple, if the Ag(I) remained in the ring, would be expected to be reversible.

vii. Polynuclear Phthalocyanines

In addition to the polynuclear metallophthalocyanines and bridged species such as PcM-X-MPc (X = O, N etc.), mentioned previosuly, other bridged species such as PcM-LL-MPc and $(\text{-LLMPc-LL-Mpc-})_n$ [125] (where LL is a bridging bifunctional ligand such as pyrazine) are known. However, solution data are not available for the last mentioned species.

Some species, such as the binuclear $EtMeO(6)[MTrNPc(-2)]_2$ [65] and the the tetranuclear spiro linked $[MTrNPc(-2)]_4$ [65, 114] show no mixed-valence behavior with cobalt(II) (Table 7) but do with Zn(II) (Table 9). Such differences may arise through the presence or absence of axial ligands, respectively inhibiting or facilitating the close approach of phthalocyanine rings.

viii. Mixed Valence Behavior

Cobalt phthalocyanine complexes of binuclear or tetranuclear phthalocyanines having flexible bridging units, do not exhibit any measureable electrochemical interaction between cobalt centers, that is, mixed-valence behavior is not observed. Rigid systems such as the anthracene and naphthalene bridged binuclear species [126] and the so-called (-1)bridge species [119] do, however, exhibit additional redox waves associated with mixed-valence species (Figure 11). These may be of the metal-centered type, such as $[Co(II)Pc(-2)]_2/[Co(I)Pc(-2).Co(II)Pc(-2)]^-$ or of the ring-centered type, such as $[Co(II)Pc(-1)]_2^{2+}/[Co(II)Pc(-1).Co(II)Pc(-2)]^+$. The splitting of a given redox process due to formation of a stable mixed-valence intermediate, is a measure of the equilibrium (comproportionation) constant, K_c, for a reaction such as:-$[M(II)Pc(-1)]_2$ + $[M(II)Pc(-2)]_2$ ---> 2 $[M(II)Pc(-1).M(II)Pc(-2)]$ (22) where the mixed-valence splitting, ΔE is related to K_c, via:

$$\Delta E = (RT/nF)Ln(K_c) \qquad (23)$$

The values of K_c so obtained for a series of mixed-valence phthalocyanine complexes of cobalt, zinc, aluminum, iron, and silicon are collected in Table 10.

They are seen to range from 20 to 3 x 10^8 (-8 to -49 kJ/mol), thereby showing a wide range of stability. The least strongly coupled systems include some zinc complexes of binuclear phthalocyanines with flexible bridging links where the cobalt analogues do not, in fact, exhibit mixed-valence behavior at all. The rigid bridged systems such as the anthracene, naphthalene and (-1)bridge species exhibit a range of mixed-valence complexes for both zinc and cobalt (Table 10), although in this last case, (-1)bridge, for cobalt, the waves were not well resolved and K_c values could not be accurately defined.

Generally speaking, mixed-valence species, for a given phthalocyanine, involving the metal ion, such as Co(II)/Co(I) and Co(III)/Co(II) were more stable than those involving the ring Pc(-2)/Pc(-1), and Co(II)/Co(I) species were more stable than Co(III)/Co(II) species.

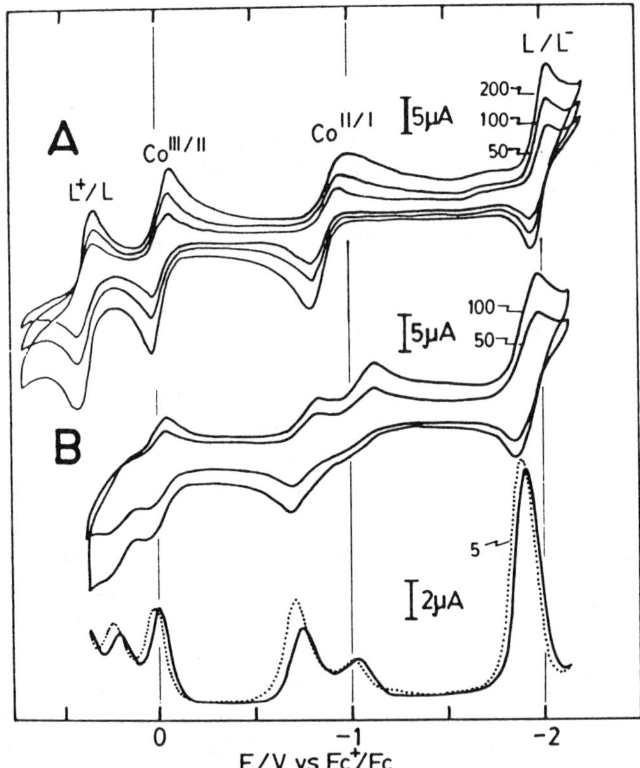

Figure 11 Cyclic and differential pulse voltammograms of (A) Co(II)[TNPc(-2)] and (B) Ant[CoTrNPc]$_2$ in DMF at a Pt electrode. Numbers indicated are scan rates. In the differential pulse studies, the solid and dotted lines indicate negative going and positive going scans respectively. Reproduced with permission from Ref. [126].

However, the strongest mixed-valence interactions were seen in ring mixed-valence RSiPc-O-PcSiR species; these exhibit both mixed-valence ring oxidation and mixed-valence ring reduction. The large K_c values for these silicon species probably reflect the shorter Pc..Pc contacts (≈. 3.3Å) than in the bridged species (≈ 4.3 Å). Oligomeric SiPc species have been prepared (Table 3) in that 2, 3 or 4 (or more) OSiPc units are strung together. The mixed-valence species of successively oligomerized species show declining stability of a given mixed-valence species with increasing chain length (Table 3), at least according to Eq.(23). However, it is probable that, due to interaction processes along the chain, this simple equation is inappropriate.

Further, the greater coupling for the Co(II)/Co(I) systems versus the Co(III)/Co(II) likely reflects interaction between the d_{z2} orbitals of each cobalt atom, directed along the inter-ring axis, there being an odd d_{z2} electron in the Co(I) species. In summary, mixed-valence behavior has been observed for the following binuclear MPc systems:

Pc(-2).Pc(-3) is observed in strongly coupled silicon oligomers, but has not been unequivocally observed with the other bridged binuclear species. The presence of an extra π^* electron repels the π-electron density in the other ring and therefore inhibits formation of these species unless they are constrained to lie close together.

Pc(-1).Pc(-2) observed with cobalt and with main group ions such as aluminum, zinc and silicon. In the case of cobalt, it is likely that both Co(II) and Co(III) mixed-valence species can be obtained by controlling the solvent (that is, availability of axial ligands). Strong coupling originates in the hole in the π orbital of one ring being capable of delocalization over the other ring.

Co(II).Co(I) is observed in inflexible binuclear species and has electronic spectroscopic characteristics, and K_c values, consistent with strong coupling between the cobalt centers (along the d_{z2} bond axis).

M(III).M(II) is observed in inflexible binuclear species linked through the phthalocyanine rings, or in PcM-O-MPc systems linked through the metal atom, especially for iron.

M(IV).M(III) is observed in PcFe-X-FePc bridged systems, X = O, N.

Some data for mixed-valence phthalocyanines containing simple bridges, such as cyanide, oxide or nitride between MPc centers [111, 127, 128] are reported in Table 10. The K_c values, for M(III)/M(II) systems are comparable with the phthalocyanine bridged systems.

The stability of the "PcFe(IV)-N-Fe(III)Pc" is extraordinary with an electrochemical splitting of 0.83 V corresponding with $K_c = 1.2 \times 10^{14}$. The "TPPFe(IV)-N-Fe(III)TPP" analogue is similarly stable (splitting 0.79 V in pyridine [111] or 1.36 V (!) in dichloroethane [129]), likely indicating that these species should best be regarded as class III fully delocalized complexes.

Table 9 Polynuclear Phthalocyanines showing Mixed Valence Behavior

Species	$E_{1/2}$/V DCB	$E_{1/2}$/V DMF
EtMeO(5)[Zn(II)TrNPc]$_2$ [65]		
[Zn(II)Pc(0)]$_2$/[Zn(II)Pc(-1)]$_2$		1.00
[Zn(II)Pc(-1)]$_2$/[Zn(II)Pc(-1).Zn(II)Pc(-2)]	0.49	0.58
[Zn(II)Pc(-1).Zn(II)Pc(-2)]/[Zn(II)Pc(-2)]$_2$	0.41	0.43
[Zn(II)Pc(-2)]$_2$/[Zn(II)Pc(-3)]$_2$	-1.19	-1.04
[Zn(II)Pc(-3)]$_2$/[Zn(II)Pc(-4)]$_2$	-1.63	-1.47
Ant[Zn(II)TrNPc]$_2$ [126]		
[Zn(II)Pc(-1)]$_2$/[Zn(II)Pc(-1).Zn(II)Pc(-2)]	0.57	0.56
[Zn(II)Pc(-1).Zn(II)Pc(-2)]/[Zn(II)Pc(-2)]$_2$	0.36	0.35
[Zn(II)Pc(-2)]$_2$/[Zn(II)Pc(-2).Zn(II)Pc(-3)]	-1.05[a]	
[Zn(II)Pc(-2)]$_2$/[Zn(II)Pc(-3)]$_2$		-1.01
[Zn(II)Pc(-2).Zn(II)Pc(-3)]/[Zn(II)Pc(-3)]$_2$	-1.31[a]	
[Zn(II)Pc(-3)]$_2$/[Zn(II)Pc(-4)]$_2$		-1.53
Nap[Zn(II)TrNPc]$_2$ [123]		
[Zn(II)Pc(0)]$_2$/[Zn(II)Pc(-1)]$_2$	1.22[a]	1.11
[Zn(II)Pc(-1)]$_2$/[Zn(II)Pc(-1).Zn(II)Pc(-2)]	0.56	0.59
[Zn(II)Pc(-1).Zn(II)Pc(-2)]/[Zn(II)Pc(-2)]$_2$	0.35	0.41
[Zn(II)Pc(-2)]$_2$/[Zn(II)Pc(-3)]$_2$	-1.18 (-1.34[b])	-1.07
[Zn(II)Pc(-3)]$_2$/[Zn(II)Pc(-4)]$_2$	-1.53	-1.45
[Zn(II)TrNPc]$_4$ [65]		
[Zn(II)Pc(-1)]$_2$/[Zn(II)Pc(-1).Zn(II)Pc(-2)]	0.53	0.58
[Zn(II)Pc(-1).Zn(II)Pc(-2)]/[Zn(II)Pc(-2)]$_2$	0.42	0.43
[Zn(II)Pc(-2)]$_2$/[Zn(II)Pc(-3)]$_2$	-1.15	-1.04
[Zn(II)Pc(-3)]$_2$/[Zn(II)Pc(-4)]$_2$	-1.64	-1.46
Nap[Co(II)TrNPc]$_2$ [123]		
[Co(III)Pc(-1).Co(III)Pc(-2)]/[Co(III)Pc(-2)]$_2$	----	0.81[d]
[Co(III)Pc(-2)]$_2$/[Co(III)Pc(-2).Co(II)Pc(-2)]	----	0.69
[Co(III)Pc(-1)]$_2$/[Co(II)Pc(-1)]$_2$ (?)	1.02	----

Species	$E_{1/2}$/V DCB	$E_{1/2}$/V DMF
[Co(II)Pc(-1)]$_2$/[Co(II)Pc(-1).Co(II)Pc(-2)]	0.63	----
[Co(II)Pc(-1).Co(II)Pc(-2)]/[Co(II)Pc(-2)]$_2$	0.49	----
[Co(III)Pc(-2).Co(II)Pc(-2)]/[Co(II)Pc(-2)]$_2$	----	0.45
[Co(II)Pc(-2)]$_2$/[Co(II)Pc(-2).Co(I)Pc(-2)]	-0.41	-0.47
[Co(II)Pc(-2).Co(I)Pc(-2)]/[Co(I)Pc(-2)]$_2$	-0.80[c,d]	-0.69[c,d]
[Co(I)Pc(-2)]$_2$/[Co(I)Pc(-3)]$_2$	-1.59	-1.57
<u>Ant[CoTrNPc]$_2$</u> [126]		
[Co(III)Pc(-2)]$_2$/[Co(III)Pc(-2).Co(II)Pc(-2)]	-----	0.62
[Co(II)Pc(-1)]$_2$/[Co(II)Pc(-1).Co(II)Pc(-2)]	0.55	-----
[Co(III)Pc(-2).Co(II)Pc(-2)]/[Co(II)Pc(-2)]$_2$	-----	0.41
[Co(II)Pc(-1).Co(II)Pc(-2)]/[Co(II)Pc(-2)]$_2$	0.38	-----
[Co(II)Pc(-2)]$_2$/[Co(II)Pc(-2).Co(I)Pc(-2)]	-0.29	-0.20
	-0.42	-----
[Co(I)Pc(-2).Co(II)Pc(-2)]/[Co(I)Pc(-2)]$_2$	-0.77[c,d]	
	-0.60[c,d]	
[Co(I)Pc(-2)]$_2$/[Co(I)Pc(-3)]$_2$	-1.58	-1.48
<u>Nap[Cu(II)TrNPc]$_2$</u> [123]		
[Cu(II)Pc(0)]$_2$/[Cu(II)Pc(-1)]$_2$	1.23[c]	
[Cu(II)Pc(-1)]$_2$/[Cu(II)Pc(-1).Cu(II)Pc(-2)]	0.76	0.81[e]
[Cu(II)Pc(-1).Cu(II)Pc(-2)]/[Cu(II)Pc(-2)]$_2$	0.56	0.51
[Cu(II)Pc(-2)]$_2$/[Cu(II)Pc(-3)]$_2$	-1.09	-0.88
[Cu(II)Pc(-3)]$_2$/[Cu(II)Pc(-4)]$_2$	-1.40	-1.32

$E_{1/2}$ values were measured by cyclic voltammetry at 200, 100, 50 and 20 mV/s. Average data $E_{1/2} = (E_{pa} + E_{pc})/2$ are reported. Hatched lines indicate that the specific redox couple is not permissible in that solvent. [a] Assignment unproven. [b] weak or broad couples of uncertain potential. [c] irreversible. [d] Data obtained from differential pulse voltammetry. [e] Weak peaks, provenance and validity uncertain. Overall charges omitted for clarity.

In general, the stability of mixed-valence phthalocyanines, as defined by the electrochemical splitting or K_c values, is somewhat larger than for the corresponding porphyrins (to the extent that such comparisons can be made).

Table 10 Comproportionation Data [65, 69, 123, 126]

	Couple[a]	Solvent	E(V)[b]	K_c[c]	K_d[d]	ΔG/ kJmol^{-1}
Ant[CoTrNPc]$_2$	I	DCB	0.17	8.4x10^{2}	1.2x10^{-3}	-17
	II	DCB	0.48	1.8x10^{8}	5.6x10^{-9}	-47
	III	DMF	0.23	9.0x10^{3}	1.1x10^{-4}	-23
	II	DMF	0.30	1.4x10^{5}	7.0x10^{-6}	-29
Nap[CoTrNPc]$_2$	I	DCB	0.14	2.8x10^{2}	2.5x10^{-3}	-14
	II	DCB	0.39	4.0x10^{6}	2.5x10^{-7}	-38
	III	DMF	0.24	1.3x10^{4}	7.5x10^{-5}	-24
	II	DMF	0.22	6.1x10^{3}	1.65x10^{-4}	-22
EtMeO(5)-	I	DCB	0.08	2.4x10^{1}	4.2x10^{-2}	-8
[ZnTrNPc]$_2$	I	DMF	0.15	3.8x10^{2}	2.6x10^{-3}	-15
[ZnTrNPc]$_4$	I	DCB	0.11	7.8x10^{1}	1.3x10^{-2}	-11
	I	DMF	0.15	3.8x10^{2}	2.6x10^{-3}	-15
Ant[ZnTrNPc]$_2$	I	DCB	0.21	4.1x10^{3}	2.45x10^{-4}	-21
	I	DMF	0.21	4.1x10^{3}	2.45x10^{-4}	-21
Nap[ZnTrNPc]$_2$	I	DCB	0.21	4.1x10^{3}	2.45x10^{-4}	-21
	I	DMF	0.18	1.2x10^{3}	8.0x10^{-4}	-18
Nap[CuTrNPc]$_2$	I	DCB	0.20	2.7x10^{3}	3.6x10^{-4}	-20
(-1)[ZnTrNPc]$_2$	I	DCB	0.26	2.95x10^{4}	3.4x10^{-5}	-25
	I	DMF	0.22	6.1x10^{3}	1.65x10^{-4}	-22
FAlPc-O-PcAlF	I	DMF	0.40	6.0x10^{6}	1.7x10^{-7}	-39
RSiPc-O-PcSiR	I[e]	CH_2Cl_2	0.49	2.7x10^{8}	3.7x10^{-9}	-49
	IV	CH_2Cl_2	0.40	7.6x10^{6}	1.3x10^{-7}	-40
FePc-O-PcFe	III[f]	Py	0.36	1.3x10^{6}	7.7x10^{-7}	-35
	V	Py	0.40	6.0x10^{6}	1.7x10^{-7}	-39
FePc-N-PcFe	III[g]	Py	0.27	3.7x10^{4}	2.7x10^{-5}	-26
	V[g]	Py	0.83	1.2x10^{14}	8.3x10^{-15}	-80

[a] I: [MPc(-1)]$_2$ + [MPc(-2)]$_2$ ---> 2[MPc(-1).MPc(-2)]. II: [Co(I)Pc(-2)]$_2$ + [Co(II)Pc(-2)]$_2$ ---> 2[Co(I)Pc(-2).Co(II)Pc(-2)]; III: [M(II)Pc(-2)]$_2$ + [M(III)Pc(-2)]$_2$ ---> 2[M(II)Pc(-2).M(III)Pc(-2)]; IV: [RSiPc(-2)]$_2$ + [RSiPc(-3)]$_2$ ---> 2[RSiPc(-2).RSiPc(-3)] (with oxygen bridge linking silicon atoms). V: [M(III)Pc(-2)]$_2$ + [M(IV)Pc(-2)]$_2$ ---> 2[M(III)Pc(-2).M(IV)Pc(-2)] (with oxygen atoms bridging the iron atoms). For the cobalt complexes, data refer to the syn isomer. [b] mixed-valence splitting energy. [c] Comproportionation constant. [d] Disproportionation constant, 1/K_c. [e] R = n-C_6H_{13}, oxygen bridges axially link the silicon atoms. [f] Oxygen bridges axially link the iron atoms. [g] nitrogen bridges axially link the iron atoms.

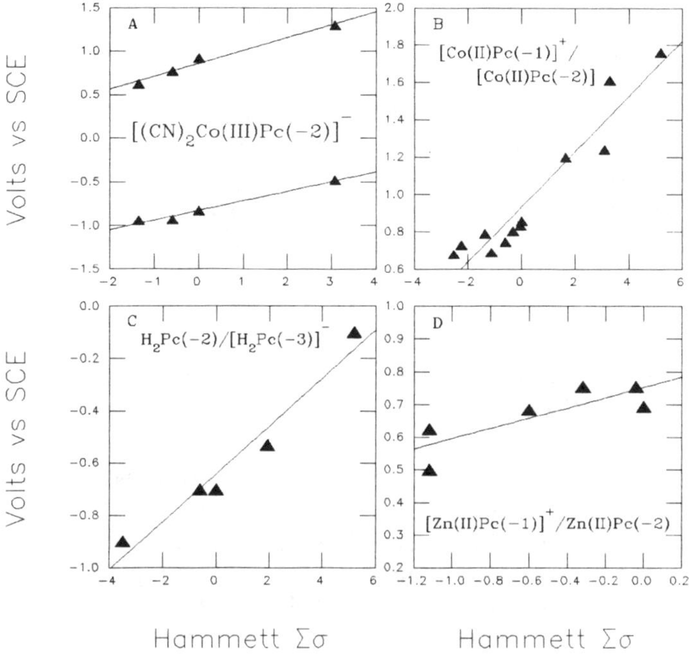

Figure 12 Hammett Sigma Plot for Substituted Phthalocyanines. Data for metal-free, zinc and cobalt species with redox processes as noted.

This is probably due to the "flatness" of the main phthalocyanine ring system, allowing for closer contact than is usually possible with the more ruffled and often highly substituted porphyrin derivatives.

ix. Hammett Relationships

Earlier brief studies [81,83,120] showed relationships betweeen the individual Hammett σ(para) parameter [135, 136]] and oxidation potentials for some tetra-and octasubstituted phthalocyanines. In Figure 12 are shown correlations for a range of redox processes for metal free, cobalt, and zinc phthalocyanines as a function of the total, $\Sigma\sigma$, value, accounting, thereby, for the number of substituents in the phthalocyanine ring [122].

Since the ring is both meta and para substituted, the use of the para parameter leads to some error. Moreover, although solvent effects in phthalocyanine redox processes are generally small, they are present; it is therefore not surprising that the data show some scatter. Overall, however, it is clear that, not surprisingly, roughly linear correlations are observed. Moreover for the limited data set shown here, the slopes do not differ greatly. Thus Figure 12 displays two oxidation data sets [Pc(-1)/Pc(-2)], both with slope 0.15 and three (first) reduction data sets, all with slope 0.10. If one may assume that these slopes apply generally to all metallophthalocyanines then it becomes possible to calculate, for example, the oxidation potential of any particular substituted metallophthalocyanine from $E(ox) = 0.15\Sigma\sigma + C$. The value of C is obtained by fitting some known experimental points for the metallophthalocyanine concerned.

F. Conclusions

This review has summarized and discussed the solution electrochemical behavior of a series of metallophthalocyanines with central ions encompassing the whole Periodic Table. The electrochemical versatility of these species, for example, the ability to tune potentials to where they might be useful in an electronic device, by changing metal ion or substituent, makes them potentially extremely valuable in the expanding field of molecular electronics. Such value is enhanced by their overall chemical and thermal stability and usual nontoxicity.

The review has also brought out major gaps in our knowledge of these systems, especially in the left-hand transition groups, and some of the heavier main group species. Greater effort should be expended in synthesizing organic solvent soluble examples of these species. The chemistry of the higher oxidation states of complexes such as molybdenum, tungsten, tin, and bismuth may prove especially illuminating.

A future article in this series will explore the surface electrochemistry and electrocatalytic properties of these fascinating materials.

Acknowledgement

ABPL is indebted to Professor C. C. Leznoff for the continuing collaboration, which has been so productive for the past 12 years, and to all those who have worked upon phthalocyanine chemistry in the Lever and Leznoff laboratories. He is also indebted to the Natural Science and Engineering Research Council (Ottawa) and the Office of Naval Research (Washington) for their continuing financial support.

Table 11 Phthalocyanine and Solvent Abbreviations used in this chapter

Mononuclear phthalocyanines:

DsPc	DisulfonylPc	TNPc	TetraneopentoxyPc
TBuPc	Tetra-t-butylPc	TAPc	TetraaminoPc
TcPc	TetracarboxyPc	TEtPc	Tetra-ethylPc
TsPc	TetrasulfonylPc	TBuPc	Tetra-t-butylPc
TMxPc	TetramethoxyPc	TNO2Pc	TetranitroPc
OCPc	OctacarboxyPc	OMPc	OctamethylPc
OBuxPc	OctabutoxyPc	OMxPc	OctamethoxyPc
$Cl_{16}Pc$	HexadecachloroPc	OCNPc	OctacyanoPc
TNO2Pc	TetranitroPc		

Binuclear phthalocyanines.

Each of the following species contains three benzene rings each with a substituted neopentoxy group (abreviated TrNPc), while the fourth ring is linked by the bridge as described. A number in parenthesis indicate the number of bridging atoms.

O(1)$[MTrNPc]_2$ an -O- ether link.
C(2)$[MTrNPc]_2$ $-CH_2-CH_2-$ link.
Cat(4)$[MTrNPc]_2$ a 1,2-catecholate link.
EtMeO(5)$[MTrNPc]_2$ a diether linkage $EtC(Me)(CH_2OPc)CH_2OPc$.
Nap$[MTrNPc]_2$ a 1,8-naphthalene direct link.
Ant$[MTrNPc]_2$ a 1,8-anthracene direct link.
(-1)$[MTrNPc]_2$ the fourth rings of each phthalocyanine are 3,4- fused.

$[MTrNPc]_4$ a $C(O)_4$ spiro tetra-ether linked (tetranuclear) phthalocyanine.

Common solvent abbreviations

ACN acetonitrile; ClN chloronaphthalene; DCB o-dichlorobenzene; DCE 1,2-dichloroethane; DMA dimethylacetamide; DMF dimethylformamide; DMSO dimethylsulfoxide; MeNp methylnaphthalene; PhCN benzonitrile; $PhNO_2$ nitrobenzene; Py pyridine; THF tetrahydrofuran.

References

1. A. B. P. Lever, Chemtech, 17 (1987) 506.
2. J. F. Myers, G. W. Rayner Canham, and A. B. P. Lever, Inorg. Chem., 14 (1975) 461
3. D. W. Clack and J. R. Yandle, Inorg. Chem., 11 (1972) 1738.
4. D. W. Clack, N. S. Hush, and I. S. Woolsey, Inorg. Chim. Acta,19 (1976) 129
5. L. D .Rollmann and R. T. Iwamoto, J. Am. Chem. Soc., 90 (1968) 1455.
6. A. B. P. Lever, S. Licoccia, K. Magnell, P. C. Minor, and B. S. Ramaswamy, Adv. Chem. Ser, 201 (1982) 237.
7. A. B. P. Lever, S. R. Pickens, P. C. Minor, S. Licoccia, B. S. Ramaswamy, and K. Magnell, J. Am. Chem. Soc., 103 (1981) 6800.
8. T. Nyokong, Z. Gasyna, and M. J. Stillman, Inorg. Chem., 26 (1987) 1087.
9. M. El Meray, A. Louati, J. Simon, A. Giraudeau, M. Gross, T. Malinski, and K. M. Kadish, Inorg. Chem., 23 (1984) 2606.
10. A. B. P. Lever, M. R. Hempstead, C. C. Leznoff, W. Liu, M. Melnik, W. A. Nevin, and P. Seymour, Pure and Appl. Chem., 58 (1986) 1467.
11. W. Liu, M. R. Hempstead, W. A. Newin, M. Melnik, A. B. P. Lever, and C. C. Leznoff, J. Chem. Soc., Dalton Trans., (1987) 2511.
12. D. Dolphin, B. R. James, A. J. Murray, and J. R. Thornback, Can. J. Chem., 58 (1980) 1125.
13. C. C. Leznoff, S. M. Marcuccio, S. Greenberg, A. B. P. Lever, and K. B. Tomer, Can. J. Chem., 63 (1985) 133.
14. J. H. Weber and D. H. Busch, Inorg. Chem., 4 (1965) 469,472.
15. F. H. Moser and A. L. Thomas, 'The Phthalocyanines', CRC Press, Boca Raton, Florida, 1983, Vol. 1,2
16. R. J. Blagrove, Austr. J. Chem., 26 (1973) 1545.
17. H. Sigel, P. Waldmeier, and B. Prijs, Inorg. Nucl. Chem. Lett., 7 (1971) 161.
18. A. B. P. Lever, Adv. Inorg. Chem. Radiochem., 7 (1965) 27; B. D. Berezin, "Coordination Compounds of Porphyrins and Phthalocyanines", Wiley, New York, 1981; F. H. Moser and A. L. Thomas, 'Phthalocyanine Compounds', Reinhold Publ. Corp., New York, 1963
19. A. J. Bard and L. J. Faulkner, "Electrochemical Methods", John Wiley, New York, 1980.
20. M. Nicholson, in this volume, Chapter 5.
21. K. M. Kadish, Progr. Inorg. Chem., 34 (1986) 435.
22. K. M. Kadish, Q. Y. Xu, and J. E. Anderson, ACS Symp. Ser. Vol. 378 (1988) 451.
23. K. M. Kadish, Chem. Rev., 88 (1988) 1121.
24. M. Gouterman in "The Porphyrins", ed. D. Dolphin, Academic Press, New

York, 1978, Vol. 1, p.1; A. M. Schaffer, M. Gouterman, and E. R. Davidson, Theor. Chim. Acta, 30 (1973) 9.
25. L. Edwards and M. Gouterman, J. Mol. Spectrosc., 33 (1970) 292.
26. M. Gouterman, G. H. Wagniere, and L. C. Snyder, Jr., J. Mol. Spectrosc., 11 (1963) 2.
27. C. Weiss, H. Kobayashi, and M. Gouterman, J. Mol. Spectrosc., 16 (1965) 415.
28. A. J. McHugh, M. Gouterman, and C. Weiss, Theor. Chim. Acta,24 (1972) 346.; A.M.Schaffer and M. Gouterman, Theor. Chim. Acta, 25 (1972) 62.
29. M. J. Stillman and T. Nyokong, in "Phthalocyanines, Properties and Applications", vol. 1, p. 133.
30. P. C. Minor, M. Gouterman, and A. B. P. Lever, Inorg. Chem., 24 (1985) 1894.
31. D. Wohrle and G. Meyer, Kontakte (Darmstadt), 3 (1985) 38.
32. N. S. Hush and A. S. Cheung, Chem. Phys. Lett., 47 (1977) 1; D. W. Clack, N. S. Hush and J. R. Yandle, Chem. Phys. Lett., 1 (1967) 157.
33. A. Louati, M. El Meray, J. J. Andre, J. Simon, K. M. Kadish, M. Gross, and A. Giraudeau, Inorg. Chem., 24 (1985) 1175.
34. D. J. E. Ingram and J. E. Bennett, J. Chem. Phys., 22 (1954) 1136.
35. F. H. Winslow, W. O. Baker, and W. A. Yager, J. Am. Chem. Soc., 77 (1955) 4751.
36. T. Fu Yen, J. G. Erdman, and A. J. Saraceno, Anal. Chem., 34 (1962) 694.
37. S. E. Harrison and J. M. Assour, J. Chem. Phys., 40 (1964) 365.
38. J. M. Assour and S. E. Harrison, J. Chem. Phys., 68 (1964) 872.
39. J. F. Boas, P. E. Fielding, and A. G. Mackay, Austr. J. Chem., 27 (1974) 7.
40. E. G. Sharoyan, N. N. Tikhomirova, and L. A. Blumenfeld, Zh. Strukt. Khim., 6 (1965) 843.
41. J. B. Raynor, M. Robson, and A. S. M. Torrens-Burton, J. Chem. Soc., Dalton Trans., (1977) 2360.
42. J. R. Harbour and M. J. Walzak, Langmuir, 2 (1986) 788.
43. A. B. P. Lever and P. C. Minor, Inorg. Chem., 20 (1981) 4015.
44. R. D. Shannon and C. T. Prewitt, Acta Crystall., 7 (1965) 27.
45. T. Nyokong, Z. Gasyna, and M. J. Stillman, Inorg. Chem., 26 (1987) 548.
46. R. O. Loutfy and Y. C. Cheng, J. Chem. Phys., 73 (1980) 2902.
47. C. C. Leznoff, S. Greenberg, S. M. Marcuccio, P. C. Minor, P. Seymour, and A. B. P. Lever, Inorg. Chem. Acta, 89 (1984) L35.
48. G. Fu, Y-S. Fu, K. Jayaraj, and A. B. P. Lever, Inorg. Chem., 29 (1990) 4090.
49. W. A. Nevin, M. R. Hempstead, W. Liu, C. C. Leznoff, and A. B. P. Lever, Inorg. Chem., 26 (1987) 570.
50. E. P. Platonova, E. Yu. Skuridin, and L. S. Degtyarev, Zh. Obshch. Khim., 54 (1984) 925.
51. S. V. Vulf'son, O. L. Kaliya, O. L. Lebedev, and E. A. Luk'yanets, Zh. Organ. Khim., 12 (1976) 123.

52. R. H. Campbell, G. A. Heath, G. T. Hefter, and R. C. S. McQueen, J. Chem. Soc., Chem. Commun., (1983) 1123.
53. R. O. Loutfy, personal communication, 1982
54. A. B. P. Lever, S. Licoccia. B. S. Ramaswamy, S. A. Kandil, and D. V. Stynes, Inorg. Chim. Acta, 51 (1981) 169.
55. A. Giraudeau, A. Louati, M. Gross, J. J. Andre, J. Simon, C. H. Su, and K. M. Kadish, J. Am. Chem. Soc., 105 (1983) 2917.
56. H. Homborg and K. S. Murray, Z. Anorg. Allg. Chem., 517 (1984) 149.
57. H. Sugimoto, T. Higashi, and M. Mori, J. Chem. Soc., Chem. Commun. (1983) 622.
58. P. Turek, J.-J. Andre, A. Giraudeau, and J. Simon, Chem. Phys. Lett., 134 (1987) 471.
59. D. Lexa and M. Reix, J. Chim. Phys. 71 (1974) 511.
60. A. Giraudeau, F.-R. F. Fau, and A. J. Bard, J. Am. Chem. Soc., 102 (1980) 5137.
61. A. B. P. Lever and J. P. Wilshire, Can. J. Chem., 54 (1976) 2514.
62. A. Wolberg and J. Manasson, J. Am. Chem. Soc., 92 (1970) 2982.
63. M. N. Golovin, P. Seymour, K. Jayaraj, Y-S. Fu, and A. B. P. Lever, Inorg. Chem., 29 (1990) 1719.
64. J. T. S. Irvine, B. R. Eggins, and J. Grimshaw, J. Electroanal. Chem., 271 (1989) 161.
65. V. Manivannan, W. A. Nevin, C. C. Leznoff, and A. B. P. Lever, J. Coord. Chem., 19 (1988) 139.
66. V. I. Gavrilov, E. A. Luk'yanets, and I. V. Shelepin, Electrokhimiya, 17 (1981) 1183.
67. V. I. Gavrilov, L. G. Tomilova, I. V. Shelepin, and E. A. Luk'yanets, Electrokhimiya, 15 (1979) 1058.
68. D. Wohrle and V. Schmidt, J. Chem. Soc., Dalton Trans., (1988) 549.
69. D. W. DeWulf, J. K. Lelend, B. L. Wheeler, A. J. Bard, D. A. Batzel, D. R. Dininny and M. E. Kenney, Inorg. Chem., 26 (1987) 266.
70. H. Li and T. F. Guarr, J. Chem. Soc., Chem. Commun., (1989) 832.
71. Y. Orihashi, M. Nishikawa, H. Ohno, E. Tsuchida, H. Matsuda, H. Nakanishi, and M. Kato, Bull. Chem. Soc., Jpn., 60 (1987) 3731.
72. C. C. Leznoff, Unpublished work
73. H. Sugimoto, M. Mori, H. Masuda, and T. Taga, J. Chem. Soc., Chem. Commun., (1986) 962.
74. A. B. Anderson, T. L. Gordon, and M. E. Kenney, J. Am. Chem. Soc., 107 (1985) 192.
75. T. M. Mezza, N. R. Armstrong, G. W. Ritter II, J. P. Iafalice, and M. E. Kenney, J. Electroanal. Chem., 137 (1982) 227.
76. B. L. Wheeler, G. Nagasubramanian, A. J. Bard, L. A. Schechtman, D. R. Dininny, and M. E. Kenney, J. Am. Chem. Soc., 106 (1984) 7404.
77. B. Simic-Glavaski, A. A. Tanaka, M. E. Kenney, and E. Yeager, J.

Electroanal. Chem., 229 (1987) 285.
78. E. Ciliberto, K. A. Doris, W. J. Pietro, G. M. Reisner, D. E. Ellis, I. Fragala, F. H. Herbstein, M. A. Ratner, and T. J. Marks, J. Am. Chem. Soc., 106 (1984) 7748.
79. E. S. Dodsworth, A. B. P. Lever, P. Seymour, and C. C. Leznoff, J. Phys. Chem. 89 (1985) 5698.
80. M. Gouterman, D. Holten, and E. Lieberman, Chem. Phys., 25 (1977) 139.
81. V. R. Shephard, Jr. and N. R. Armstrong, J. Phys. Chem., 83 (1979) 1268.
82. V. I. Gavrilov, A. P. Konstantinov, V. M. Derkacheva, E. A. Luk'yanets, and I. V. Shelepin, Zh. Fiz. Khim., 60 (1986) 1448.
83. V. I. Gavrilov, L. G. Tomilova, V. M. Derkacheva, E. V. Chernykh, N. T. Ioffe, I. V. Shelepin and E. A. Luk'yanets, Zh. Obshch. Khim., 53 (1983) 1347.
84. C. Paliteiro and J. B. Goodenough, J. Electroanal. Chem., 239 (1988) 273.
85. The rest potential is the open circuit potential, when zero current is flowing, that is, no oxidation or reduction chemistry is taking place.
86. A. B. P. Lever, Ph. D. Thesis, London, 1960.
87. T. J. Klofta, J. Danziger, P. Lee, J. Pankow, K. W. Nebesny, and N. R. Armstrong, J. Phys. Chem., 91 (1987) 5646.
88. R. Behnisch and M. Hanack, Synth. Metal, 36 (1990) 387.
89. M. Hanack and A. Datz, Chem. Ber., 119 (1986) 1281.
90. J. E. Elvidge and A. B. P. Lever, J. Chem. Soc., (1961) 1265.
91. This complex was initially identified as Cr(II)[TBuPc(-2)].H_2O but, given its electrochemical data, it is more likely HOCr(III)[TBuPc(-2)] Previously Unpublished analytical data are: Found C, 71.6; H, 6.4; N, 13.3 Calcd. for HOCr(III)[TBuPc(-2)] C, 71.53; H, 6.13; N, 13.90 %
92. a) A. B. P. Lever and J. P. Wilshire, previously unpublished data; b). T. Nyokong, G. Ferraudi, and M. Feliz, submitted to J. Photochem. Photobiol. 1991.
93. A. B. P. Lever, P. C. Minor, and J. P. Wilshire, Inorg. Chem., 20 (1981) 2550.
94. P. C. Minor and A. B. P. Lever, Inorg. Chem., 22 (1983) 826.
95. A. B. P. Lever and P. C. Minor, Adv. Mol. Relax. Proc., 18 (1980) 115.
96. D. J. Cookson, T. D. Smith, J. F. Boas, P. R. Hicks, and J. R. Pilbrow, J. Chem. Soc., Dalton Trans., (1977) 211.
97. A. B. P. Lever and J. P. Wilshire, Inorg. Chem., 17 (1978) 1145.
98. K. M. Kadish, L. A. Bottomley, and J. S. Cheng, J. Am. Chem. Soc., 100 (1978) 2731.
99. M. Mossoyan-Deneux and D. Benlian, J. Inorg. Nucl. Chem., 43 (1981) 635.
100. W. Kalz, H. Homborg, H. Kuppers, B. J. Kennedy, and K. S. Murray, Z. Naturforsch., B 39b (1984) 1478.
101. W. A. Nevin, W. Liu, M. Melnik, and A. B. P. Lever, J. Electroanal. Chem., 213 (1986) 217.
102. M. Brezina, W. Khalil, J. Koryta, and M. Musilova, J. Electroanal. Chem., 77

(1977) 237.
103. R. S. Nicholson and I. Shain, Anal. Chem., 36 (1964) 706.
104. E. Ough, T. Nyokong, K. A. M. Creber, and M. J. Stillman, Inorg. Chem., 27 (1988) 2724.
105. Y. Orihashi, H. Ohno, E. Tsuchida, H. Matsuda, H. Nakanishi, and M. Kato, Chem.Lett., (1987) 601.
106. L. A. Bottomley, C. Ercolani, J. -N. Gorce, G. Pennesi, and G. Rossi, Inorg. Chem., 25 (1986) 2338.
107. M. Hanack, A. Lange, M. Rein, R. Behnisch, G. Renz, and A. Leverenz, Synth. Metal, 29 (1989) F1.
108. G. Ferraudi, S. Oishi, and S. Muraldiharan, J. Phys. Chem., 88 (1984) 5261.
109. Y. H. Tse, L. Persaud, P. Seymour, and A. B. P. Lever, work in progress.
110. J. Ouyang, K. Shigehara, A. Yamada, and F. A. Anson, J. Electroanal. Chem., 297 (1991) 489.
111. L. A. Bottomley, J. -N. Gorce, V. L. Goedken, and C. Ercolani, Inorg. Chem., 24 (1985) 3733.
112. C. Ercolani and B. Floris, this volume, Chapter 1.
113. G. Ferraudi and D. R. Prasard, J. Chem. Soc., Dalton Trans., (1984) 2137.
114. W. A. Nevin, W. Liu, S. Greenberg, M. R. Hempstead, S. M. Marcuccio, M. Melnik, C. C. Leznoff, and A. B. P. Lever, Inorg. Chem., 26 (1987) 891.
115. P. A. Bernstein and A. B. P. Lever, Inorg. Chem., 29 (1990) 608.
116. K. Kusuda, K. Shiraki, and H. Yamaguchi, Ber. Bunsenges. Phys. Chem., 92 (1988) 725.
117. V. I. Gavrilov, L. G. Tomilova, E. V. Chernykh, O. L. Kalija, I. V. Shelepin, and E. A. Luk'yanetz, Zh. Obshch. Khim., 50 (1980) 2143.
118. E. B. Fleischer, S. Jacobs, and L. Mestichelli, J. Am. Chem. Soc., 90 (1968) 2527.
119. C. C. Leznoff, H. Lam, S. M. Marcuccio, W. A. Nevin, P. Janda, N. Kobayashi and A. B. P. Lever, J. Chem. Soc. Chem. Commun., (1987) 699.
120. Y. Orihashi, H. Ohno, and E. Tsuchida, Mol. Cryst. Liq. Cryst., 160 (1988) 139.
121. Y. Orihashi, N. Kobayashi, E. Tsuchida, H. Matsuda, H. Nakanishi, and M. Kato, Chem. Lett., (1985) 1617.
122. A. B. P. Lever, unpublished observations.
123. C. C. Leznoff, H. Lam, W. A. Nevin, N. Kobayashi, P. Janda, and A. B. P. Lever, Angew. Chem. Int. Ed., 26 (1987) 1021.
124. H. C. Brown and Y. Okamoto, J. Am. Chem. Soc., 80 (1958) 4979.
125. M. Hanack, C. Deger, and A. Lange, Coord. Chem. Rev.,83 (1988) 115.
126. N. Kobayashi, H. Lam, W. A. Nevin, P. Janda, C. C. Leznoff, and A. B. P. Lever, Inorg. Chem., 29 (1990) 3415.
127. M. Hanack, and A. Hirsch, Synth. Metal., 19 (1988) 139.
128. G. Rossi, M. Gardini, G. Pennesi, C. Ercolani, and V. L. Goedkin, J. Chem. Soc., Dalton Trans., (1989) 193.

129. K. M. Kadish, R. K. Rhodes, L. A. Bottomley and H. M. Goff, Inorg. Chem., 20 (1981) 3195.
130. C. Ercolani, F. Monacelli, S. Dzugan, V. L. Goedken, G. Pennesi, and G. Rossi, J. Chem. Soc. Dalton Trans., (1991) 1309.
131. A. Beck, M. Hanack, and K. M. Mangold, Chem.Ber., 124 (1991) 2315.
132. A. Leverenz, Ph. D. Thesis, Tubingen, 1990
133. A. MacGragh, C. B. Storm, and W. S. Koski, J. Am. Chem. Soc., 87 (1965) 1470.
134. T. G. Brown and B. M. Hoffman, Mol. Phys., 39 (1980) 103.
135. C. D. Ritchie, "Physical Organic Chemistry", 2nd Edn., Marcel Dekker, NY, 1990.
136. H. C. Brown and Y. Okamoto, J. Am. Chem. Soc., 80 (1958) 4979.
137. M. Hanack, personal communication, 1992.
138. L. F. LeBlevenec and J. G. Gaudiello, J. Electroanal. Chem., 312 (1991) 97.

2

Electrochromism and Display Devices

M. M. Nicholson

A. INTRODUCTION

Electrochromism was originally defined by Platt as the change in an absorption spectrum due to the presence of an electric field [1]. Such an effect ordinarily is much too small to cause a visible change, and it disappears when the field is removed. Usage of the term *electrochromism* was later extended to include changes in color or optical contrast brought about by application of a voltage to a material — usually a solid — in an electrochemical cell. These larger effects result from faradaic reactions, with the passage of direct current, closely akin to those in battery electrodes. The spectral change persists after the signal is removed but can be reversed by applying a voltage that causes current to flow in the opposite direction. Faradaic electrochromic systems that produce high optical contrast and can be electrically cycled many times without deterioration are of interest for electronic information displays.

Sandwich-type diphthalocyanine complexes of the lanthanide elements are among the most promising electrochromic display materials, offering not only high contrast, but virtually a full spectrum of colors from a single compound. This chapter first discusses fundamental aspects of the faradaic electrochromism in phthalocyanines, with emphasis on the lanthanide diphthalocyanines and closely related materials. It then describes the structures and performances of experimental phthalocyanine display devices and identifies areas in which further development is needed to make this technology available for a wide range of applications.

B. ELECTROCHROMISM

Color changes due to successive electron-transfer reactions of dissolved phthalocyanines at electrode surfaces have been known for many years. For example, Clack and Yandle reported stepwise reductions of several divalent-metal phthalocyanines and a single-ring aluminum compound involving up to four or five electrons per molecule in tetrahydrofuran [2]. Pronounced color changes were noted, but since the faradaic nature of these processes was recognized at the outset and solid films were not involved, there was no precedent for calling them electrochromic.

The first description of electrochemically generated multicolor responses in a solid film of a rare-earth diphthalocyanine was given by Moskalev and Kirin in 1970 [3]. They observed that lutetium diphthalocyanine, on a transparent semiconductive tin oxide substrate, developed colors ranging from blue, through green, to "cinnamon" in aqueous 0.1 *M* KCl as the potential was varied from –0.9 to +0.9 V with respect to a saturated calomel electrode. This behavior was designated, from an operational viewpoint, as electrochromism. Several absorption spectra were given, and similar results were noted for the diphthalocyanines of neodymium, europium, and yttrium. In 1972, the same investigators reported additional data for the lutetium compound and proposed certain chemical reactions to explain the color changes [4]. These early interpretations excluded direct faradaic processes of the dye material, invoking, instead, field-induced solid-phase ionization and complexation with electron acceptors such as oxygen.

Several years later, starting from the observations of Moskalev and Kirin, Nicholson and Galiardi began the evaluation of lanthanide diphthalocyanines for use in electronic information displays [5,6]. On low-resistance tin oxide, lutetium diphthalocyanine films of good color intensity were switchable in less than 50 ms at energy densities considerably lower than those of the monochrome tungsten oxide electrochromic, which had been studied extensively. The report of favorable switching parameters, coupled with unique multicolor capability, prompted a number of industrial laboratories to enter this research area.

Much work on electrochromic diphthalocyanines continues at this time. Soviet investigators tend to emphasize the basic aspects of synthesis, characterization, physical chemistry, and electrochemistry, while the U.S., British, French, and Japanese work, though often fundamentally significant, has been motivated primarily by the prospect of display applications.

Section B surveys the occurrence of visible-range electrochromism in phthalocyanine films and comments briefly on synthesis and electrode fabrication for rare-earth diphthalocyanine materials. The principal methods of investigation used for these systems are indicated, representative electrochromic spectra are given, and sources of more extensive spectroscopic data are listed. The diphthalocyanine electrode processes are then discussed in terms of reaction stoichiometrics, mechanisms, equilibrium relationships, kinetics, and the role of oxygen.

i. Occurrence in Phthalocyanines

Solid-state electrochromism has been observed in phthalocyanine compounds of more than thirty metals, predominantly among the sandwich-type diphthalocyanines (bisphthalocyanines) of the lanthanide and actinide rare earths and the related Group III elements lanthanum, yttrium [3], and scandium [4]. It also occurs in diphthalocyanines of the Group IV elements tin [7], zirconium [8], and hafnium [8]. Single-ring electrochromic phthalocyanines include the metal-free [9] and magnesium [10] compounds and those of transition elements, including divalent iron, cobalt, nickel, copper, zinc, and molybdenum [7,11]. Observation of the color changes is more difficult with low-conductivity materials.

Many electrochemical studies bearing on electrochromism have been done with solutions of phthalocyanines having various groups substituted on the rings to increase the solubilities in selected media. Such groups include sulfonic acid [12], *t*-butyl [13], and dodecyl [14]. Substituted phthalocyanines — even some with large groups on the rings — can also exhibit electrochromism when examined as films in contact with nonsolvent media. Among these are lanthanide diphthalocyanines substituted with *t*-butyl or propoxy groups [15], with cyano [16] or with octyloxymethyl or dodecyloxymethyl groups [17]. Significant improvement in the electrochromic behavior was reported recently with a crown-ether-substituted lutetium diphthalocyanine [18]. Also electrochromic are films of the single-ring octacyanophthalocyanines of hydrogen, zinc, and copper(II) [16,19].

With the ability to exist in several oxidation states corresponding to different visible spectra, all phthalocyanines are potentially electrochromic. Kinetically, however, the occurrence of faradaic electrochromism in a film some tens of molecules thick requires electrical conduction in each of the color forms. Moreover, the film must provide transport paths for both electrons and ions, so that it remains electrically neutral when it switches. Although each color phase does not have to conduct by both mechanisms, an insulating layer must not develop in the course of switching, and ions essential for charge compensation must not become trapped out of contact with the electrolyte, unable to escape.

Charge-transport characteristics vary widely within the family of phthalocyanine materials. The electrochromism of greatest practical interest is that capable of many switching cycles. Some applications demand millions of cycles without significant loss of color intensity or quality. The electrode film must remain intact during this treatment. Hence, its microstructure should not vary too much with the insertion and removal of ions. The desired features of multiple colors, solid-state conductivity, and dimensional stability are inherent properties of diphthalocyanine compounds. Electrochromism observed in the less conductive single-ring phthalocyanines is typically limited to two colors.

The diphthalocyanines comprise a large group of materials. Kirin and Moskalev listed twenty elements that were known by 1971 to have diphthalocyanines and predicted that a total of forty-one elements could form such complexes [20]. The criterion was that the radius of the metal atom should exceed the coordination space of 1.35 Å available in the phthalocyanine ligand. The metal ion can then be stably situated between the two macrocycles. Smaller ions are coordinated into the ring, preferentially forming monophthalocyanines, in which electrochromism is not readily observed.

ii. Preparation of Diphthalocyanine Electrochromic Materials

An extensive discussion of phthalocyanine synthesis is beyond the scope of this chapter. However, because materials preparation is so essential to electrochromics research, methods used to synthesize, purify, and fabricate films of lanthanide diphthalocyanines are briefly reviewed below.

A crude diphthalocyanine product is prepared by heating a mixture of the rare-earth acetate or chloride with *o*-phthalonitrile to approximately 300° C [21]. It is advisable to recrystallize the phthalonitrile if a practical grade material is used. Once it begins, the exothermic ring-formation reaction is completed in a short time, leaving a dark solid mixture that contains the diphthalocyanine $LnPc_2$, a monophthalocyanine LnXPc, where X is acetate or chloride, some H_2Pc, and, with certain rare earths, a triphthalocyanine Ln_2Pc_3 [22–24]. When a small amount of the crude product is rubbed onto conductive tin oxide glass, it is usually found to be electrochromic, despite the presence of unswitchable components. The solid reaction product is ground to a fine powder and then washed successively with several organic solvents. The sequence acetic anhydride, acetone, dimethylformamide is appropriate for lutetium diphthalocyanine [21].

The further purification that is needed depends on the rare-earth element. For the lutetium compound, vacuum sublimation of the desired component at ~10^{-6} torr is sufficient, about 60% of the solid charge in the evaporator being recoverable as diphthalocyanine [5]. For the lighter elements, especially those from lanthanum through neodymium, the separation is more difficult and is often done by solution chromatography on an alumina column [22,25]. Because Ln_2Pc_3 has an absorption spectrum closely resembling that of a diphthalocyanine species in a solvent such as dichloromethane [24], care must be taken to identify the chromatographic fractions. Unless the triphthalocyanine is removed prior to sublimation, a large amount of H_2Pc appears in the sublimation product [24,26]. M'Sadek et al. attributed this to hydrolysis of Ln_2Pc_3 yielding $LnHPc_2$ and H_2Pc [24]. Results of Clarisse and Riou for synthesis products of Dy, Gd, and Nd appear to confirm this, although the role of a triphthalocyanine was not discussed [27].

A different separation method was devised for the actinide diphthalocyanines [28]. Moskalev, Shapkin, and Darovskikh isolated polycrystalline americium diphthalocyanine by anodic deposition from a solution of the reaction product in dimethylformamide containing 1% hydrazine hydrate and demonstrated that the same technique yielded the diphthalocyanines of Nd, Eu, Ho, and Lu [29]. The hydrazine hydrate apparently served as the electrolyte and as a reagent to keep the diphthalocyanine in its more soluble reduced form until it contacted the platinum anode.

Uniform diphthalocyanine films for use in electrochromics research or device development are easily produced in a few minutes by vacuum sublimation if the troublesome impurities noted above are absent. Adhesion of the film to an oxide substrate may be improved by vacuum annealing, which can be done with an infrared lamp [5]. Vacuum deposition with plasma polymerization also has been investigated for ytterbium diphthalocyanine films [30].

Results of Ando et al. for neodymium diphthalocyanine [31], and of Moskalev and Shapkin for several lanthanide compounds [32], imply that electrodeposition offers another general method for obtaining uniform thin films. Occasionally, films have been cast from solution manually [16,19], by spinning [18], or by the Langmuir-Blodgett technique [15,33]. The simple expedient of rubbing the solid dye material onto the substrate [4] can be very useful in qualitative experiments, although the uniformity and adhesion of such films are predictably poor. Information on films modified by organic additives or binders is given in Section C on displays.

iii. Methods of Investigation

Many techniques have been employed in the quest for mechanistic understanding and performance improvement in the rather complex diphthalocyanine electrochromic systems. Electrochemical data and optical absorption spectra are obtained with cells as simple as that in Fig. 1. The cell may be filled and sealed in an inert atmosphere or provided with a gas inlet and outlet for deaeration. The conductive substrate beneath the dye film is preferably semiconductive tin oxide or indium-tin oxide with a sheet resistance less than 25Ω/square. Semitransparent gold or another inert metal substrate is sometimes used, but these have the disadvantages of lower transparency and smaller usable voltage ranges in aqueous electrolytes. The electrolyte often contains chloride ion because in most cases this anion participates readily in the electrochromic reactions, and it permits the convenient use of silver-silver chloride counter and reference electrodes without separators. The dye electrode is usually controlled potentiostatically, although galvanostatic measurements have some advantages in kinetics studies. For in situ recording of spectra, a cell of this type is placed in the sample beam of a spectrophotometer, and a matching blank cell that includes a conductive glass plate without the dye film is placed in the reference beam.

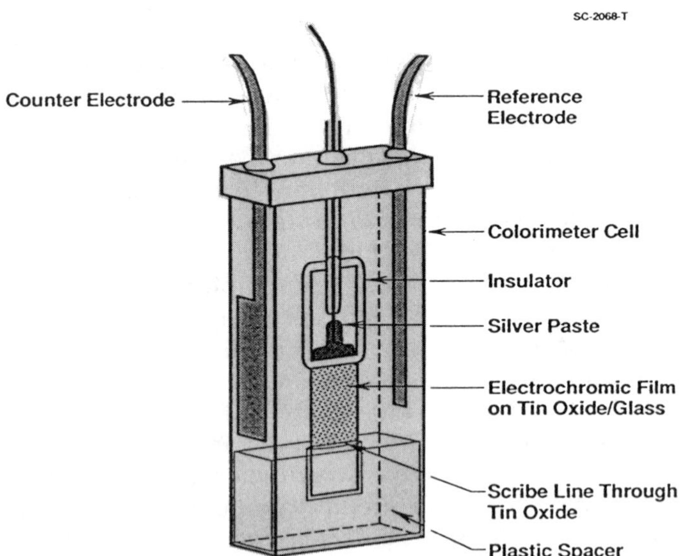

Figure 1. Typical experimental cell.

Supplementing the research by direct spectroelectrochemical approaches are studies involving solid-state conductivity, electron paramagnetic resonance, magnetic susceptibility, magnetic circular dichroism, mass spectroscopy, x-ray crystallography, electron microscopy, radio-tracer measurements, chemical redox reactions, and acid-base equilibria.

iv. Electrochromic Spectra and Oxidation States

The electronic spectroscopy of phthalocyanines is discussed fundamentally in other chapters of Volumes 1 and 2. The present section surveys the spectra of electrochromic phthalocyanines from an experimental viewpoint and offers a concise guide to the rather extensive literature on that subject. It also correlates diphthalocyanine spectra with oxidation states and indicates approaches by which the respective colors can be modified. Although spectral shifts with oxidation state occur in the ultraviolet and near-infrared regions as well as the visible, this discussion emphasizes the visible wavelength range, essentially 400 to 700 nm, which is most pertinent to display applications.

Table 1 gives most of the sources of solid and solution spectral data for electrochromic phthalocyanines *without ring substitution* that were published through 1990. Many of these references include data on the original, reduced, and oxidized forms. This tabulation is limited to compounds for which electrochromism has been observed in the solid state.

a. Solid-film spectra

Because lutetium diphthalocyanine provides a full range of colors from orange or reddish-orange to violet and the films are easily prepared by vacuum sublimation, this compound has been the most extensively studied among the lanthanide complexes. This is evident in Table 1a. Sources of solid-film spectra for electrochromic phthalocyanines outside the lanthanide series are given in Table 1b.

Moskalev and Kirin were the first to report the lutetium diphthalocyanine electrochromic spectra, using an aqueous 0.1 *M* KCl electrolyte [3]. Typical curves for the cycled green, light blue, and orange forms in 1 *M* KCl are shown in Fig. 2. Other spectra, representing further reduction of the film to dark purple-blue and violet, are given in [48]. These forms are discussed in the section on electrode processes.

A lutetium diphthalocyanine film prepared by vacuum deposition is nearly always green, with a prominent *Q* band characteristic of phthalocyanines near 665 nm. Because the lanthanide elements are predominantly

Table 1 Sources of UV-Visible Spectral Data on Electrochromic Phthalocyanines Without Ring Substitution[a]

a. Lanthanide Diphthalocyanines

Ln	Solid Films	Solutions: Dimethyl-formamide	Solutions: Dichloro-methane	Solutions: o-Dichloro-benzene
La			34[b]	
Pr	35	36		35
Nd	13,29,37	28[c],29[c]	24,27,34[b]	13
Sm				13
Eu	29	28[c],29[c]	24	
Gd	37,38	36	27	
Tb			27	
Dy	26	36	24,26,27,39	
Er	38,40,41,42	21		13
Tm	43	39		
Yb	33,38,44		27,39	
Lu	3,4,5,7,29, 37,44,45,46 47,48,49	29[c],47,50	14,24,27,51 52,53	54

b. Other Electrochromic Phthalocyanines

Compound	Solid Film[d]	Solutions: Dimethyl-formamide	Solutions: o-Dichloro-benzene	Solutions: Dichloro-methane
H_2Pc	9			
ZnPc	11			
$ScPc_2$	37	20,43		55
YPc_2	37	21	34[b]	
$ZrPc_2$	56		8	
$HfPc_2$		43	8	
MoOPc	56			
UPc_2	55			55
$AmPc_2$	29	28[c],29[c]		

[a]Numbers are for references at the end of this chapter.

[b]With methanol.

[c]With hydrazine hydrate.

[d]Electrochromism also reported in films of MoPc and CoPc [7].

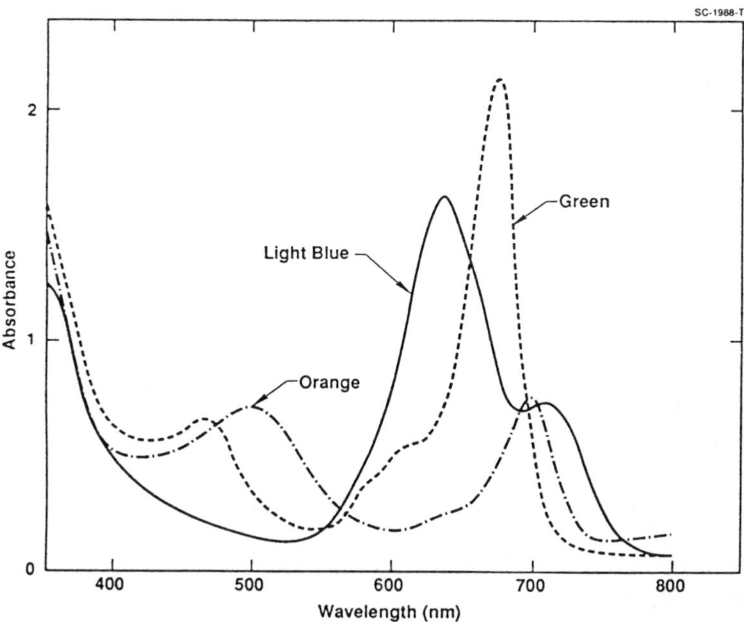

Figure 2. Absorption spectra of a lutetium diphthalocyanine film in aqueous 1 *M* KCl.

trivalent, the formula of this compound was originally assumed to be $LuHPc_2$, with trivalent lutetium and one labile hydrogen in the molecule. When Corker et al. found that the green form had an unpaired electron, they suggested the formula $LuHPc_2^+A^-$, where A^- was an unidentified anion [50]. Other investigators later proposed that the anion was superoxide, O_2^- [7,57]. Meanwhile, the paramagnetism was confirmed [44,51], and the mass spectrum was found to be that of $LuPc_2$, rather than $LuHPc_2$, when determined by the conventional electron-impact method [7,51]. Most investigators now accept the radical formula $LuPc_2$ for the green species and $LuHPc_2$ for the diamagnetic one-electron reduced blue. The position of the odd electron in $LuPc_2$ has been a subject of speculation [44,51,58,59]. Markovitsi et al. have assigned it to one of the phthalocyanine rings, primarily on the basis of a 1382-nm charge-transfer band in the near-infrared spectrum [39]. Figure 3 represents the $LnPc_2$ type of molecular structure [60].

The reduced form $LuHPc_2$ has been found in vacuum-deposited films [7] and in some bulk preparations [26,61] by field-desorption mass spectroscopy. Although its identity was obscured for some time, pure $LuHPc_2$ was eventually isolated by Moussavi et al., who characterized its detailed molecular structure by x-ray crystallography [61]. The double form of the *Q* band in the film spectrum of this reduction product varies

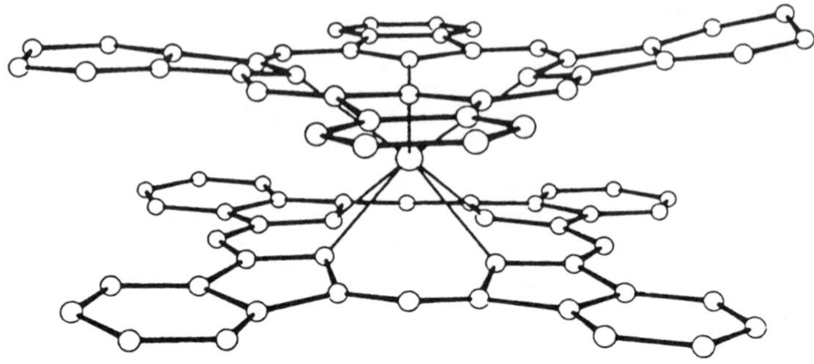

Figure 3 Molecular structure of a lanthanide diphthalocyanine in the green state. Reproduced with permission from [60] (Fig. 2). Copyright (1980) American Chemical Society.

with *p*H in aqueous media, the height of the 710–nm peak increasing above *p*H 4 [48]. Within the general redox scheme, the one-electron oxidized species is represented as $LuPc_2^+$, and a solid form containing an anion, as $LuPc_2^+A^-$.

With diphthalocyanines of tetravalent metals such as uranium, the color sequences are offset one step from those of the lanthanide complexes in terms of the net electrical charges; UPc_2 is blue, UPc_2^+ is green, and so forth [55]. This is consistent with the organic ring valences as the main color-controlling factor. The presence of two Pc^{2-} rings in the sandwich structure results in a blue color, while a Pc^{2-} in combination with a Pc^- produces a characteristic green.

The film spectra of lutetium diphthalocyanine show some variations with deposition conditions [47,62]. Changes on exposure to air provide further evidence that the green material originally deposited as $LuPc_2$ may be converted to $LuHPc_2^+O_2^-$ when it is removed from the vacuum system [47]. Spectra determined before and after contact with air are shown in Fig. 4. The broader peak of curve b, which lacks the vibrational structure to the left of the Q band, apparently results from interactions between closely packed $LuPc_2$ molecules. If the substrate is heated during the dye deposition, the peak is further broadened, and the reactivity toward air is lost or diminished. Hydration of the solid may also modify the spectrum, and the absorption peaks will be flattened if cracks develop in the film [48]. Still other perturbations can arise from changes in crystal structure or molecular orientation. Hence, the fabrication and treatment of the dye film should be rather closely controlled when good reproducibility of the spectral peak heights is important.

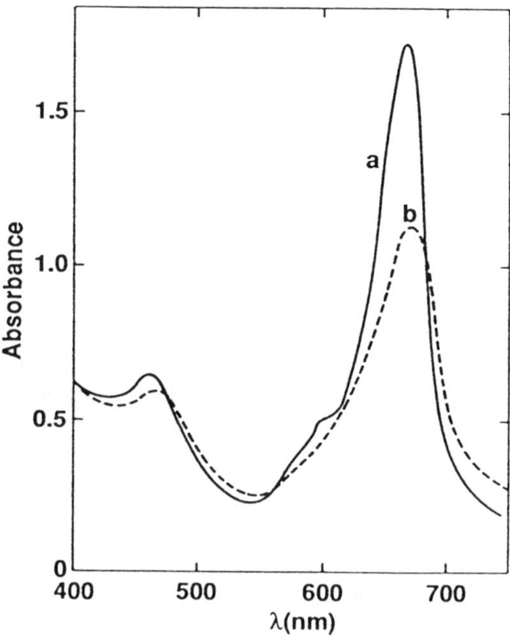

Figure 4 Spectrum of a vacuum-sublimed lutetium diphthalocyanine film on glass. (a) After exposure to air; (b) before exposure to air. Reproduced from [47] (Fig. 1) by permission of the publisher, The Electrochemical Society, Inc.

b. Solution spectra

Many influences that complicate the solid-film spectra are absent in solution. Accordingly, most of the systematic studies relating the spectra of diphthalocyanines to the nature of the rare-earth ion and to various ring substituents have been done on solutions. Solvents often used are dimethylformamide (DMF), dichlorobenzene, and dichloromethane. One caution applies to DMF: Although the dye materials are readily soluble, the green forms can be reduced by an amine impurity that is commonly present, even in the reagent-grade solvent [47,50]. For that reason, other solvents are preferred unless the DMF is specially purified [50].

Table 2 summarizes trends observed in the most prominent visible-range peaks of some diphthalocyanine solution spectra. These compounds undergo major spectral and color changes on oxidation or reduction, but the effects on the peak position λ_{max} due to the various substituents are relatively small. However, there are some very significant differences in the molar extinction coefficient ε, and correspondingly in absorption-band width, within each oxidation state. For example, with the green state in

Table 2 Some Trends in Lanthanide Diphthalocyanine Solution Spectra[a,b]

Ln	Ring substituents[c]	1e Reduced (blue)		Original (green)		1e Oxidized (orange or red)			
		λ_{max}	log ε	λ_{max}	log ε	λ_{max}	log ε	λ_{max}	log ε
a. Effects of lanthanide element; spectra in dichlorobenzene [13][d]									
Nd	--	634	4.91	674	4.76	718	4.41	494	4.52
Sm	--	628	4.93	670	4.78	712	4.42	488	4.50
Er	--	620	4.87	662	5.03	696	4.50	480	4.52
Lu	--	624	4.76	658	5.01	684	4.22	474	4.24
b. Effects of ring substituents; spectra in dichlorobenzene [54][d]									
Lu	4-t-butyl	622	5.10	662	5.21	696	4.69	482	4.70
Lu	4-phenoxy	624	5.19	668	5.30	702	4.72	466	4.69
Lu	4-phenyl	632	5.22	674	5.32	708	4.77	484	4.75
Lu	3-bromo-5-t-butyl	662	4.69	688	4.99	732	4.46	548(?)	4.24
c. Effects of long-chain alkyl substituents; spectra in dichloromethane [14][e]									
Lu	--	620	5.10	660	5.19	690	4.67	472	4.67
Lu	octyl	627	5.02	668	5.08	701	4.69	488	4.65
Lu	dodecyl	631	5.02	671	5.08	704	4.69	492	4.65

[a]Predominant peaks; references include additional data.

[b]λ_{max} in nm; ε in l $mole^{-1}$ cm^{-1}. Logarithms to base 10.

[c]Eight groups per molecule in B; sixteen in C.

[d]With 0.2 M Bu_4NClO_4.

[e]With 0.1 M i-Pr_4NPF_6.

dichlorobenzene, ε for $NdPc_2$ exceeds that for $LuPc_2$ by a factor of 1.8. The substitution of eight *t*-butyl groups on the lutetium diphthalocyanine molecule increases ε in dichloromethane — for the blue form by a factor of 2.1, for the green by 1.6, and for the two peaks of the orange by 2.9 and 3.0. Reference [63] gives more extensive data on octa-*t*-butyldiphthalocyanines, including those of nine lanthanide complexes in three oxidation states. In contrast, the bulkier C_8 and C_{12} alkyl chains have only slight effects on the spectra in dichloromethane (Fig. 5), even when sixteen groups per molecule are present [14].

Dissolution of the diphthalocyanines in any of these low-polarity solvents greatly sharpens the spectral peaks, in comparison with those of the solid films. This is illustrated by Fig. 6 for the green form of dysprosium diphthalocyanine [26]. Such an effect, which is typical of phthalocyanine spectra [64], is discussed in Volume 1, Chapter 3.

v. Electrode Processes of the Lanthanide Diphthalocyanines

The color transitions in rare-earth diphthalocyanine electrode films are unusually fast for layer thicknesses on the order of 1000 to 2000 Å. On a low-resistance substrate, potential-step switching typically occurs in less than 50 ms. This led early investigators to question whether the electro-

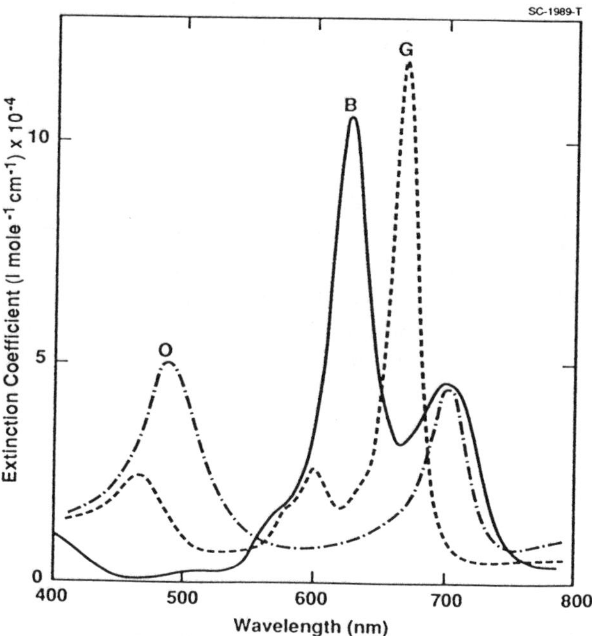

Figure 5 Spectra of $[(C_{12})_8Pc]_2Lu$ in dichloromethane. (G) Initial green form; (O) oxidized orange form; (B) reduced blue form. Reproduced with permission from [14] (Fig. 2), Pergamon Press PLC.

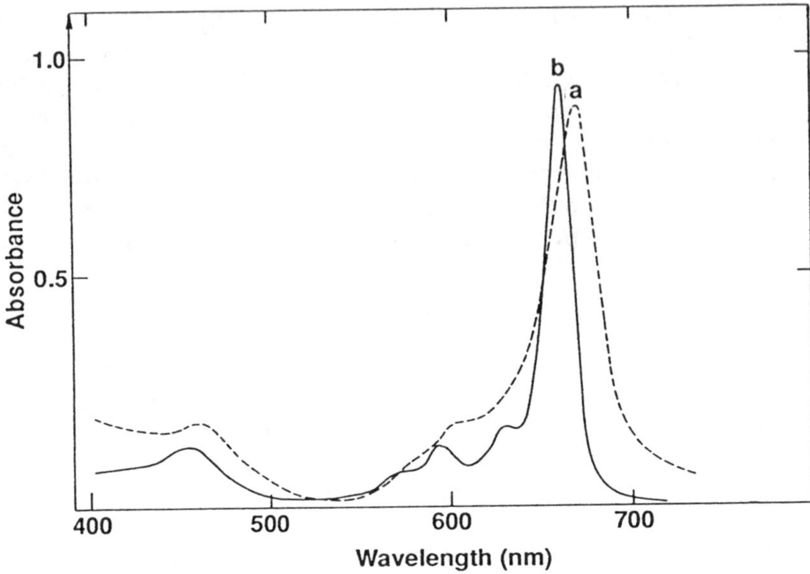

Figure 6 Spectra of dysprosium diphthalocyanine. (a) Sublimed film; (b) solution in dichloromethane. Reproduced with permission from [26] (Fig. 2), Elsevier Sequoia S.A.

chromism involved direct faradaic reactions of the phthalocyanine or was due to other processes. Consequently, there was great interest in determining the charges transferred per molecule and identifying the rate-controlling factors. From work in many laboratories, a coherent, though rather complex, picture of these systems has emerged. Kinetic processes have been modeled quantitatively in several instances, and some approximate equilibrium *p*H–potential diagrams have been developed. Complications arise from background reactions of the electrolyte, changing morphology of the electrode film, and certain influences of oxygen that are not fully understood.

Lanthanide diphthalocyanine electrode processes in the presence of liquid electrolytes will be discussed here. Information on cells with solid or semisolid electrolytes, which pertain largely to applications, is given in Section C on display devices.

a. *n* values and faradaic reactions

The first determination of *n*, the number of electrons transferred per molecule, for a diphthalocyanine electrode reaction probably was that published by Moskalev and Shapkin in 1978 [32]. A known weight of diphthalocyanine was anodically deposited on platinum from a solution of the reduced form in dimethylformamide containing hydrazine hydrate.

The deposit was then cathodically stripped with a linear voltage sweep, and the charge was measured by integration of the current–voltage curve. In a variation of this technique, the electrode bearing the anodic deposit was removed from the original cell and treated cathodically in aqueous potassium chloride, where it remained insoluble. Several lanthanide diphthalocyanines were investigated this way. One-step reductions, with *n* close to unity, were observed in all cases for what must have been green-to-blue transitions. The authors represented the oxidized forms as $PcLnPc_{ox}$, where Pc_{ox} signifies the radical group Pc^-. The soluble cathodic product in DMF was written $PcLnPc^-$, and PcLnPcH was proposed for the insoluble reduced film. Although later information would indicate K^+ rather than H^+ as the counterion for such conditions, the Soviet workers' early insight on the stable radical form $LnPc_2$ is noteworthy.

Corker, Grant, and Clecak used electron spin resonance (ESR) to investigate the electrochemical reactions of lutetium diphthalocyanine dissolved in purified dimethylformamide [50]. Starting with the green form, they prepared violet, blue, and yellow-red species. The electrolyte ions were those of tetrabutylammonium tetrafluoroborate. The solid green material and the green solution were paramagnetic, as was the violet reduction product, while the blue reduced form and the yellow-red oxidation product were ESR silent. These results indicated the following scheme for solutions in the aprotic DMF medium:

$$\text{Violet} \xleftarrow{+1e} \text{Blue} \xleftarrow{+1e} \text{Green} \xrightarrow{-1e} \text{Yellow-Red} \qquad (1)$$

Corresponding formulas were proposed for the redox species on the assumption that the paramagnetic green was a salt $[LuHPc_2]^+A^-$ containing an unidentified anion A^-.

Later ESR measurements by Walton, Ely, and Elliott confirmed the paramagnetism of lutetium diphthalocyanine in the green state but revealed some complexities in oxidation of the green films [44]. The ESR signal was lost on oxidation in contact with aqueous potassium chloride, consistently with a one-electron step, but was retained in the corresponding process with sodium sulfate. Walton et al. suggested that either *n* was 2 for the sulfate product or that the uptake of oxygen, indicated from other types of experiments [65], might account for the unexpected paramagnetism of the orange sulfate form. Surprisingly, ytterbium diphthalocyanine showed no ESR signal in the green or the orange state.

Chang and Marchon determined magnetic susceptibilities of bulk preparations of lutetium diphthalocyanine and several of its chemically formed oxidation and reduction products [5]. They proposed the scheme

$$\underset{\text{Violet}}{LuPc_2^{2-}} \xleftarrow{+1e} \underset{\text{Blue}}{LuPc_2^{-}} \xleftarrow{+1e} \underset{\text{Green}}{LuPc_2} \xrightarrow{-1e} \underset{\text{Reddish-Brown}}{LuPc_2^{+}} \xrightarrow{-1e} \underset{\text{Red}}{LuPc_2^{2+}} \qquad (2)$$

and presented mass spectral data confirming the radical formula $LuPc_2$ for the paramagnetic green.

Further electrochemical *n* values became available from moving-boundary experiments [66,67] and from an incremental voltage-step procedure carried out under near-equilibrium conditions [48]. The boundary-propagation experiments yielded *n* values of roughly 2 for the oxidation of uncycled lutetium diphthalocyanine films in both chloride and sulfate electrolytes. The cathodic measurements (Subsection b) were more difficult, but one-electron reductions of the green to light blue were confirmed in some instances. The incremental voltage-step method, also applied to films, gave an *n* value of 1 for the green/orange, blue/green, and blue/violet transitions, and in acidic solutions, a value of 2 for the blue/dark-purple-blue process. Both of these investigations [66,67,48] will be more fully discussed in the following.

By 1982, many formulas had been proposed for species within the lutetium diphthalocyanine redox system [57]. Still at issue were the possibility of more than one green form, the extent of *p*H dependence, and the sometimes elusive role of oxygen.

More recent work by Gavrilov et al. on a special blue form of lutetium octa-4-*t*-butyldiphthalocyanine illustrates the complex nature of these materials [68,69]. This blue solid was produced in the customary chemical synthesis and separated chromatographically from the green fraction. In *o*-dichlorobenzene solution with a neutral electrolyte, the blue produced six well-resolved pairs of voltammetric peaks at a gold electrode, while the green material from the same synthesis produced only four of those pairs. These unusual results were explained on the basis that the chemically synthesized blue material was a dimeric form. The two additional pairs of peaks were for one-electron processes of the dimer; in those cases, *n* was 0.5 electron per lutetium atom. The solid blue dimer was convertible to the green monomer $[(t\text{-Bu})_4Pc]_2Lu$ by exposure to strongly anodic potentials or by grinding the solid material [68].

b. Ion-insertion mechanisms

When a solid film undergoes an electrochemical reaction, the electrons gained or lost by the film are compensated by transfer of ions to or from the electrolyte. This is necessary to preserve electrical neutrality of the film. The ion-transport processes are remarkably fast in lanthanide diphthalocyanine films, even with polyatomic counter ions such as sulfate or acetate. Accordingly, research on the electrochromic mechanisms has been concerned with ionic mobilities in the films, the structural order or porosity of the organic solids, the effects of ionic radii and solvation, and the possible formation of adducts with electrolyte acids or salts. Information on the ion-insertion mechanisms obtained by several experimental approaches is summarized in this section.

Nicholson and Pizzarello investigated the faradaic reactions of lutetium diphthalocyanine with the solid-state moving-boundary technique illustrated in Fig. 7 [65–67,70]. The film was sublimed onto a nonconductive substrate such as sapphire or Mylar. A metal contact was applied to the upper end of the film, and the lower end was placed in the electrolyte. Under a constant applied current, the color change began at the electrolyte interface and was propagated upward with a well-defined boundary. The total voltage and the distance traveled by the boundary were monitored as functions of time. In this configuration, the system was ohmically controlled and could be represented as two resistors, of lengths x and $\ell - x$, in series. The sheet resistivities of the two color phases were resolved by raising the electrolyte level in known increments and measuring the resulting voltage drops. These experiments yielded carrier mobilities in the ionic phases, as well as approximate n values and bulk film resistivities. A qualitative version of this technique was used independently by Yamana to examine the reaction sites in erbium diphthalocyanine films [41].

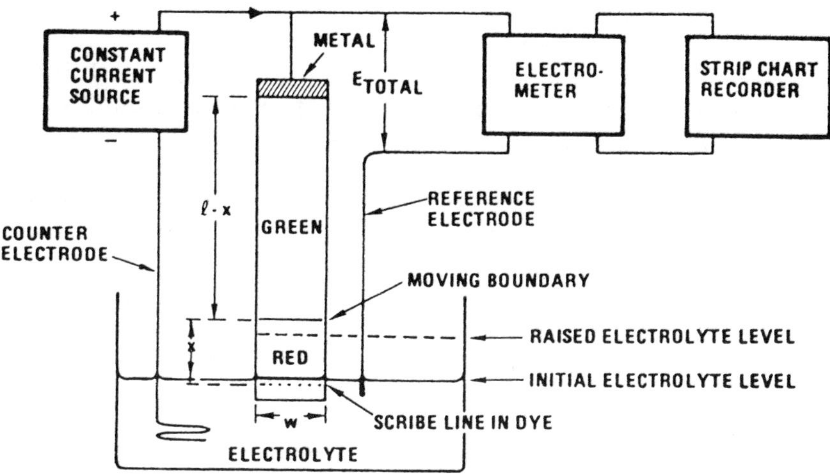

Figure 7 Experimental arrangement for moving-boundary measurements. Reproduced from [66] (Fig. 2) by permission of the publisher, The Electrochemical Society, Inc.

Some of the moving-boundary results for the anodic and cathodic electrochromic processes of lutetium diphthalocyanine are given in Table 3. The mobilities were calculated on the assumption of one predominant charge carrier in the color-converted phase. Mobilities near 4×10^{-6} cm^2/V s were determined for the anions in the red (orange) oxidation products. This value is comparable to 1.2×10^{-5} cm^2/V s for Ag^+ in single-crystal silver β-alumina [71]. The cathodic product formed in the moving-boundary experiments depended on the electrolyte. Neutral solutions of alkali metal salts produced light blue materials, while the strongly acidic 1 *M* HCl yielded a dark violet product with a cation mobility of 8×10^{-7} cm^2/V s. The *p*H effects will be discussed in more detail in Subsection C.

Table 3 Mobilities of Counter Ions in Lutetium Diphthalocyanine Electrode Films Determined by the Moving Boundary Technique

Color State	Electrolyte	Counter Ion	Mobility (cm^2/V s)x10^6	Ref.
Orange	1 M KCl	Cl^-	4	66
Orange	1 M Na_2SO_4	SO_4^{2-}	4	66
Violet	1.2 M HCl	H^+	0.8	67

The presence of chloride and sulfate ions in the respective oxidation products prepared by boundary propagation was confirmed directly by autoradiographs, with ^{36}Cl and ^{35}S as radiotracers, and by energy-dispersive spectroscopy on the same types of specimens, performed in the scanning electron microscope [70].

Using x-ray photoelectron spectroscopy, Collins and Schiffrin also found the expected counteranions in platinum-supported lutetium diphthalocyanine films that had been electrolyzed in aqueous KCl or KF [7]. Smaller proportions of the counterions were also present in the cycled green films.

Cyclic voltammetry provides further evidence of the difference between fresh and cycled films [7,49,72]. Figure 8, from the work of Castaneda et al., illustrates a "break-in" effect that is commonly observed, whether the initial scan is anodic or cathodic. The first scan typically requires several tenths of a volt more than later scans to reach the current peak. This behavior, which is characteristic of organic electroactive films, is thought to indicate an increase in porosity of the film as it is penetrated by the solvent.

Castaneda and Plichon made a voltammetric study of lutetium diphthalocyanine in aqueous electrolytes, including twenty cation–anion combinations [72]. To explain some asymmetries in the current–voltage curves, they proposed that an electrolyte MA might be present in the

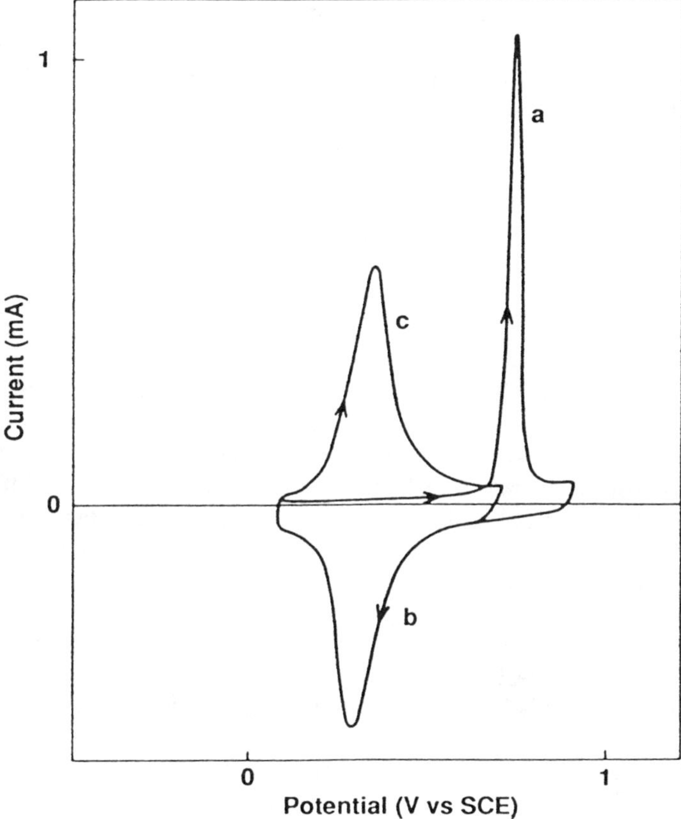

Figure 8 Voltammograms of lutetium diphthalocyanine film on platinum in 0.5 *M* H_2SO_4, scanned at 100 mV/s. (a) Initial scan; (b) steady-state cycling. Reproduced with permission from [49] (Fig. 2), Elsevier Sequoia S.A.

green, blue, and red forms of a cycled film, and that its transport *as a neutral entity* into or out of the film could be a slow step that hinders the overall charge-compensation process. An example of such a sequence is:

$$\underset{\text{Film}}{LuPc_2{}^+A^-,MA} \longrightarrow \underset{\text{Film}}{LuPc_2{}^+A^-} + \underset{\text{Soln}}{MA} \qquad \text{Slow} \qquad (3)$$

$$\underset{\text{Film}}{LuPc_2{}^+A^-} + e + \underset{\text{Soln}}{M^+} \longrightarrow \underset{\text{Film}}{LuPc_2,MA} \qquad \text{Fast} \qquad (4)$$

Besbes et al. in the same laboratory investigated substituted lutetium diphthalocyanines with 16 long-chain alkoxymethyl groups in each molecule [17]. These materials also behave as liquid crystals [73]. Some unexpected anion effects on the green/orange system were found, and the major difference between the initial and later scans illustrated by Fig. 8 was not observed.

Bardin, Plichon, et al. greatly modified the cation specificity of lutetium diphthalocyanine by substituting crown-ether groups of the type 15-crown-5 at all of the 3,4 positions on the phthalocyanine rings [18]. Films of the substituted compound were investigated by cyclic voltammetry in several aqueous electrolytes. Figure 9 gives examples of the effects found in alkali-metal salt solutions. With $KClO_4$, $NaClO_4$, or KCl, there were two pairs of nearly symmetrical current peaks, showing reversibility of the green/orange and one-electron blue/green processes. As expected, both K^+ and Na^+ were easily inserted in the film with the crown substituents present. Sulfate ion was somewhat more difficult to insert than chloride, as indicated by the flatter green/orange peaks in Fig. 9(b). With $LiClO_4$, both oxidation and reduction of the film were severely hindered, as Fig. 9(c) shows. This was further evidence that the oxidation involved more than a simple incorporation of the anion. The behavior of the film toward acidic electrolytes was even more strikingly changed by the crown-ether groups. In HCl and $HClO_4$, no voltammetric responses were observed until a salt such as KCl or $KClO_4$ was added.

c. Equilibrium and *p*H–potential relationships

Formal electrode potentials for lanthanide diphthalocyanine redox couples in several organic solvents have been determined or estimated from solution voltammograms [14,52,53,74]. Most of the values pertain to the one-electron oxidation or one-electron reduction of the $LnPc_2$ form. For example, with aprotic solutions of the $LuPc_2$ system in dichloromethane, L'Her et al. found an $E°$ of +0.03 V versus Fc/Fc^+ for the $LuPc_2/LuPc_2^+$ couple and –0.45 V versus Fc/Fc^+ for the $LuPc_2^-/LuPc_2$ [52], where Fc/Fc^+ represents the standard ferrocene/ferrocinium system [75,76]. These values correspond, approximately, to +0.21 and –0.27 V versus Ag/AgCl,Cl^- (aqueous). Castaneda et al. found that the respective $E°$s with C_8 and C_{12} alkyl-substituted lutetium compounds differed from these by less than 0.1 V [14].

More recently, Konami et al. determined $E°$s for diphthalocyanines of twelve lanthanide elements and yttrium in aprotic *o*-dichlorobenzene solutions containing tetrabutylammonium perchlorate [74]. Both linear-sweep and potential-pulse voltammetry techniques were used. The results included $E°$s for one-electron steps involving species from $LnPc_2^+$ to $LnPc_2^{3-}$. If $LuPc_2$ is represented by the symbol D, the $E°$ values were 0.547 V versus Ag/AgCl(sat. KCl) for D/D^+, 0.094 V for D^-/D, –0.989 V for D^{2-}/D^-, and

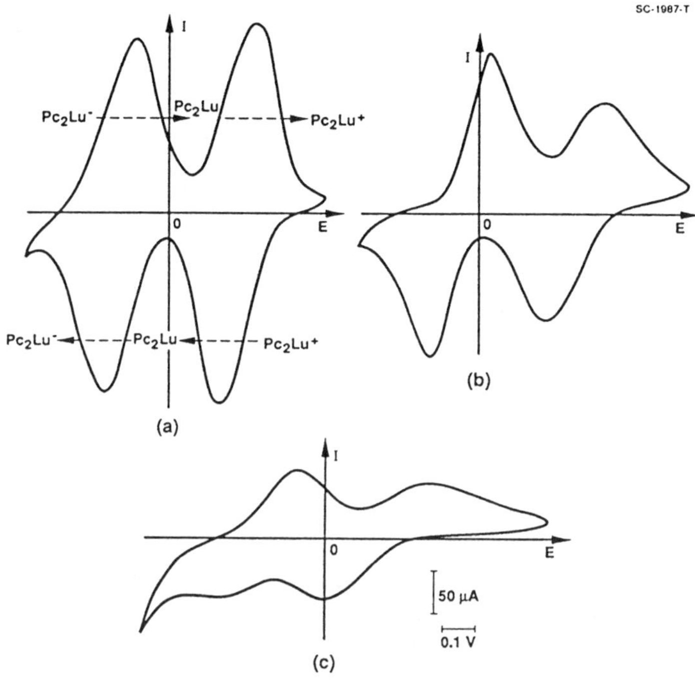

Figure 9 Voltammograms of crown-ether substituted lutetium diphthalocyanine film on SnO_2 substrate in contact with neutral aqueous electrolytes, scanned at 20 mV/s. (a) Sat. $KClO_4$, 1 *M* $NaClO_4$ or 1 *M* KCl; (b) sat. Na_2SO_4 or K_2SO_4 or 1 *M* NaCl (c) 1 *M* $LiClO_4$. Reproduced with permission from [18] (Fig. 1), Elsevier Sequoia S.A.

–1.351 V for D^{3-}/D^{2-}. Although these were soluble materials, the clear resolution of four aprotic redox steps involving five oxidation states is of interest in relation to the electrochromism of solid diphthalocyanine films.

Table 4 gives formal potentials for the one-electron oxidation step (nominally green/orange) in solid films of $LuPc_2$ and three of its ring-substituted forms. The voltammetric method involved averaging the peak potentials for the forward and reverse voltage sweeps. This is valid for fast electrochemical processes. In the spectroelectrochemical method, the potential at zero current was determined for the condition in which half of the spectral change for the transition had occurred in the film. The equilibrium potentials $E°_{G/O}$ for these couples, of the type $LuPc_2/LuPc_2^+$, are independent of *p*H. Substitution of eight propoxy groups into the molecule had essentially no effect on $E°_{G/O}$, but eight *t*-butyl groups apparently hindered the insertion of chloride ions to the extent that an

Table 4 Formal Potentials of Green/Orange Transition in Films of Unsubstituted and Substituted Lutetium Diphthalocyanine in Aqueous Electrolytes[a]

Green Form	Electrolyte	$E°_{G/O}$ (V vs Ag/AgCl,Cl$^-$)	Experimental Method	Ref.
$LuPc_2$	0.1 M HCl	0.56	Spectro-electrochemical	77
$LuPc_2$	0.1 M TBACl	0.60	Spectro-electrochemical	77
$Lu(C_3H_7O)_8Pc_2$	0.1 M HCl +0.1 M KCl	0.57	Voltammetry	15
$Lu(t\text{-}C_4H_9)_8Pc_2$[b]	0.1 M HCl +0.1 M KCl	0.75	Voltammetry	15
$Lu(15\text{-crown-}5)_8Pc_2$[b]	Satd $KClO_4$ +$HClO_4$ 3.3 < pH ≤ 6.0	0.23	Voltammetry	18

[a]E° is given for the electrolyte indicated, not for unit counter ion activity in solution.
[b]Langmuir-Blodgett film.

additional 0.19 V was required thermodynamically to oxidize the film. The ion selectivities introduced by the crown-ether groups are complex, as noted previously. Though much larger than *t*-Bu, the 15-crown-5 groups facilitated the oxidation; with a perchlorate counterion, $E°_{G/O}$ was 0.33 V lower than for the unsubstituted compound in chloride solutions.

Some other processes in the diphthalocyanine electrochemical systems are very sensitive to acids. A *p*H dependence would be expected from the behavior of $LuHPc_2$ as a weak acid, but several effects less obvious than this are important. In 1971, Moskalev and Kirin investigated the acid–base properties of tetrasulfonated diphthalocyanines in aqueous solutions by spectrophotometric titrations [12]. The sulfonic acid groups were fully ionized, but the labile protons in the substituted M(III)HPc_2 forms were removable only in alkaline solutions. The pK_a values of 8.7, 8.6, and 10.2 were determined for the sulfonated blue forms of gadolinium, yttrium, and lutetium diphthalocyanines, respectively, at 20°C. In the 1989 study of crown-ether-substituted lutetium diphthalocyanine complexes, a pK_A of 3.8 was estimated for the form LuH(CR)Pc_2 [18].

Nicholson and Weismuller reported the approximate *p*H–potential, or Pourbaix, diagram of Fig. 10 for lutetium diphthalocyanine in 1982 [48]. The experimental points correspond to equal concentrations of the oxidized and reduced species in film electrodes, as judged from the spectral peaks, at essentially zero current. The *n* values were measured by incremental potential-step coulometry. This diagram is somewhat approximate; more data are needed in the *p*H 3 to 4 region, and a low concentration of phthalate or phosphate buffer was present in the *p*H 3 to 7 range, in addition to the 1 *M* KCl. The disappearance of the green form of the diphthalocyanine near *p*H 0 is clearly evident, however.

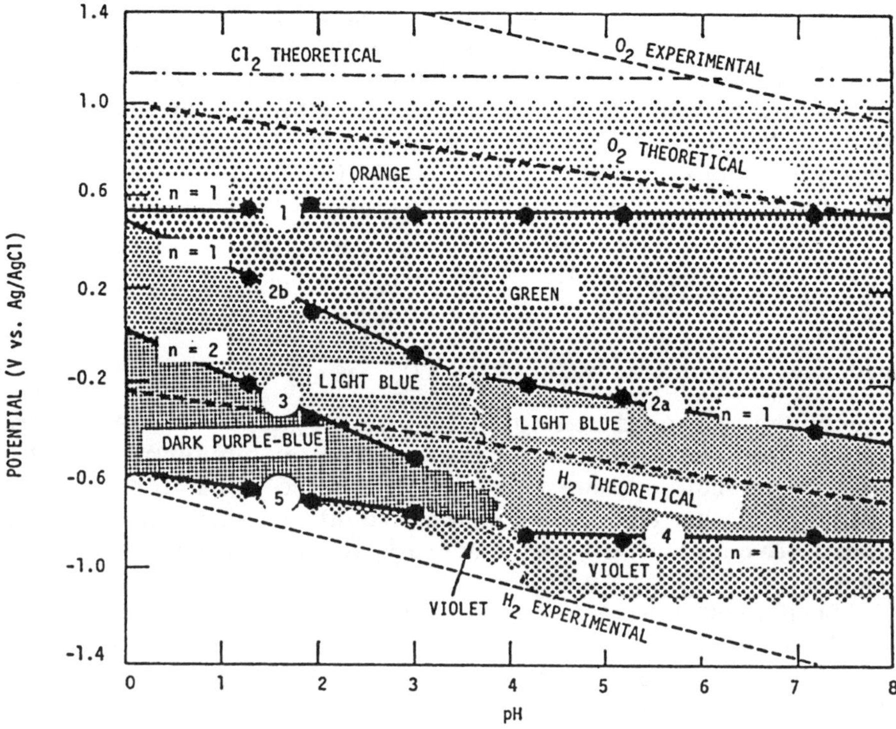

Figure 10 pH–potential diagram of a lutetium diphthalocyanine film on tin oxide. Reproduced with permission from [48] (Fig. 5).

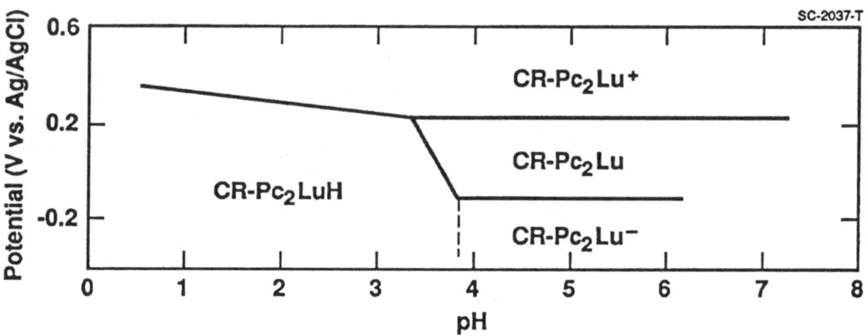

Figure 11 pH–potential diagram of crown-ether-substituted lutetium diphthalocyanine in saturated $KClO_4$ + $HClO_4$ solutions. Adapted with permission from [18] (Fig. 4), Elsevier Sequoia S.A.

In 1985, L'Her et al. investigated the effect of adding trifluoracetic acid to solutions of lutetium diphthalocyanine in dichloromethane [53]. They observed the disproportionation of the green form to orange and blue at an acid concentration near 10^{-3} M and demonstrated by rotating-disk voltammetry that the one-electron waves of the green/orange and blue/green couples merged into a single two-electron wave, for the blue/orange, under those conditions. The electrode reaction was then given as

$$\underset{\text{Blue}}{LuPc_2H_y{}^{(y-1)+}} \rightleftharpoons \underset{\substack{\text{Orange}\\ \text{(Brown)}}}{LuPc_2{}^{+}} + yH^{+} + 2e \qquad (5)$$

(The symbol y is used here to distinguish the number in the chemical formula from the faradaic n.) Thus more than one proton might add to $LuPc_2^-$ in dichloromethane solution. If y were not equal to the faradaic n value of 2, this electrode potential would not show the "normal" dependence on pH (59 mV/unit) that nearly always occurs in organic redox systems.

Such unusual pH dependence had also been found in the diagram of Fig. 10 for Line 2b (light-blue/green) and Line 3 (light-blue/dark-purple-blue). Those abnormally steep slopes were attributed to formation of acid adducts with the solid reduced forms of lutetium diphthalocyanine [48].

In 1989, Castaneda et al., apparently unaware of [48], sought evidence of the disproportionation of lutetium diphthalocyanine films contacted by aqueous acids, using cyclic voltammetry [49]. They observed a blue/orange two-electron peak and concluded that the disproportionation point was near pH 0. A schematic pH–potential diagram showing that feature and four oxidation states of lutetium diphthalocyanine was given.

The diagram of Fig. 11 was determined by Bardin et al. for the crown-ether-substituted lutetium diphthalocyanine in perchlorate solutions [18]. There are several differences between this diagram and Fig. 10 for the unsubstituted lutetium system: (a) The crown-ether green form is more easily oxidized, with the green/orange line occurring at 0.23 V versus Ag/AgCl, compared to 0.56 V for the unsubstituted system: (b) The green crown-ether compound disproportionates at pH 3.3, rather than pH 0. (c) The blue/orange line could be determined for the crown-ether system at pH values to the left of the disproportionation point; that line in Fig. 11 has a slope of –47 mV. Both diagrams show evidence of acid ionization processes in the vicinity of pH 3.5 to 4.0. Bardin identified this reaction in the crown-ether system as

$$LuH(CR)Pc_2 \rightleftharpoons Lu(CR)Pc_2{}^{-} + H^{+} \qquad (6)$$

For the two light blue phases in Fig. 10, this transitional *p*H region probably includes the steps [48]

$$LuPc_2H_3^{2+} \rightleftarrows LuPc_2H_2^{+} + H^{+} \quad (7)$$

$$LuPc_2H_2^{+} \rightleftarrows LuPc_2H + H^{+} \quad (8)$$

In Eqs. (6)–(8) equivalent amounts of anions are understood to be present in the solids along with the various cationic reduction products.

Pourbaix diagrams have not been reported for diphthalocyanines of tetravalent metals. However, Corbeau et al. found that UPc_2 does not disproportionate in the presence of acid and inferred that, with four bonds from the metal atom to the phthalocyanine rings, this molecule cannot be protonated [55]. The tetravalent-metal systems may be entirely independent of *p*H.

d. Kinetics

Several efforts have been made to develop quantitative kinetic models for the lanthanide diphthalocyanine electrode processes [7,38,78]. It is recognized that the rate-controlling step may depend on the current density and therefore may vary with the experimental technique. Thus, a process that appears fast and electrochemically reversible in a slow voltammetric scan may become rate-limited under a large imposed current or a potential step. Moreover, with ion-insertion electrodes, there can be very pronounced kinetic differences between the reactions of fresh and cycled films [49,72]. Figure 8 is a good example of such behavior.

Nicholson and Pizzarello investigated the oxidation of precycled lutetium diphthalocyanine films from the green to the orange (red) state in aqueous 1 *M* KCl by the galvanostatic-transient technique [78]. The potential of the film on a tin oxide substrate was measured as a function of time under applied current densities ranging from 0.3 to 265 mA/cm^2. The transition time for completion of the electrode reaction, which was found from optical, as well as electrical transients, ranged from 10 s to 7 ms. Several kinetic models were considered, but the principal data, from 0.3 to 6 mA/cm^2, were most consistent with a two-layer system in which the oxidation began at the electrolyte interface and moved through the film, retaining a boundary between the green and orange phases. In several plotting forms, the transient data fit the equations for an expanding space charge within the orange ionically conductive phase.

A parameter $\kappa\mu_i$ was evaluated, where μ_i is the ionic mobility in the oxidized phase and κ is its dielectric constant. This product was 4×10^{-5}

cm^2/V s. Taking the mobility of 4×10^{-6} cm^2/V s from the moving-boundary experiments reported in Table 3, a dielectric constant of 10 was estimated for the oxidized film. This appears to be the right order of magnitude, although κ could exceed 10 with water and ions present. The effective μ_i would then be correspondingly lower.

Collins and Schiffrin sought a kinetic model for the $LuPc_2$ oxidation process to fit their linear-sweep voltammograms in aqueous 1 *M* NaF and 1 *M* KCl [7]. They too ruled out a simple reversible process, using instead, a classical electrode kinetics treatment with a rate constant and transfer coefficient. The concentrations of oxidized and reduced species were assumed to be uniform throughout the film. Voltammograms calculated for this model could be made to fit the experimental ones very well up to the position of the maximum current, but the two curves diverged beyond that point. The same work included potential-step measurements that, for times less than 1 s, were treated with a purely diffusional model. A diffusion coefficient of ~10^{-11} cm^2/s, presumed to be that of the chloride ion, was evaluated. This is much lower than the diffusion coefficient of 1.1×10^{-7} cm^2/s, which corresponds to the chloride mobility of 4×10^{-6} cm^2/V s found by the moving-boundary technique.

A later paper by Collins and Schiffrin discussed galvanostatic-transient measurements on the oxidation of lanthanide diphthalocyanine films in ethylene glycol containing chloride or fluoride salts [38]. Noting that the diffusion coefficient of 1.1×10^{-7} cm^2/s is exceptionally large for an organic film, they preferred a modified Nernstian approach rather than a space-charge model for the electrode. The analysis used assumes discrete redox centers throughout the film, each with its own standard potential [79,80]. The apparent film capacitance is then attributed to the spread in redox potentials. This treatment gave the expected shape for the galvanostatic transient, but, as the authors recognized, a model with so many adjustable parameters has limitations for the assignment of a physical mechanism to the process described.

Clearly, much remains to be learned about the kinetics of diphthalocyanine electrodes and the solid-state transport processes involved. The film thickness probably increases considerably with uptake of solvent and ions. The forward and reverse processes are sometimes very asymmetrical. In developing a model, it is convenient, and often sufficient, to consider only one type of charge carrier. However, the conduction can be mixed in electrochromic systems. The diffusion of ions may be greatly enhanced by concurrent electron or hole transport [81]. When this occurs, an ionic diffusion coefficient may not have a unique value in a given chemical phase of the electrode.

Electronic conduction in the diphthalocyanine materials is generally understood to occur by electron hopping between sites of different oxidation states. Figure 12 shows evidence of this in the sudden increase of conductivity due to current reversal in a dual moving-boundary experiment

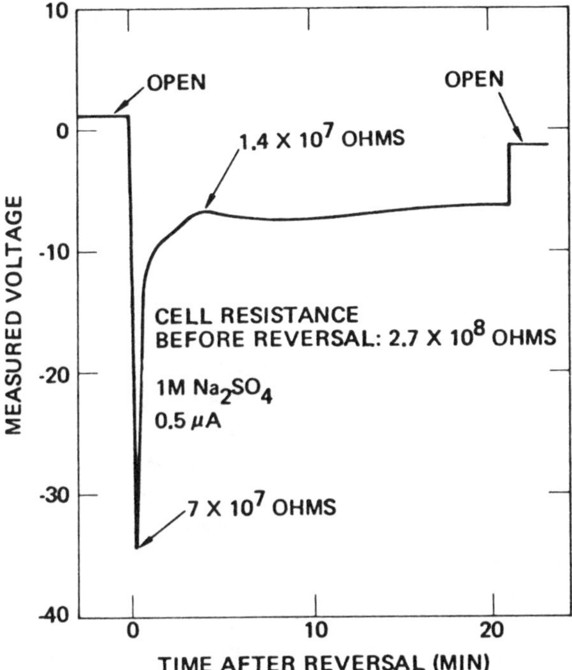

Figure 12 Voltage–time relationship following current reversal in a dual-boundary experiment with a lutetium diphthalocyanine film [82].

[82]. The lutetium diphthalocyanine film was on a sapphire substrate, arranged horizontally with each end contacted by the electrolyte. The orange oxidized and blue reduced phases had been propagated from the respective ends of the film at an applied current of 5 μA. During that process, the cell resistance rose to 2.7×10^8 Ω. Immediately after current reversal, the resistance dropped to 7×10^7 Ω, and 4 min later it was 1.7×10^7 Ω. Only a small percentage of the film had been reconverted to green, but the resistance had dropped by a factor of 16 due to rapid penetration of the ionic phases by electronic charge carriers. Similar effects were observed in single-boundary current reversal.

e. The role of oxygen

A discussion of electrochromic diphthalocyanines would be incomplete without comment on the pronounced effects that oxygen can have on these systems. Some observations and mechanisms proposed to explain them are noted in the following. It is apparent that further research is needed to clarify the role of oxygen in the electrode reactions and especially in their kinetics.

Collins and Schiffrin found oxygen in vacuum-deposited lutetium diphthalocyanine films at levels comparable to that of lutetium and suggested that it could be present as hydroxide or superoxide [7]. A difference in the absorption spectra of sublimed films before and after exposure to air (Fig. 4) was mentioned earlier [47]. M'Sadak et al. also determined oxygen in original vacuum-deposited films and showed that the amount increased on application of anodic potentials in aqueous electrolytes [26].

Oxygen seems to promote the electrochromic switching process and may be essential under some conditions to initiate the conversion from green to orange. This usually requires a large anodic overpotential, as illustrated by Fig. 8. Nicholson and Pizzarello observed that ambient oxygen was necessary to propagate the anodic green/orange reaction boundary in a lutetium diphthalocyanine film contacted by a sulfate electrolyte [65]. With a chloride solution, however, the boundary could advance in the absence of oxygen. Walton et al. found differences in ESR spectra of the oxidation products formed in sulfate and chloride electrolytes, which may have indicated a difference in oxygen uptake by the respective oxidized forms [44].

Faradaic *n* values approaching 2 often are observed in the first oxidation of green lutetium diphthalocyanine films. Using related information from solution spectra, Nicholson and Weismuller suggested that these larger *n*s could be due to the presence of $LuHPc_2^+O_2^-$ in the films [47]. Since the superoxide ion O_2^- is a fairly strong reducing agent, this complex could undergo a two-electron oxidation initially. With $LuPc_2$ as the corrected formula for the oxygen-free green form, equations for the electrode reactions [47] become:

$$\underset{\text{Initial Green}}{LuHPc_2^+O_2^-} + X^- \longrightarrow \underset{\text{Orange}}{LuPc_2^+X^-} + H^+ + O_2 + 2e \tag{9}$$

$$\underset{\text{Cycled Green}}{LuPc_2} + X^- \longrightarrow \underset{\text{Orange}}{LuPc_2^+X^-} + 1e \tag{10}$$

In accordance with Eq. (10), the reversible potential for the green/orange transition in cycled films is independent of *p*H and shows the expected Nernstian relationship to the anion concentration [77]. It would be very interesting to examine the possible influences of anion and hydrogen-ion concentrations, as well as oxygen pressure, on the initial oxidation process.

C. DISPLAY DEVICES

The various oxidation states of the rare-earth diphthalocyanines correspond to colors ranging throughout the visible spectrum. Figure 10 defines conditions for producing violet, dark blue, light blue, green, and orange film colors by completing the indicated faradaic reactions of lutetium diphthalocyanine. Still other, intermediate colors result from partial conversion of the film from one oxidation state to another. Yellow is produced by partial oxidation of a green film to orange, and blue-green by its partial reduction to blue. Many pleasing contrasts are obtained when selected pairs of colors are formed on adjacent coplanar electrode areas. Thus, a patterned electrode plate with segments connected to independent electrical leads can be used to display alterable numerals, letters, or graphic designs.

Before the multicolor electrochromism of lanthanide diphthalocyanines was generally recognized, electrochromic display development had been limited to monochrome systems, primarily the blue/white tungsten oxide and the magenta/white viologens. The availability of multiple colors, even electrically "tunable" colors, along with other favorable performance characteristics, generated renewed engineering interest in electrochromics as display materials.

i. Device Structures and Display Characteristics

A segmented electrochromic display cell is illustrated schematically in Fig. 13. Although only two segments are shown, many may be present, and all are electrically addressable by leads to an external power supply. The color of a given segment or group of segments is controlled in most cases by selecting the voltage to be applied between that electrode area and a common counter electrode. After the faradaic reaction is completed, the switched electrodes are preferably left on open circuit to avoid possible effects of background current. Such a display is said to have nonvolatile memory. Should fading occur, in minutes or hours, for example, the image can be refreshed by reapplying the original drive voltage.

Figure 13 shows an electrolyte cavity filled with a porous white material such as titanium dioxide powder. In this configuration, the color display is viewed by reflected light, as one views a printed page. The counter electrode then can be a single plate covering the entire back area of the cell. This electrode should have a coulombic capacity relatively large compared to that of the electrochromic layer, so that it maintains a fixed electrochemical potential during the switching process. Its cycle life should be at least as great as that of the display electrode. Reversible couples such as Ag/AgCl, Pb/$PbSO_4$, and Pb/PbF_2 are used as counter

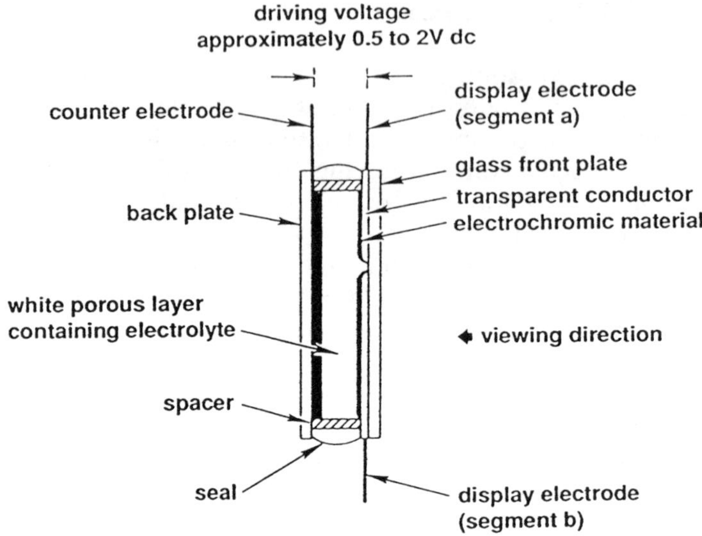

Figure 13 Schematic design of electrochromic display cell. Reproduced with permission from [83] (Fig. 1), Pergamon Press PLC.

electrodes in demonstration electrochromic cells and would be suitable for some practical display devices.

A back-lighted display is preferred for some applications. The device then is, in effect, a variable color transparency, which may be viewed directly or projected on a screen. In a back-lighted display, the white porous material is omitted, and the counter electrode is placed outside the field of view, usually in the shape of a frame around the viewing area. Alternately, the counter electrode can be a semitransparent screen or a transparent system with the desired faradaic response. Transparent faradaic counter electrodes would need to be further developed experimentally for use in combination with organic electrochromics. The special requirements of electrochromic matrix displays are considered in Subsection iv.

Many performance features desired in flat-panel display materials are inherent to the electrochromic diphthalocyanines [5,6,84]. The range of color and the fast response, typically less than 50 ms, distinguish these systems from other electrochromics, including tungsten oxide. Surprisingly, the rapid switching continues with lutetium diphthalocyanine at temperatures as low as –50°C in a concentrated calcium chloride electrolyte, and it has been observed in preliminary experiments up to 100°C [6]. The switching charge density of 1 to 3 mC/cm^2 is several times lower than those of inorganic electrochromics. This advantage results from the very high molar extinction coefficients of phthalocyanines.

The switching energy for the $LuPc_2$ system usually falls within the range of 1 to 4 mJ/cm^2. Hence, *if the display is switched only once per second,* the average power to the area that is actually switched is on the order of 2 mW/cm^2. With less frequent switching, the power consumption is correspondingly lower. In special applications, therefore, the electrochromic would require less power than a liquid-crystal display, with which it is most often compared. Although its power consumption is very low, the liquid crystal must be activated continuously to maintain the image. More important advantages of the electrochromic over liquid crystals are wide viewing angle, noncritical electrode spacing, greater temperature range, and coincident color, that is, the ability to produce the full color range in the same film area without the use of a tricolor light-filter system. "Shades of gray," or different intensities of the same color, are not directly available with multicolor electrochromics as they are with liquid crystals or with a monochromic system such as tungsten oxide.

Still to be improved for the diphthalocyanine displays are the cycle life, the hue of the red color, and the technology for matrix addressing. Progress in each of these areas is reviewed in this section. Despite the success of liquid crystals, most display users prefer a solid device. All of the components need not be true solids; gels and other semisolid materials are acceptable if they do not release liquid or vapor under ordinary conditions of use or storage. Examples of experimental solid and semisolid diphthalocyanine electrochromic cells are included in the discussion of cycle life.

ii. Cycle Life

A thoroughly sealed electrochromic cell can have a shelf life of several years. No single fundamental cause limits the cycle life of such devices, but many factors can contribute to the deterioration of a diphthalocyanine film or its inability to continue switching at an acceptable rate. One failure mechanism often predominates with a given cell composition and switching routine. Even without extensive development, however, the lanthanide diphthalocyanine electrodes have surpassed typical secondary battery electrodes in cycle life by two or three orders of magnitude. The thin-film structure of the electrochromic appears to be the main reason for its greater cycling ability.

a. Liquid electrolytes

The *p*H–potential diagram of Fig. 10 is useful in defining safe operating voltage ranges for the lutetium diphthalocyanine electrodes in aqueous electrolytes. The heavy broken lines (theoretical) indicate thermodynamic limits imposed by the electrolytic formation of hydrogen and oxygen under the assumption that each gas is produced at a pressure of

1 atm. Gas evolution obviously would tend to disrupt a thin organic film. In practice, both theoretical limits can be exceeded because of substantial overvoltages for hydrogen and oxygen formation on the semiconductive tin oxide substrate. The potentials for rapid onset of these background currents are shown by the experimental broken lines in Fig. 10. Those lines represent approximate outer limits for switching the electrochromic material; the practical limits may differ in the presence of the phthalocyanine film.

Table 5a summarizes life data for green/orange cycling of lutetium diphthalocyanine in liquid electrolytes. Corresponding information for cycling to the blue and purple states is given in Table 5b. Most life testing in liquid electrolytes has been done in three-electrode cells of the type illustrated by Fig. 1. The input signal usually consists of potentiostatic stepping between two voltage levels. Each level is maintained, typically, for 1 s, although several variations have been used, including shorter voltage steps with off periods between [38], and a four-level potentiostatic cycle approximating that routine [5,84].

The progress of a cycling experiment is monitored visually, by measuring charge and absorbance transients, and by occasionally recording complete spectra at constant potentials. Different criteria, some only qualitative, have been used in various laboratories to define the end of the cycle life. A 50% drop in absorbance or switching charge relative to early-stage cycling may be taken as a practical end of the electrode life [85].

Three failure modes have been recognized for lutetium diphthalocyanine electrodes [5,85]: (a) Fading, or decrease of color intensity, which is thought due to chemical attack of the dye material, (b) peeling or less obvious loss of adhesion between the dye film and the conductive substrate, and (c) relatively abrupt cessation of switching due to corrosion of the electronic contact to the substrate.

The life of the lutetium diphthalocyanine film in green/orange cycling is much greater in aqueous acids than in neutral solutions [7,8,85]. There is some evidence that fading in the neutral electrolytes is due to attack of the oxidized phthalocyanine complex by hydroxide ions [7]. Figure 10 shows that anodically formed oxygen also could be involved, especially with a platinum or gold substrate instead of tin oxide. A Nafion binder produced a three–fold life improvement in KCl but eventually led to peeling of the film as the polymer swelled [85]. Collins and Schiffrin found that the green/orange (or green/yellow-tan) cycle life could be extended by three orders of magnitude by using salt solutions in ethylene glycol instead of water [38]. They also reported that a modified input signal consisting of a current-step reduction followed by a potential–step oxidation was beneficial.

Unfortunately, the blue and purple colors are not attainable with film electrodes in ethylene glycol and some other organic media because

Table 5 Life Data for Lutetium Diphthalocyanine Electrodes in Liquid Electrolytes

a. Green/Orange or Blue-Green/Orange Cycling

Substrate	Electrolyte[a]	Life (Cycles)	Ref.
Pt	KCl, KBr or KF	10^4 to 10^5	7
Au	0.5 M $LiClO_4$	$>10^6$ [b]	17
Pt	TEACl, LiCl or TEAF in ethylene glycol[c]	$>4 \times 10^6$	38
SnO_2	1 M KCl	5×10^4	85
SnO_2	1 M KCl + 0.2 M NaAMPS[d]	8×10^4	85
SnO_2 (Nafion Binder)	1 M KCl	2×10^5	85
Au	H_2SO_4+$(NH_4)_2SO_4$	$>10^7$	86
Au	1 M H_2SO_4	2×10^7	37
SnO_2	1 M KCl + 0.33 M HCl (pH 2)	2×10^5	85
SnO_2	1 M KCl + 0.24 M AMPS (pH 0.6)	3×10^6	85

b. Cycling to Blue or Purple States

Substrate	Electrolyte	Colors	Life (Cycles)	Ref.
Au	H_2SO_4	Purple/Green	$\sim 10^5$	86
Au	1 M H_2SO_4	Purple/Orange	$>10^5$	37
SnO_2	1 M KCl + 0.005 M NaAMPS	Blue/Orange	4×10^4	85
SnO_2	1 M KCl + 0.25 M AMPS	Blue/Orange	6×10^3	85
SnO_2	1 M KCl + 0.2 M HCl	Blue/Orange	3×10^4	85

[a]Aqueous unless noted.

[b]Octyloxymethyl substituted.

[c]TEA represents tetraethylammonium.

[d]AMPS represents 2-acrylamido-2-methylpropanesulfonic acid.

those forms tend to dissolve in organic solvents [38,67]. Through further research, it should be possible to find combinations of diphthalocyanine materials and organic media that will provide these film colors. Aprotic electrolytes are especially desirable to prevent electrolytic hydrogen formation. Work with lutetium diphthalocyanine solutions in dimethylformamide [50] and dichlorobenzene [51,74] strongly suggests that protons are not necessary to form the dark blue and purple, although H^+ is the usual counterion associated with these reduced states in aqueous electrolytes.

Figure 10 implies that the blue colors of lutetium diphthalocyanine should be produced in aqueous electrolytes below *p*H 3 with less risk of hydrogen formation than at *p*H 3 to 7. This potential benefit results from the positions of Lines 2b and 3 in the figure. Nevertheless, the cycle life involving the reduced forms has generally been 10^4 to 10^5 cycles, while 10^6 and 10^7 cycles are obtained in green/orange or blue-green/orange switching. The main problem with the strongly reduced films appears to be cracking or loss of adhesion. This may be caused by hydrogen formation, even with careful control of the drive signal, or by expansion of the film as it complexes with acids. Peeling also occurs when the dark blue film is exposed to oxygen. It is more severe, therefore, in a poorly sealed cell.

Several approaches to cycle-life improvement involve chemical or physical modification of the diphthalocyanine film. Besbes et al. demonstrated a life greater than 10^6 cycles for the green/orange transition of octyloxymethyl–substituted lutetium diphthalocyanine in aqueous $LiClO_4$ [17]. Because this material has liquid-crystal character, it may be able to retain a more ordered structure during the cycling process. The research of Bardin et al. on crown-ether-substituted lutetium diphthalocyanine was discussed in Section B,v [18]. The voltammogram of Fig. 9(a) shows that cation insertion was facilitated by the crown-ether groups with $KClO_4$, $NaClO_4$, and KCl electrolytes. The one-electron blue/green reaction then became almost as reversible as the green/orange. Such reversibility implies long cycle life; if the electrochemical reaction is fast enough, there will be no need to drive the electrode into regions where gas formation can occur. The *p*H–potential diagram of Fig. 11 indicates conditions under which three colors are obtainable with the crown-ether material.

Other modifications described by Liu et al. showed promising results in voltammetry and initial life testing [15]. Lutetium and erbium diphthalocyanines octasubstituted with propoxy and or *t*-butyl groups were prepared. In addition to the uniformly substituted compounds, there was one type with mixed propoxy and *t*-butyl substituents on both rings and another with these two groups incorporated asymmetrically, one kind of group in each ring. The electrochromic materials were examined as solution-cast films and as highly structured Langmuir–Blodgett films, in which the macrocycles were shown to stand on edge. The "break-in" effect

(Fig. 8) typical of most diphthalocyanines was absent with the Langmuir films, and improved colors were observed in 1500 voltammetric cycles.

b. Solid or semisolid electrolytes

Work with solid diphthalocyanine electrochromic cells has been somewhat exploratory. Spatial limitations make it inconvenient to use a reference electrode, and many of these cells have also lacked a well-poised counter electrode. Significant cycling capabilities have been demonstrated, nevertheless, as noted in Table 6.

Generally, both mobile cations and mobile anions are needed for the electrochromic material to exhibit its full span of colors. This precludes the use of classical solid electrolytes having only one mobile ionic species. With lead fluoride, for example, only the green form and the orange or reddish oxidation product of the diphthalocyanine would be expected. Electrolyte formulations containing polymers with a small amount of water or an organic plasticizer appear more promising because they can provide both types of mobile ions, as well as higher conductivity.

The cell described by Yamamoto et al. contained a potassium nitrate electrolyte solidified with an epoxy resin and had erbium diphthalocyanine on the counter electrode as well as the display plate [88]. With a white titanium oxide pigment in the electrolyte layer, only the front dye film was visible by reflected light. The life data (Table 6) are comparable to those in liquid-electrolyte cells.

Table 6 Life Data for Cells with Solid or Semisolid Electrolytes

Cell	Input Signal (V)	Colors	Life (Cycles)	Ref.
ITO/$LuPc_2$/WO_3/Hydron-S[a] + H_2O/Au	2		10^4	87
ITO/$LuPc_2$/PbF_2/Au	1 to 2	Green/Orange	10^5	45
ITO/$ErPc_2$/ KNO_3 + TiO_2[b] +Epoxy + H_2O / $ErPc_2$/ITO	2 to -2	Green/Orange Blue/Orange	10^6 7×10^4	88
ITO/$ErPc_2$/ PVA - Salt[c] + H_2O Vapor / ITO	3	Green/Orange	$>10^3$	89

[a]Poly(2-hydroxyethylmethacrylic acid).

[b]Opaque white electrolyte layer.

[c]KBr, KCl, KI or $CaCl_2$.

Another solid cell making effective use of polymers is that of Sammells and Pujare [90]:

SnO_2	$LuPc_2$	PolyAMPS + Salt	Nafion	PolyAMPS + $CeCl_3$ + Salt	SnO_2

In this cell, the electrolyte was a salt, either 0.1 *M* Na_2SO_4 or 0.1 *M* KCl, incorporated in a solid polymeric form of AMPS (2-acrylamido-2-methylpropanesulfonic acid). A Nafion film served as a separator, and ceric chloride provided a faradaic counter-electrode reactant. Reversible multicolor cycling was observed.

iii. Colors

Qualitatively, the transmission maximum, rather than the absorption peak, provides a first approximation to the hue of a color, although the shapes and widths of the absorption bands are very important. Quantitative color characteristics of lutetium diphthalocyanine films are documented in [48]. Detailed procedures for converting a visible spectrum to a set of three color coordinates are given in the reference work on color science by Wyszecki and Stiles [91]. Such coordinates may be expressed in the CIE (Commission Internationale d'Eclairage) system, which is generally used in the display industry, or in the Munsell system, for which standard color samples are readily available [92]. The samples can be very useful in describing a nonemissive display because photographic reproduction of colors is seldom adequate. Reference [48] gives colors in both coordinate systems for each of the lutetium diphthalocyanine oxidation states represented in Fig. 10 and for many intermediate, or blended, colors corresponding to partial completion of the faradaic reactions.

Two other color effects may be correlated with the absorption spectra [48]. (a) Light scattering, or translucency, usually develops to some extent as the electrochromic film is cycled. This is indicated by an elevated baseline in the spectrum. (b) Film discontinuities in the form of microcracks may result from gas formation on the electrode or from film expansion or thermal mismatch with the substrate. Such discontinuities alter the shape of the spectrum from the typical peaks and valleys toward a flat wavy line, and the visual appearance, toward a neutral gray. Both effects are useful in the characterization of cycled films.

Aesthetically, the green color of lutetium diphthalocyanine and the blue and violet colors of its reduced forms are very satisfactory. Some improvement is needed, however, to achieve consistently a true red.

Although the one-electron oxidized form of $LuPc_2$ is sometimes perceived as red, its usual appearance in a solid film is orange. Fortunately, the $LnPc_2^+$ species is more susceptible than the green or blue form to color modification by chemical substitution (Section B,iv). The neodymium compound, for example, forms a purplish-red, rather than an orange, oxidation product [85]. Thus, color improvement by the use of mixed lanthanide diphthalocyanines is possible.

Frampton et al. obtained positive results with a variation of this approach [93]. Instead of combining the diphthalocyanines of two rare earths, they used a mixed rare-earth oxide powder as the original starting material for the phthalocyanine synthesis. This oxide mixture included 30% Dy_2O_3, 25% Er_2O_3, and more than five other oxides. Enhanced colors and extended cycle life were reported.

iv. Matrix Addressing

In a matrix display, the panel is covered with an array of closely spaced dots, all of which are switchable, so that virtually any form of image can be produced. A real-time picture is attainable if the entire field can be changed rapidly enough to avoid a visible flicker (~25 ms). Other uses of matrix displays require versatility in imaging but are far less demanding in terms of speed and cycle life.

Selectivity of the dots addressed is a primary concern for any matrix display. Engineers prefer to use directly multiplexed circuitry when this is practical. All display dots in a given column are then connected to a common lead. The column positions may be considered x coordinates. A perpendicular array of opposing row electrodes provides the y coordinates. The selected dot is activated by applying a signal between the corre-sponding row and column leads. Because direct multiplexing requires a steep threshold in the response of the display medium to the input signal, this addressing method can be limited to small panels.

A major advance in flat-panel display technology, originated by T. P. Brody, eliminates the problems due to inadequate threshold and interactions between neighboring dots [94]. Known as active-matrix addressing, this concept uses a nearly invisible network of transistors and capacitors built directly onto the display panel. Each pixel (picture element) then has, in effect, its own independent switch to the power supply. As a result, large panels can function as effectively as small ones.

The rapidly developing liquid-crystal video technology is based on the active-matrix principle. Full color is achieved with a triad system, in which the switchable panel is overlaid with a nonswitchable color-matrix filter. This filter consists of groups of red, blue, and green dots, which are in registry with liquid-crystal electrode dots. The display is backlighted with white light, and the liquid crystal acts as a shutter to select the light transmitted to the viewer. Thus, a red dot is seen if light is blocked at the adjoining blue and green dots, and yellow, if only the blue light is blocked. The resolution of such a display is limited by the fact that each pixel contains three dots, with the attendant active-matrix elements.

Because a multicolor electrochromic material provides a full range of colors in each dot, it offers the prospect of higher resolution than a liquid-crystal matrix. The electrochromic requires no added filter and can function well in a reflective or a transmissive mode. Reference [95] describes an experimental feasibility study on multicolor diphthalocyanine matrix displays. It was shown that a display dot may be represented electrically as a voltage-dependent resistor and capacitor in series. Typical parameters for the green/orange switching of lutetium diphthalocyanine are 25 Ω cm^2 and 1000 $\mu F/cm^2$. The large area capacitance is nearly all pseudocapacitance of the electrochemical reaction. These values correspond to a time constant of 25 ms regardless of the dot size. This time is inherent in the electrochromic system and does not include any delay introduced by the resistance of the substrate or the drive circuit.

The various faradaic reactions have voltage thresholds that could permit multiplexed addressing. However, the valuable memory feature of the display would be lost through galvanic action if groups of dots were placed on a common electronic conductor; with lutetium diphthalocyanine, a blue (reduced) dot and an orange (oxidized) dot, connected together and in contact with the same electrolyte layer, would self-discharge to green. Direct multiplexing must be avoided for this reason.

Active-matrix addressing offers a solution to the self-discharge problem of electrochromics. The principal adaptations required are (a) a means for handling the relatively high instantaneous switching current density and (b) a physical structure that does not corrode when used in close proximity to an electrolyte. These are engineering and fabrication tasks. Proprietary design concepts already have been developed to provide the switching currents, even at video rates, and the fabrication of a corrosion-resistant network should be feasible with existing process technology. Pattern definition in the diphthalocyanine layer itself can be accomplished by photolithography with lift-off etching [45,95].

v. Potential Applications

Multicolor electrochromic displays are a potential competitor for a significant share of the very large liquid-crystal display market. Liquid crystals have received intensive development for many years. This effort continues, especially for flat panels to replace the cathode ray tube. Consequently, the less mature electrochromic technology might be expected to find its first applications where liquid crystals are inadequate. This would include displays requiring open-circuit memory or extremely low long-term average power combined with infrequent switching. The attractive colors and good legibility in reflected light make the electrochromics a candidate for many small displays even in segmented formats. Such applications include electronic timepieces, calculators, automotive instrument panels, games, and novelties. Simple, unpatterned electrochromic cells also are of interest as variable light filters.

The development of active-matrix drive for electrochromics would enhance the foregoing display applications and create significant opportunities for uses in larger panels such as status boards, variable maps, and large-screen projection displays in which the feature of coincident color could be very important.

ACKNOWLEDGMENTS

The author's research on phthalocyanine electrochromic systems was supported by U.S. Government contracts with the Air Force Office of Scientific Research, the Naval Air Development Center, and the Office of Naval Research. Funding for complementary effort was provided by Rockwell International Corporation. Assistance of the Rockwell International Science Center in preparation of the manuscript for this chapter is gratefully acknowledged.

REFERENCES

1. J. R. Platt, *J. Chem. Phys.*, 34 (1961) 862.
2. D. W. Clack and J. R. Yandle, *Inorg. Chem.*, 11 (1972) 1738.
3. P. N. Moskalev and I. S. Kirin, *Opt. Spektrosk.*, 29 (1970) 414.
4. P. N. Moskalev and I. S. Kirin, *Russ. J. Phys. Chem.*, 46 (1972) 1019.
5. M. M. Nicholson and R. V. Galiardi, "Investigation of Lutetium Diphthalocyanine as an Electrochromic Display Material," Final Report, Contract N62269–76–C–0574, C77–215/501, NADC–76283–30, AD–A039756, May 1977, Rockwell International, Anaheim, California.
6. M. M. Nicholson and R. V. Galiardi, *SID Int. Symp. Dig.*, IX (1978) 24.
7. G. C. S. Collins and D. J. Schiffrin, *J. Electroanal. Chem.*, 139 (1982) 335.
8. L. G. Tomilova, N. A. Ovchinnikova, and E. A. Luk'yanets, *Zh. Obshch. Khim.*, 57 (1987) 2100.

9. T. Sugiyama, T. Okamoto, and H. Yamamoto, *Japan Appl. Phys. Symp.*, Autumn (1984) 14p–D–15.

10. J. L. Kahl, L. R. Faulkner, K. Dwarakanath, and H. Tachikawa, *J. Am. Chem. Soc.*, 108 (1986) 5434.

11. J. M. Green and L. R. Faulkner, *J. Am. Chem. Soc.*, 105 (1983) 2950.

12. P. N. Moskalev and I. S. Kirin, *Russ. J. Inorg. Chem.*, 16 (1971) 57.

13. L. G. Tomilova, E. V. Chernykh, N. T. Ioffe, and E. A. Luk'yanets, *Zh. Obshch. Khim.*, 53 (1983) 2594.

14. F. Castaneda, C. Piechocki, V. Plichon, J. Simon, and J. Vaxiviere, *Electrochim. Acta*, 31 (1986) 131.

15. Y. Liu, K. Shigehara, M. Hara, and A. Yamada, *J. Am. Chem. Soc.*, 113 (1991) 440.

16. B. Schumann, D. Wöhrle, and N. I. Jaeger, *J. Electrochem. Soc.*, 132 (1985) 2144.

17. S. Besbes, V. Plichon, J. Simon, and J. Vaxiviere, *J. Electroanal. Chem.*, 237 (1987) 61.

18. M. Bardin, E. Bertounesque, V. Pilchon, J. Simon, V. Ahsen, and O. Berkaroglu, *J. Electroanal. Chem.*, 271 (1989) 173.

19. D. Wöhrle, H. Kaune, and B. Schumann, *Makromol. Chem.*, 187 (1986) 2947.

20. I. S. Kirin and P. N. Moskalev, *Russ. J. Inorg. Chem.*, 16 (1971) 1687.

21. P. N. Moskalev and I.S. Kirin, *Russ. J. Inorg. Chem.*, 15 (1970) 7.

22. I. S. Kirin, P. N. Moskalev, and N. V. Ivannikova, *Russ. J. Inorg. Chem.*, 12 (1967) 497.

23. K. Kasuga, M. Ando, H. Morimoto, and M. Isa, *Chem. Lett.*, (1986) 1095.

24. M. M'Sadak, J. Roncali, and F. Garnier, *J. Chim. Phys.*, 83 (1986) 211.

25. I. S. Kirin, P. N. Moskalev, and Y. A. Makashev, *Russ. J. Inorg. Chem.*, 10 (1965) 1065.

26. M. M'Sadak, J. Roncali, and F. Garnier, *J. Electroanal. Chem.*, 189 (1985) 99.

27. C. Clarisse and M. T. Riou, *Inorg. Chim. Acta*, 130 (1987) 139.

28. P. N. Moskalev and G. N. Shapkin, *Radiokhimiya*, 19 (1977) 356.

29. P. N. Moskalev, G. N. Shapkin, and A. N. Darovskikh, *Russ. J. Inorg. Chem.*, 24 (1979) 188.

30. M. Yamana, K. Kanda, N. Kashiwazaki, M. Yamamoto, T. Nakano, and C. Walton, *Japan J. Appl. Phys.*, 28 (1989) L1592.

31. Y. Ando, H. Mizuguchi, and K. Kawanobe, *Japan Kokai* 77 98,688 (1977); *Chem. Abst.* 89:68608x.

32. P. N. Moskalev and G. N. Shapkin, *Electrokhimiya*, 14 (1978) 574.

33. M. Petty, D. R. Lovett, P. Townsend, J. M. O'Connor, and J. Silver, *J. Phys. D: Appl. Phys.*, 22 (1989) 1604.

34. K. Kasuga, M. Ando, and H. Morimoto, *Inorg. Chim. Acta*, 112 (1986) 99.

35. H. Isago and R. Hasegawa, *Chem. Express*, 4 (1989) 233.

36. M. Yamana, *Tokyo Denki Univ. Res. Rept.*, 25 (26) 1977) 39.

37. M. T. Riou and C. Clarisse, *J. Electroanal. Chem.*, 249 (1988) 181.

38. G. C. S. Collins and D. J. Schiffrin, *J. Electrochem. Soc.*, 132 (1985) 1835.

39. D. Markovitsi, T. H. Tran-Thi, R. Even, and J. Simon, *Chem. Phys. Lett.*, 137 (1987) 107.

40. H. Yamamoto, M. Noguchi, and M. Tanaka, *Japan J. Appl. Phys.*, 23 (1984) L221.

41. M. Yamana, *Ohyo Butsuri*, 48 (1979) 441.

42. M. T. Noguchi, H. Sugiyama, H. Yamamoto, and S. Tanaka, *Japan Appl. Phys. Symp.*, Spring (1984) 6a–Y–9.

43. P. N. Moskalev and N. I. Alimova, *Russ. J. Inorg. Chem.*, 20 (1975) 1474.

44. D. Walton, B. Ely, and G. Elliott, *J. Electrochem. Soc.*, 128 (1981) 2479.

45. N. Egashira and H. Kokado, *Japan J. Appl. Phys.*, 25 (1986) L462.

46. F. A. Pizzarello and M. M. Nicholson, *J. Electrochem. Soc.*, 128 (1981) 1288.

47. M. M. Nicholson and T. P. Weismuller, *J. Electrochem. Soc.*, 131 (1984) 2311.

48. M. M. Nicholson and T. P. Weismuller, "A Study of Colors in Lutetium Diphthalocyanine Electrochromic Displays," Final Report, Contract N00014–81–C–0264, C82–268/201, AD–A12083, October 1982, Rockwell International, Anaheim, California.

49. F. Castaneda, V. Plichon, C. Clarisse, and M. T. Riou, *J. Electroanal. Chem.*, 233 (1987) 77.

50. G. A. Corker, B. Grant, and N. J. Clecak, *J. Electrochem. Soc.*, 126 (1979) 1339.

51. A. T. Chang and J. C. Marchon, *Inorg. Chim. Acta*, 53 (1981) L241.

52. M. L'Her, Y. Cozien, and J. Courtot-Coupez, *J. Electroanal. Chem.*, 157 (1983) 183.

53. M. L'Her, J. Cozien, and J. Courtot-Coupez, *Compt. Rend. Acad. Sci. Paris*, 300, II (1985) 487.

54. L. G. Tomilova, E. V. Chernykh, V. I. Gavrilov, I. V. Shelepin, V. M. Derkacheva, and E. A. Luk'yanets, *Zh. Obshch. Khim.*, 52 (1982) 2606.

55. P. Corbeau, M. T. Riou, C. Clarisse, M. Bardin, and V. Plichon, *J. Electroanal. Chem.*, 274 (1989) 107.

56. J. Silver, P. J. Lukes, P. K. Hey, and J. M. O'Connor, Polyhedron, 8 (1989) 1631.

57. M. M. Nicholson, Ind. Eng. Chem. Prod. R&D, 21 (1982) 261.

58. J. C. Marchon, J. Electrochem. Soc., 129 (1982) 1377.

59. A. De Cian, M. Moussavi, J. Fischer, and R. Weiss, Inorg. Chem., 24 (1985) 3162.

60. K. Kasuga, M. Tsutsui, R. C. Petterson, K. Tatsumi, N. Van Opdenbosch, G. Pepe, and E. F. Meyer, Jr., J. Am. Chem. Soc., 102 (1980) 4835.

61. M. Moussavi, A. De Cian, J. Fischer, and R. Weiss, Inorg. Chem., 27 (1988) 1287.

62. A. N. Sidorov and P. N. Moskalev, Zh. Fiz. Khim., 62 (1988) 3015.

63. L. G. Tomilova, E. V. Chernykh, and E. A. Luk'yanets, Zh. Obshch. Khim., 57 (1987) 2368.

64. B. R. Hollebone and M. J. Stillman, Faraday Trans. II, Phys. Chem., 7 (1987) 2107.

65. M. M. Nicholson and F. A. Pizzarello, J. Electrochem. Soc., 127 (1980) 2617.

66. M. M. Nicholson and F. A. Pizzarello, J. Electrochem. Soc., 126 (1979) 1490.

67. M. M. Nicholson and F. A. Pizzarello, J. Electrochem. Soc., 128 (1981) 1740.

68. V. I. Gavrilov, A. P. Konstantinov, E. A. Luk'yanets, and I. V. Shelepin, Elektrokhimiya, 22 (1986) 1667.

69. V. I. Gavrilov, V. A. Vazhnina, A. P. Konstantinov, and V. I. Shelepin, Elektrokhimiya, 22 (1986) 1112.

70. F. A. Pizzarello and M. M. Nicholson, J. Electron. Mat., 9 (2) (1980)231.

71. M. S. Whittingham and R. H. Huggins, J. Electrochem. Soc., 118 (1971) 1.

72. F. Castaneda and V. Plichon, J. Electroanal. Chem., 236 (1987) 163.

73. C. Piechocki, J. Simon, J. J. André, D. Guillon, P. Petit, A. Skoulious, and P. Weber, Chem. Phys. Lett., 122 (1985) 124.

74. H. Konami, M. Hatano, N. Kobayashi, and T. Osa, Chem. Phys. Lett., 165 (1990) 397.

75. H. M. Koepp, H. Wendt, and H. Strehlow, Z. Elektrochem., 64 (1960) 483.

76. R. R. Gagne, C. A. Koval, and G. Lisensky, Inorg. Chem., 19 (1980) 2854.

77. M. M. Nicholson and T. P. Weismuller, "Electrochemistry of Rare-Earth Diphthalocyanines," Final Report, Contract F49620–83–C–0088, SC5383.1FR, AD–A145813/2/GAR, July 1984, Rockwell International, Anaheim, California.

78. M. M. Nicholson and F. A. Pizzarello, J. Electrochem. Soc., 127 (1980) 821.

79. W. J. Albery, M. G. Boutelle, P. J. Colby, and A. R. Hillman, J. Electroanal. Chem., 133 (1982) 135.

80. P. J. Peerce and A. J. Bard, J. Electroanal. Chem., 114 (1980) 89.

81. W. P. Weppner and R. A. Huggins, J. Electrochem Soc., 124 (1977) 1569.

82. M. M. Nicholson and T. P. Weismuller, unpublished work, Contract N00014–77–C–0636.

83. M. M. Nicholson, "Electrochromic Materials," in Encyclopedia of Materials Science and Engineering, M. B. Bever, Editor (1986) 1406.

84. M. M. Nicholson and T. P. Weismuller, *Proc. IEEE 1983 National Aerospace and Electronics Conf.*, 1 (1983) 368.

85. M. M. Nicholson, S. Lewkowitz, and F. A. Pizzarello, "Multicolor Electrochromic Displays. Exploratory Development." Final Report, Contract N00014–85–C–0415, SC5433.FR, July 1989, Rockwell International, Thousand Oaks, California.

86. Y. Bessonat, G. Gerard, G. Leroy, and M. LeContellec, *SID Int. Symp. Dig.*, XIII (1982) 102.

87. T. Hoshino, H. Megawa, and K. Miyashita, *1980 Gen. Mtg. Elec. Sci. Soc.*, Northern Branch, Japan, Abst. D–35.

88. H. Yamamoto, Y. Ushioda, M. Tanaka, S. Yamaguchi, and H. Enjoji, *Proc. SID*, 28 (1987) 247.

89. M. Starke, I. Androsch, and C. Hamann, *Phys. Stat. Sol.*, 120 (1990) K95.

90. S. F. Sammells and N. U. Pujare, *J. Electrochem. Soc.*, 133 (1986) 1065.

91. G. Wyszecki and W. S. Stiles, *Color Science: Concepts and Methods, Quantitative Data and Formulae*, Wiley–Interscience, New York, 1982.

92. *Munsell Book of Color, Neighboring Hues Edition, Matte Finish Collection*, Macbeth Color and Photometry Division, Kollmorgen Corporation, New York, 1976.

93. C. S. Frampton, J. M. O'Connor, J. Peterson, and J. Silver, *Displays*, October (1988) 174.

94. T. P. Brody and P. R. Malmberg, *Int. J. Hybrid Microelec.*, II (1979) 29.

95. M. M. Nicholson, F. A. Pizzarello, and T. J. La Chapelle, "Multicolor Electrochromic Dot-Matrix Display Investigation," Final Report, Contract N00014–79–C–0434, ONR–CR339–005–1F, June 1980, Rockwell International, Anaheim, California.

3

Phthalocyanine-Based Molecular Electronic Devices

Branimir Simic-Glavaski

The past several years have witnessed an upsurge of interest in the field of molecular electronics, the field that uses molecules to sense, generate, and process information.This field is truly interdisciplinary and is motivated by the increasing demands and challenges presented by information, sensing, and processing systems. It is based on recent advances in science and technology that are rapidly expanding in the areas of communications and computing. Developments in the area of molecular electronic devices have been presented at many international conferences [1-6], beginning with the first workshop held in Washington D.C. in 1981, organized by the late Forest Carter.

Molecular electronic devices using phthalocyanine molecules were discussed at this first workshop, which was not surprising to those familiar with the versatility of phthalocyanine molecules.

A. FUNDAMENTALS:COMPUTERS, MOLECULAR ELECTRONIC DEVICES, AND PHTHALOCYANINES

i. Computers

Intelligence, both natural and artificial, presents one of the most challenging frontiers in science and technology. Information and communication processes that form intelligence are the subject of interdisciplinary research efforts in many laboratories.

The purpose of this chapter is to clarify and explain why molecular electronic devices may play a significant role in generation of information and information processing and the role of phthalocyanine molecules in molecular electronics.

Generation of information and information processing systems are closely interrelated to the problems originating in materials, devices that use novel concepts in materials and applications of devices in various complex systems that may process information.

At this point one may ask how to generate and process information. One of the obvious answers would be related to studies of abstract computer systems. On the other hand, computation, whether it is performed by mechanical, electronic, or biological systems such as neural networks or the brain, is a physical process that requires a physical system. This physical system performs mathematical work and obeys the laws of physics.

In 1962, Brillouin discovered and established the relation between information and energy [7] and concluded that energy dissipation must increase with the reliability of the measurement, the relation being $kTln(1/\eta)$, where η , k and T are the probability of measurement error, Boltzmann's constant, and the temperature, respectively.

This discovery was the cornerstone that provides the thermodynamic foundations for information in an energy-time domain. A definition of a computer as a system that transforms energy into mathematical work establishes the fundamental limit of the minimum energy per bit of information. Two other important factors are related to the speed of operation: temporal resolution, or how long must it take to generate a signal or bit of information and to the spatial density of elementary processing elements, that is, what is the minimum size of a processing element in a computer.

A computer as a system may be considered from several points of view. One major classification is, however, whether the system is reversible or irreversible, depending on whether the system is energy nondissipative or dissipative. From a thermodynamic point of view, the information system is irreversible when information is lost during the process of computation. In other words, that information has to be converted into heat, which is a dissipative effect. According to their modes of operation, computers may also be classified as analog or digital. One may say that the analog mode of operation imitates natural dynamics more reliably, while the digital mode of operation is more suitable for number-crunching operations, because such systems have higher accuracy.

Further functional division can be into serial and parallel, step-by-step modes of operation (including Turing machines), random access machines, cellular automata, and tree automata, all of which are capable of simulating one another. These divisions also include thermodynamically reversible computers, such as the ballistic models of Fredkin and Toffoli [8] and the Brownian model of Bennett [9]. These computers can compute at finite speed with zero or close to zero energy dissipation, respectively. Brownian computers may be considered as the closest ones to molecular electronic devices, as they allow thermal noise to influence their performance. For example, chemical reactions are reversible and are controlled via Brownian motion, which accomplishes the forward reactions, bringing product molecules together, pushing them backward through the transition state, and letting them emerge as reactant molecules.

The required energies are of the order of about $100kT$ and correspond to a covalent bond which prevents spontaneous molecular rearrangements at room temperature. Such systems already exist in natural systems and may be exemplified by RNA polymerase, the enzyme that synthesizes a complementary RNA copy. This chemical process may represent a reversible Turing machine [10].

The size of a computer or processing element may be yet another classifying factor.

Feyman suggested [11] that an atom may be used as a computer where an atomic ground state is considered as a logic "zero" and its excited state as logic "one." However, he left open the question of how such states can be achieved and controlled; later we shall see that such controls can be obtained.

The practical realization of a computing system can be developed using several methods based on mechanical, electrical, electro-optical and optical foundations.

a. Mechanical computers

Mechanical models of computers are based on modeling dynamics using Newtonian mechanics and approximate force fields that allow the implementation of mechanical rod logic [12]. This concept has produced methods for signal transmission, logic gates, data storage registers, and power supplies.

b. Electrical computers

In electrical computers models, the information is carried by current and potential. These types of computers are the most practical and popular today.

The principal processing elements are about a micrometer in size and contain multiple transistors and other components. The processing elements can be fitted into integrated circuits and onto a planar array. The construction of such systems is limited by lithography problems. Three dimensional processors are unlikely to be built in the near future because of construction difficulties related to interconnecting wires. One measure of computer merit, which is often used to describe the power of a computer chip, is gates hertz per centimeter squared (gate Hz/cm^2), where the gates are processing or memory elements switching with a frequency expressed in hertz. Advanced integrated circuits based on electricity as a carrier of information are of the order of $10^{14\text{-}15}$ gate Hz/cm^2. We may also find another definition of a chip power expressed in gate interconnects operations/second cm^2.

c. Electro-optical computers

Electro-optical computers are based on electro-optical processing elements, where an optical output may be modulated using electrical methods, for example, systems that use Bragg cells or microchannel plate devices. Microchannel plate devices [13-15] are popular in artificial neural systems that are used in the simulation of functions of natural neural networks. The plates are currently limited to sizes of approximately 1000 x 1000 processing elements and speeds at about 100 μs.

d. Optical computers

In optical computing, an optical signal is a carrier of information. This computing methodology has become a promising alternative to electrical methods in computing since the discovery of the first lasers in early sixties. Discovery of optical bistability was used to formulate optical logic and its applications in optical computing [16]. This type of computing has a number of potential advantages: transfer of information is performed at the speed of light; very high bandwidth and optical wiring are especially attractive for parallel processing. Lenses and prisms can transfer images consisting of millions of resolvable spots. Optical beams can cross each other without interaction. The other advantage is related to optical information storage capacities which is very high in the forms of surface and volume holograms as discussed by van Heerden [17] and Gaylord (18). Van Heerden demonstrated that the maximum storage densities of thin film and volume holograms are

$$S_{2D} = 1/\lambda^2$$

$$S_{3D} = n^3/\lambda^3$$

bits per unit or for $n = 1.5$ and $\lambda = 514.5$ nm $S_{2D} = 3.78 \times 10^8$ bit/cm^2 and $S_{3D} = 2.48 \times 10^{13}$ bit/cm^3. Additional discussions on optical computing and its advantages can be found in numerous reports of Huang [19], Caulfield [20], Casasent [21], Neff [22], Smith [23], Goodman [24], Keyes [25] and others.

ii. Fundamental Limits

In considering the question as to which technology is best for computational purposes, one realizes that several factors may be critical. However, it is obvious that energy, time, and size are the most fundamental, that is, the minimum energy required per bit of information, shortest switching time, and minimum size of the processing element.

Von Neumann [26] argued that an irreversible computer operating at temperature T must dissipate at least $kTln2$ per elementary act of information. This amount of energy corresponds to the energy change of a hydrogen bond which is altered at room temperature and is about 3×10^{-21} J. For example, DNA replication, transcription and protein synthesis has a high energy efficiency, dissipating only (20-100)kT per nucleotide or amino acid inserted under physiological conditions.

The other fundamental limiting physical factor associated with computation is related to the physical transfer of information through space. In the classical study of Wigner [27], it was established that the time limits are related to the Heisenberg uncertainty principle, which limits the information processing rate of the computer processing element.

The size of a computing processing element can in theory be reduced to the size of an individual atoms as proposed earlier by Feyman [11]. The possible realization of molecules or atoms as computer elements can be achieved using the latest developments in tunneling microscopy as discussed in numerous reports presented at the conference on Molecular Electronics-Science and Technology [5]. However, size reduction can be downscaled even to a single electron, as proposed by Likharev [28, 29]. In an isolated tunnel junction with small capacitance, the transfer of a single electron is able to change the voltage across the barrier. This effect, frequently called single electronics, may be used in the construction of a reversible Brownian computer based on Josephson devices [30].

iii. Processing Elements

Studies of the physics of communications and information processing in computers have focused on individual devices and their fundamental limits in the energy time domain of a computational process.

a. Electronic processing elements

In the late nineteenth century electrical communication systems consisted of electrical generators, transmission lines, and simple mechanical devices which could switch electrical currents ***on*** and *off*. Such primitive electrical systems required simple materials: iron to provide magnetic fields and copper to conduct electricity. The invention of the vacuum tube in 1906 provided methods to control electricity electronically and much later led to the realization of the electronic digital computer in the 1940s. Until the 1950s vacuum tubes were essential parts of modern communications and computing technologies. The switching and amplifying elements required a lot of power, which eventually finished as unwanted dissipative heat. The invention of the transistor in 1949, based on semiconducting materials has revolutionized research and development in digital device technology. The integrated silicon chip, invented, in 1958-1959, has rapidly reduced power dissipation and production cost and at the same time increased the speed of operation and reliability.

Modern electronic computers are digital, contain many logic gates, and still dissipate vast amounts of energy compared to kT. Digital and especially binary modes of operation provide a high accuracy of calculation, and digital computers are usually constructed in a serial configuration. The principal building blocks are transistors constructed from various types of semiconducting materials.

Silicon oxides are mainly used in field effect transistors **(FET),** which are the principal component in logic gates [31]. Higher electron and hole mobilities in germanium [32] were used to improve the speed of device operation; however, this material is inferior with regard to stability and dielectric properties. Gallium arsenide, with impurities which form electron states far below the conduction band as deep traps, has a higher frequency of oscillation and has been suggested as a competitor for silicon [33]. GaAs devices are usually smaller than the silicon ones and can be constructed with additional effort. The Josephson tunneling effect in superconductors has been investigated in many laboratories, using lead alloys and niobium as superconductors [34]. Josephson tunneling devices are driven with lower voltages and therefore use less power. Transistor threshold logic (TTL) devices are also frequently used as elementary devices in computer systems [35]; however, TTL devices dissipate energy continuously, while complementary metal oxide semiconductor (CMOS) and Josephson junction devices dissipate energy only when they are switched. All these discussed devices are macroscopic in nature and dissipate about $10^{11}kT$ per bit of information. A similar energy dissipation of about $10^{11}kT$ but at much slower switching rates, is typical of a natural signal processor like a neuron.

b. Optical processing elements

Optical computing as a method is seriously constrained by the absence of suitable optical processing and memory elements, although many optical processing elements have been demonstrated and discussed in the literature. In light-emitting or laser diodes (LED) the principal action is based on a laser operation [36] and can reach rates up to 1 GHz [37]. $LiNbO_3$ devices were constructed for waveguide modulators to monitor interchip signals in very high-speed integrated circuits (VHSIC) [38]. The main problem is that these devices are not cascadable and cannot be used as the input to another device. Other devices are based on phase modulation [39], the acousto-optical effects [40] and second-harmonic generation [41]. The discovery of optical bistable devices [42], with their potential of being cascadable, has improved the prospects for optical computing. Low-power integrated optics, light switches, and low-energy integrated optical bistable devices capable of performing optical logic functions have been reported [43,44] with potential applications in optical computing [45,46]. Another class of bistable devices is based on the electronic operation of $Ga_xAl_{1-x}As$ and involves multiple quantum wells so as to function as an electroabsorption modulator. Such a device is known as a self-electrooptical effect device (**SEED**) [47,48]. These devices operate in a reflective mode and require about a picojoule per bit of operation. Their switching time is about a nanosecond, and they have relatively low contrast (> 100) a property much needed for a wide dynamic range [49]. It is obvious that the availability of optical switches, optical bistable devices, spatial light modulators, and optical processing elements is influenced by the advancement of appropriate materials to be utilized, in optical hardware [50]. Organic materials may provide the ideal medium for optical processing elements [51]. The data in Table 1 show required switching energies and speeds for some optical processing elements. A comparative energy-time diagram for processing elements is shown in Fig. 1.

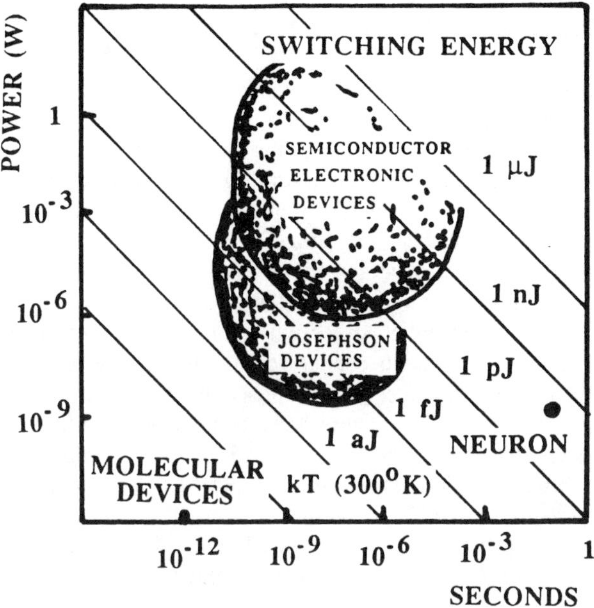

Figure 1. *A comparative switching energy-time diagram for some technologies.*

TABLE 1. Bistable Resonators

	Switching time (sec)	**Switching energy** (J)
CS_2	5×10^{-10}	1.5×10^{-4}
GaAs	4×10^{-8}	8×10^{-9}
$LiNbO_3$	5×10^{-8}	5×10^{-13}
Bistable liquid crystal	4×10^{-2}	2×10^{-8}
Glass-CS_2	2×10^{-12}	4×10^{-7}
Rhodopsin	1×10^{-9}	2×10^{-12}

B. MOLECULAR ELECTRONIC DEVICES

The concept that molecular electronic devices can be used as processing elements in computing has been suggested as an attractive alternative. The main problem in electronic and optical processing elements has been the question of operational energy and the switching time as shown in Fig.1. Molecular devices permit substantial improvements: operational energy requirements can be reduced to about 100*kT*; the switching speed may also be reduced because its fundamental operations may be based on different principles; and the density of processing elements can be reduced to the size of molecules and atoms. The proposed models of such devices range from the highly conceptual to those which have been demonstrated in the laboratory as being feasible . In an earlier report on the physical limits of heat generation in computation, Landauer [52] proposed that a typical potential well with two minima may describe bistability, where *X* is a generalized coordinate representing the quantity that is switched (see Fig.2), and could be applied to ferrites, ferroelectrics, and thin magnetic films. This principle was elaborated theoretically in the model of Aviram and Seiden [53]. The model involves organic molecules such as ferrocene, ferrocenium, quinones, or pyridinium. Digital memory is based on mixed-valence quantum wells similar to the schematic presentation in Fig. 2.

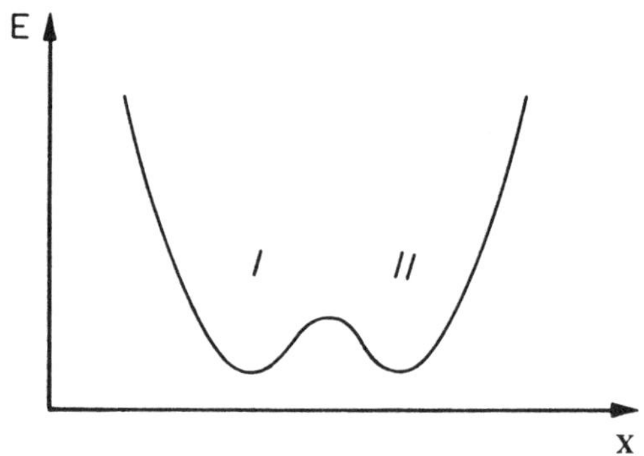

Figure 2. *One-dimensional double well with memory states I and II. E is the energy and X is the distance coordinate.*

According to Aviram and Seiden, the quantum wells are believed to be characteristic of the π–bonding system. In the example of hemiquinone (see Fig. 3), the hydrogen bond and the proton transfer from one potential well to another may be considered as dynamic molecular information storage. The process of proton transfer can be achieved by electrical or optical methods. Proton transfer requires energies of the order of 100-1000*kT* per molecule or per bit of information. A distinguishable position for protons may be achieved by inserting CH_2 chains between quinone molecules.

Figure 3. *An example of hemiquinone as a molecular switch. The position of protons indicate a molecular memory state [53].*

Recent developments in the field of conducting polymers have shown that polyacetylenes $(CH)_n$ have the potential of being applied in molecular electronic devices. Polyacetylene molecules have poor conductivity ($\sigma < 10^{-5}\ \Omega^{-1}cm^{-1}$). However, the conductivity of doped polyacetylenes as *n* or *p* type [54] can be dramatically changed and may approach the conductivity of copper. The concept of the soliton in polyacetylene chains has been extensively employed in numerous conceptual examples of molecular electronic devices presented by Carter (see [1-3]).

Another class of molecules for potential use in molecular electronics is tetracyanoquinodimethane (TCNQ) and its derivatives. Potember et al. [116] deposited thin films of TCNQ on copper or silver substrates. In a current-controlled system it was possible to form salts of Ag-TCNQ or Cu-TCNQ which provide bistable low-high impedance. The switching speeds are on the order of 1 ns, and the precise reaction energy is not known. Bistability is achieved in the presence of light, most probably through redox couples of metallo- and nonmetallo-TCNQ sandwiched between silver or copper electrodes.

Proteins as molecular devices have been proposed by Ulmer [55], where the main problem is related to a crystallization, modification of the crystal lattices, and self-assembly of protein subunits. It is expected that newly developed techniques for site-directed mutagenesis would be useful in the engineering the amino acid sequence of any cloned protein. In the relatively simple system of bacteriorhodopsin embedded in purple membranes, Sasabe et al. [56] analyzed the conversion of light into the *pH* gradient across the cell membrane and suggested that these photoelectric responses might be applicable to the construction of biochips. Aizawa et al. [57] were concerned with the reversible electron transfer between an adsorbed enzyme and the electrode and the modulation of enzyme activity. The role of the bacteriorhodopsin in converting light energy into a hydrogen ion gradient which chemiosmotically drives the synthesis of ATP has been of great research interest to Birg and his coworkers [58]. An optically coupled molecular electronic **NAND** gate based on the substituted porphyrin system is shown in Fig. 4.

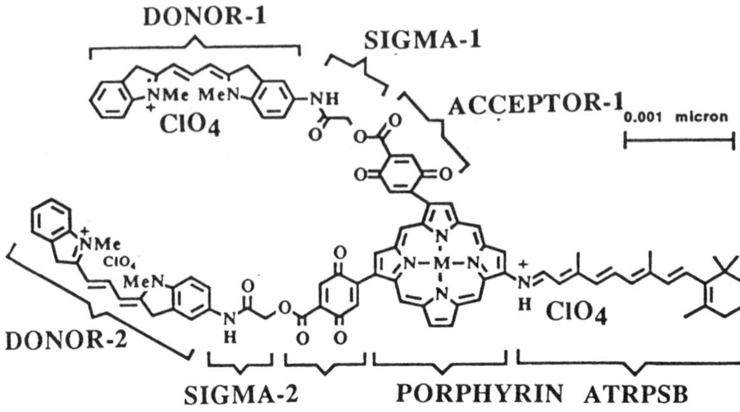

Figure 4. *Molecular electronic NAND gate [58].*

Two optically coupled inputs are available by using π–σ–π, donor (cyanine)-sigma(-CH_2-CO-CH_2-)-acceptor (quinone) complexes. The quinone is attached to the porphyrin (P) molecule that serves as a charge integrator and stabilizes a retinyl protonated Schiff base (ATRPSB). If both cyanine inputs are excited, then the absorption band of the output ATRPSB shifts from ~520 to ~590 nm. It is also assumed that it is unrealistic to use a single molecule to represent one bit of information in a digital environment, and it is necessary to use ensemble averaging, where optical coupling is the most convenient method of communicating with large ensembles of molecules. **NAND** gates, using rhodopsin complexes operating at a practical level, require about a picojoule at 77 K with switching times of about 1 ns. The applications of porphyrin molecules in molecular devices have been discussed in numerous reports (see [1-6]), most notable is one that describes an electronic shift register memory at the molecular level [59]. The shift register is constructed using zinc or palladium porphyrin and a chain (polymer) of electron transfer active species. The polymer consists of three molecular groups per repeat unit. Unidirectional electron transfer is caused by altering the redox state of the porphyrin molecule and requires about $50kT$ per bit of information. This energy level is similar and consistent with the value reported earlier [60].

Phthalocyanine and other metallo macrocycle molecules have been widely studied and applied in various systems. Phthalocyanine molecules are *p* type semiconductors with a typical energy gap of about 1.8 eV [61]. The semiconducting and structural properties of phthalocyanines resemble those of doped polyacetylenes; thus one can avoid the laborious difficulties of doping and cross linking polyacetylene chains. From the structural point of view, phthalocyanine molecules may form a variety of configurations. The most characteristic forms are monomer or aggregated units, ring stacked [62] or polymeric sheet type [63] shown in Fig. 5.

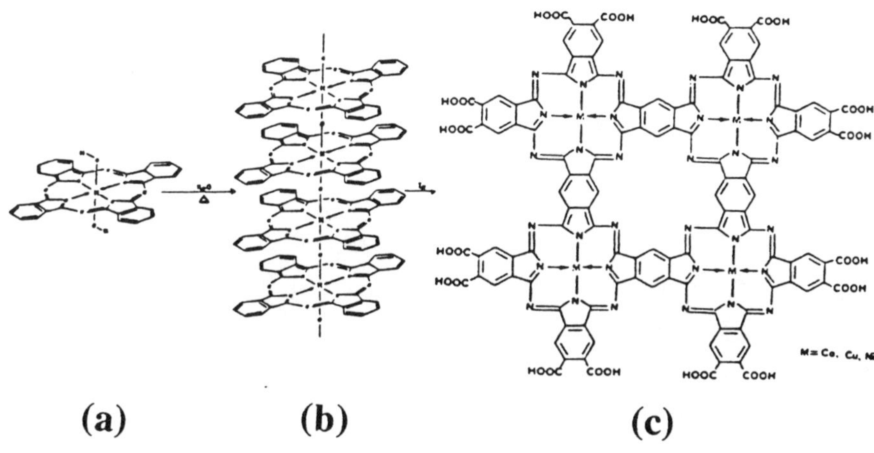

Figure 5. *Schematic representation of different phthalocyanine structures. (a) Monomer, (b) ring stacked and (c) polymer sheet .*

The electrical and optical properties of phthalocyanine molecules are strongly dependent on structural configuration. The general properties of phthalocyanine molecules have been presented and discussed in other reports, however, some fundamental and relevant properties are presented again here. The basic 18π electron structure in metal free and other metallo phthalocyanine molecules has very rich molecular orbitals and available energy levels. The populated ground HOMO and SHOMO levels are $a_{1u}(\pi)$ and $a_{2u}(\pi)$ and the first excited LUMO level is $e_g(\pi^*)$. The electronic transitions $a_{1u}(\pi)$---> $e_g(\pi^*)$ and $a_{2u}(\pi)$---> $e_g(\pi^*)$ are responsible for the intense visible absorption *Q* and ultraviolet Soret bands, positioned at about 660 and 330 nm respectively. The *Q* absorption band splits into *Q(0,0)* and *Q(1,0)* as a result of vibrational conditions. An example of the absorption spectrum obtained from cobalt tetrasulfonated phthalocyanine (Co-TsPc) is shown in Fig. 6.

The distributions of molecular charge is also presented in Fig. 6 with the most representative intermolecular charge transfers. The absorption spectrum is characteristic of phthalocyanine molecules, regardless of whether it is obtained from water soluble forms, like tetrasulfonated forms, or from ring-stacked or polymeric configurations.

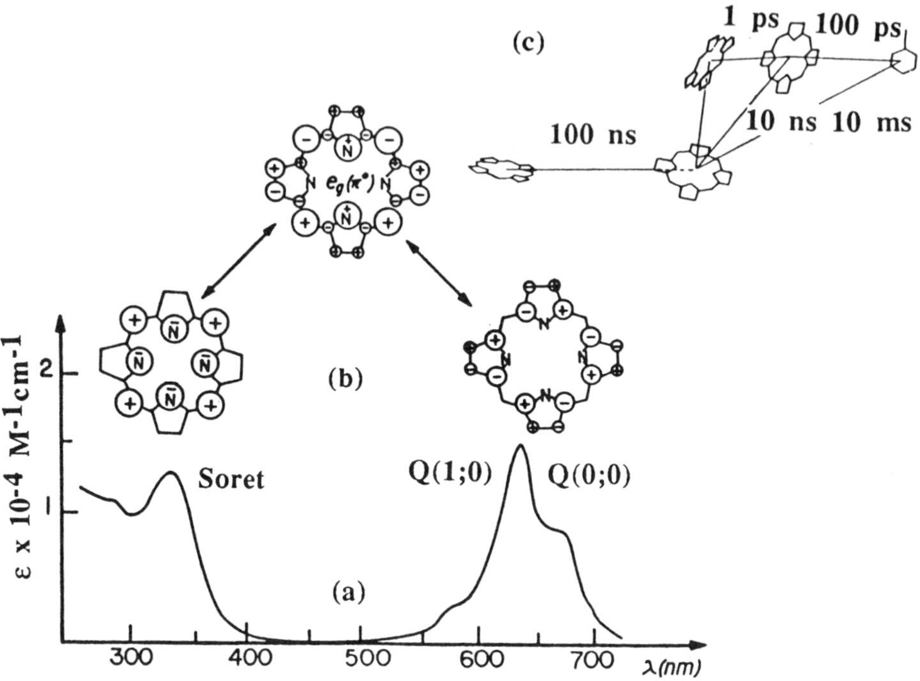

Figure 6. *(a) A typical UV visible absorption spectrum obtained from 10^{-5}* ***M*** *CoTsPc in water, pH~7, (b) distribution of electrons responsible for the absorption Soret and Q bands; (c) electronic transitions for various macrocycle structures.*

The first suggestion that phthalocyanine molecules might be used as **NOR** gates was proposed by Carter [64]. Gallium and nickel phthalocyanines in a ring-stacked configuration were proposed to act as a molecular analogue to a **NOR** gate. The system is conceptual and is shown in Fig. 7. It is assumed that the structure in Fig. 7(a) is an analogue to the electronic switch in Fig. 7(b). Modified gallium phthalocyanines are bridged with fluorine to form a ring-stacked configuration that starts and finishes with nickel phthalocyanines linked axially with $-(SN)_n$ and -S-$(CH)_n$ chains, respectively. The linkage groups are proposed to make contacts with metal electrodes. Two A and B inputs imagined through $-CH_2-SO_3^-$ groups are attached to the modified gallium phthalocyanines, and the output is through a $-(SN)_n$ chain attached to the top nickel phthalocyanine. The model assumes that the periodic potential is necessary for tunneling can be switched ***on*** or ***off***. This proposed conceptual model of the molecular **NOR** gate is very attractive; however, the methods of construction and control are not resolved.

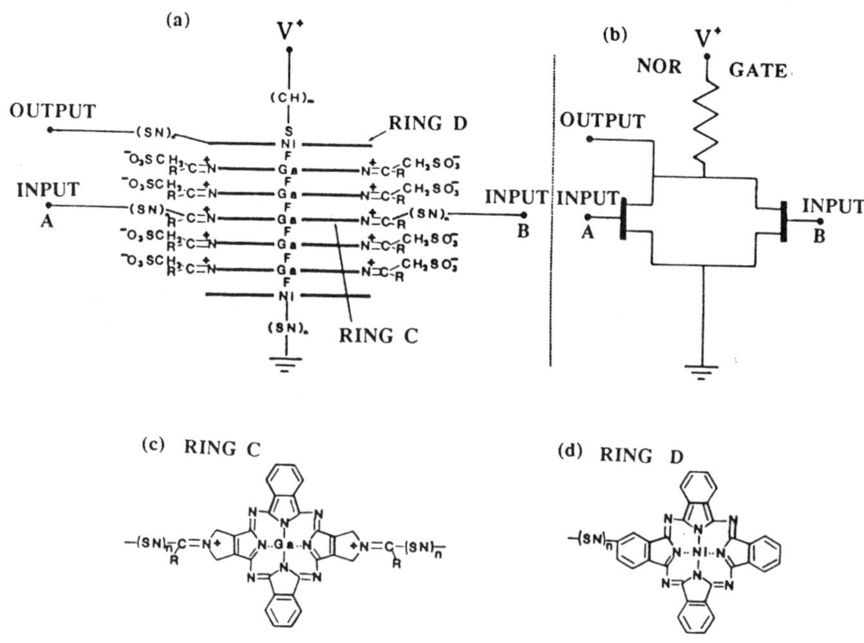

Figure 7. *A molecular analogue to a **NOR** gate based on phthalocyanine molecules. After [64].*

Marks and co-worker [65] have assembled face to face phthalocyanines into a stacked configuration with about 100 subunits or more. The stacked metallo phthalocyanines (MPc) are connected through a linkage X, where M = Si, Ge, Sn and linkage X = O. The "one dimensional wire" of $(OMPc)_n$ has high chemical and thermal stability. In order to make a stacked system electrically conductive, it is necessary that individual phthalocyanine units are positioned in close spatial proximity i.e. between 0.3 and 0.4 nm, that the conduction pathway has a minimum of energetic "hills" and "valleys", and that the central metals must be in different oxidation states. Partial oxidation of the metal ions can be obtained by doping stacked phthalocyanines with halogen elements. Molecular iodine is a good oxidant and is accommodated in ion channels within the crystal lattice. In this way the conductivity increases from about 3 x 10^{-8} to about 5000 $\Omega^{-1}cm^{-1}$ for (Ni-Pc)I at 25K . The charge carriers in (Ni-Pc)I are associated with delocalized π orbitals along the stack of the macrocycles. The conductivity of about 700 $\Omega^{-1}cm^{-1}$ for $(H_2\text{-Pc})I$ illustrates that it is not related only to metal ions.

A novel method for the construction of cofacial metallo phthalocyanines suggests that such structures may be superconductive at room temperatures [66].

Baker et al. [67] demonstrated that metal free phthalocyanines may be deposited by the Langmuir-Blodgett technique onto a common substrate. Thin films show a linear current-voltage (I-V) response proportional to the thickness of the deposited phthalocyanine material. In a subsequent report [68], it was shown that deposited copper phthalocyanine changes conductivity when the film is exposed to nitrogen dioxide.

Langmuir-Blodgett films of tetra 4 tert butyl, chloride bridged, silicon phthalocyanine (ttb $Pc\text{-}SiCl_2$) show similar linear I-V characteristics. Stacked (ttb $PcSiCl_2$) are edge on deposited on the substrate with the molecular faces having their orientation perpendicular to the substrate [69].

It was also shown that such phthalocyanines deposited on ITO substrates and sandwiched between aluminum electrodes, exhibit a photovoltage as a function of the number of Langmuir-Blodgett layers [70]. Electroluminescent and electrical effects were observed from Au/Cu-Pc/GaP structures. Deposition of Cu-Pc was achieved by the Langmuir-Blodgett method [71]. The electroluminescence conversion efficiency is dependent on the number of LB layers, showing a maximum for 5.6-nm-thick films. The effect of electroluminescence is explained in terms of an electron tunneling mechanism through the phthalocyanine film.

Phthalocyanines are light-sensitive molecules, and their conductivity can be modulated with circularly polarized light [72]. The sensitivity occurs in electron donor-acceptor complexes. In an oxidizing medium and in the presence of an electron-accepting oxygen molecule, Pc acts as a donor. Circularly polarized light interaction with the Pc in an oxygen or electron-accepting atmosphere is possible from two points of view. On the one hand, magnetic circular dichroism studies of the metal Pc indicate a resonance interaction with the central metal and the red light of the visible spectrum. On the other hand, light interaction with the oxygen adsorbed on the Pc ring occurs over most of the visible spectrum. The conductivity of metal-free phthalocyanine changes as a function of changes in light polarization, and indicates that light interacts with electron -accepting oxygen rather than with metal *d* electrons. Barrett et al. suggest that such an effect offers the possibility of fast light-induced switching.

Metal-free phthalocyanine is a molecule that undergoes photochemical hole burning. In this way it is possible to cause photoinduced tautomerization involving the reorientation of two protons inside the central ring. Tautomerization in the phthalocyanine system most likely involves a long-lived state of the lowest-lying triplet state and is limited to the triplet lifetime. Holes of 120 MHz width may be burned in metal-free phthalocyanine in polyethylene at 1.4 K. About 10^4 holes may be stored in the frequency domain with time burning on the order of approximately

100 μs. Phthalocyanines have a relatively high hole-burning quantum efficiency, which is about 10^{-3} making them along with some porphyrin molecules very attractive molecular systems [73].

C. PHTHALOCYANINE MOLECULES AS MOLECULAR DEVICES

i. Fundamental Concepts

Phthalocyanine molecules as materials for processing elements have been studied and applied in molecular electronic devices by Simic-Glavaski [60, 74 - 81]. It was recognized that fast ($<10^{-12}$ s) intramolecular charge transfer can serve as a carrier of information that requires a small amount of energy, of the order of only a few 10kT! Intense optical signals can be obtained from molecular assemblies that contain only 10^4-10^5 molecules. Multiple redox states serve as multiple quantum wells and also as multivalued memory states. In addition, sulfonated phthalocyanines are biocompatible and can serve as read-out and also as write-in systems that may interface large macromolecules such as DNA and natural biological tissue such as neurons or cells [79, 82, 83]. Structural configurations have interesting electro-optical responses and properties that can be used to provide some special functions that are applicable in complex systems such as artificial neural network. Furthermore, phthalocyanine-based molecular electronic devices can operate as either analog or digital devices. The generation of information is based on intramolecular charge transfer and can easily be controlled and obtained using electrical and optical methods that involve standard electrochemical and optical spectroscopic experimental methods. Intramolecular charge transfer causes a change of electronic densities at a particular atomic location, as shown in Fig. 8.

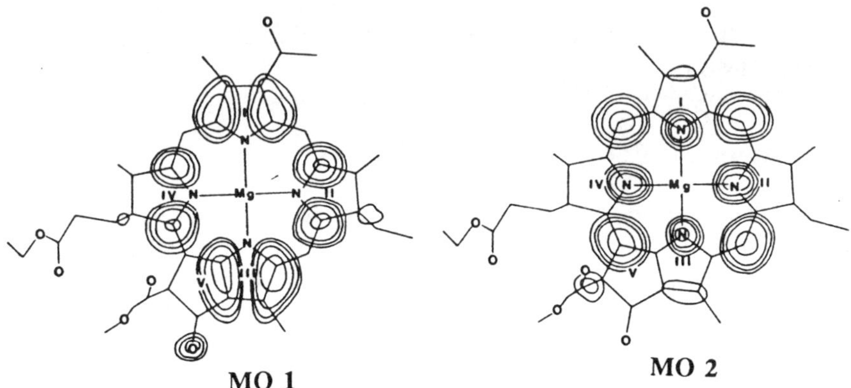

Figure 8. *Schematic electron densities for ground states of molecular orbitals MO of a macrocycle molecule [1].*

ii. Switching

Switching in phthalocyanine molecules is obtained using intramolecular or interorbital charge transfers. The transfer of an electron from its ground *m* to a final *n* state constitutes an act of switching similar to the concept proposed by Feyman [11]. Ground *m* and final *n* states may represent logic low and high states, respectively. A schematic transition of an electron from its ground to its final state is shown in Fig. 9, and it is performed via a virtual or excited state. Short transition times (~10^{-16} s) are indicated together with a much longer (~10^{-13} s) lifetime of the virtual state. Therefore, the switching time is determined by the lifetime of the virtual state.

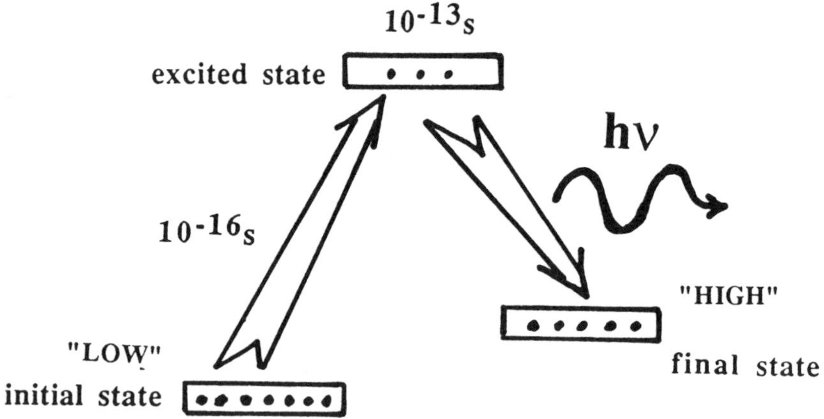

***Figure 9.** A diagram for an electronic transition in Raman scattering representing a switch.*

iii. Memory

The multiple redox states of phthalocyanine molecules may be considered as memory states and
may be altered rapidly by electrical or optical methods.

iv. Methods

Phthalocyanine molecules H_2, Co, Fe, and Cu in sulfonated (-TsPc) and regular (Pc) form have shown [75, 88, 89] a strong electro-optical coupling effects. These effects are based on the fast intramolecular electron transfer associated with an appropriately controlled molecular redox state. A change of redox state in turn alters the polarizability of a particular bond and, therefore, the Raman scattering signal for that bond. Such a modulated optical signal has been used as a carrier of information.

The modulation of optical signals from phthalocyanine molecules has been achieved using relatively simple methods. In an analog electro-optical mode of operation, optical signals from phthalocyanines adsorbed on some electrode interfaces can be controlled by electrochemical cyclic voltammetry and interrogated by surface enhanced Raman spectroscopy. However, if the control of redox states is achieved by optical methods, using light at frequencies ω_2 and ω_3 which coincide with the wavelength of the absorption maxima in the Soret and *Q*-bands, then the optical output at the frequency $\omega_1+\Delta\omega$ behaves in a digital form, where $\Delta\omega$ is a frequency shift of the appropriate Raman line. In the digital operational mode one employs a writingin laser operating at two frequencies ω_2 and ω_3 and a readout laser operating at frequency ω_1 in the nonabsorbing window at about 500 nm, as shown schematically in Fig. 10.

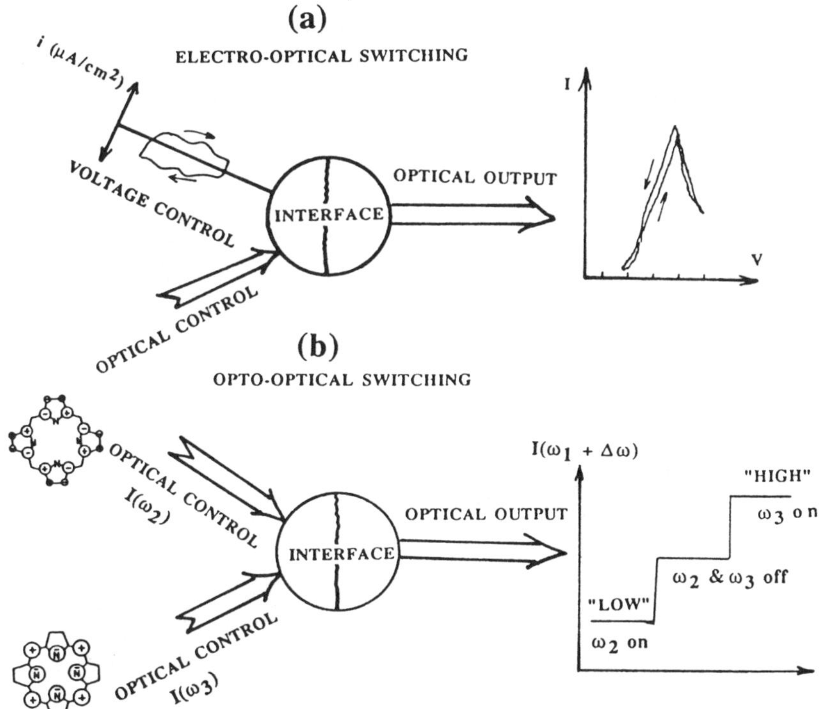

Figure 10. *Schematic diagram for optical switching. (a) in the electro-optical switching the interface of phthalocyanine molecules adsorbed on a solid substrate such as Ag or Au is controlled by electrical voltage input and optical control input at a frequency* ω_1*, the intensity of the optical output at a frequency* $\omega_1 + \Delta\omega$*, is in an analog form ; (b) in an opto-optical mode of switching, the intensity I of the optical digital output at frequency* $\omega_1+\Delta\omega$ *is controlled with two optical inputs at frequencies* ω_2 *and* ω_3*.*

Phthalocyanine molecules are commercially available, but their purity varies with supplier. Less available are water-soluble tetrasulfonated forms and more exotic structures such as stacked- ring [62] and polymeric sheet [63] types which have to be synthesized. Water-soluble tetrasulfonated phthalocyanines are suitable for electrochemical and biological systems which involve aqueous media. The molecules can be synthesized by the method of Weber and Busch [90]. Phthalocyanine molecules have been deposited on various electrode interfaces [91] and electrochemical cyclic voltammetry has been used to analyze the redox states of phthalocyanine molecules. A cyclic voltammogram obtained from Fe-TsPc adsorbed on pyrolytic graphite is shown in Fig. 11. The voltammogram was obtained in an alkaline medium saturated with Ar gas. The four characteristic redox peaks indicate Faradaic process. The calculated coverage shows a value of about 10^{-10} M/cm^2, which corresponds to a monolayer of adsorbed Fe-TsPc [91]. The energy levels and assignment of redox peaks are also shown in Fig. 11 with a clear indication of four quantum wells.

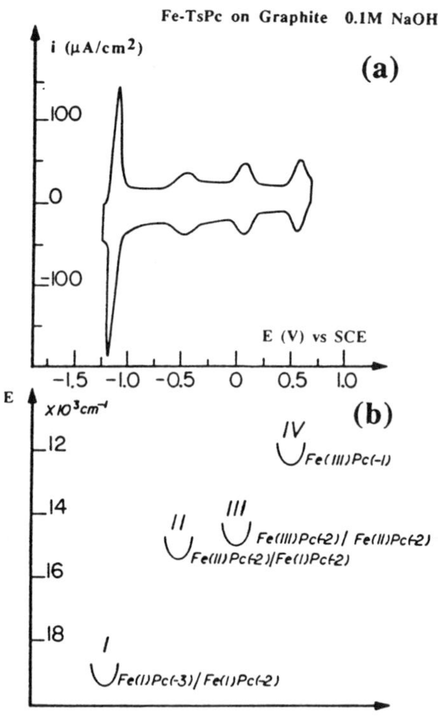

Figure 11. *(a) Cyclic voltammogram obtained from adsorbed Fe-TsPc on a graphite electrode. Scan rate 200 mV/s; (b) multiple quantum wells corresponding to the different redox states of the adsorbed Fe-TsPc on a graphite electrode.*

A change of molecular state from quantum well I to quantum well II corresponds to approximately $30kT$ and that would be the energy required to generate a bit of information.

Phthalocyanine molecules can be deposited on silver (poly- or monocrystals) or gold substrates using electrochemical or simple spreading techniques; as discussed in prior work [88, 91]. Cyclic voltammograms obtained for phthalocyanine molecules adsorbed on a silver or a gold substrates are limited between the potentials of appropriate hydrogen evolution and silver/gold oxidations. The electrode coated with macrocycle molecules is placed in a spectro-electrochemical cell that provides a means for electrical and optical control. All measured potentials were performed using a saturated calomel electrode (SCE).

Optical signals originating from phthalocyanine molecules can be modulated using combined electrochemical and optical methods.

If molecules are illuminated with a light radiation of intensity $I(\omega_i)$ at incident light frequency ω_i then the induced electric moment μ is proportional to the electric field $\mathbf{E_i}$ of the incident light radiation (84)

$$\mu = \alpha \mathbf{E_i} \cos \omega_i t. \tag{1}$$

The emitted light I_s from an induced dipole moment is given by

$$I_s = |\mu|^2 = |\alpha \mathbf{E_i}|^2 \tag{2}$$

In general, the intensity is a function of the polarizability α, and therefore the intensity changes are influenced by changes in α.

Raman scattering in terms of classical theory provides the relationship between the total intensity of a Raman band scattered over a solid angle 4π by a randomly oriented molecule undergoing transitions from an initial state *m* to a final state *n* and is given by

$$I_{mn} = (2^3\pi)/(3^2c^4)I_i(\omega_i-\omega_v)^4\Sigma|(\alpha_{\mathbf{i};\mathbf{j}})_{mn}|^2, \tag{3}$$

where ω_v is the frequency shift of the Raman line, $\alpha_{\mathbf{ij}}$ is the *i, j*th element of the scattering tensor, and $i, j = x, y, z$.

The general form for the *i, j*th component of the scattering tensor is given by the following sum over all vibronic states of the molecule:

$$\alpha_{i,j} = \frac{1}{\hbar} \sum_r \left[\frac{\mu_{rn}\mu_{mr}}{\omega_{rm} - \omega_i + \frac{1}{2}\Gamma_\nu} + \frac{\mu_{mr}\mu_{rn}}{\omega_{rn} + \omega_i + \frac{1}{2}\Gamma_\nu} \right] \tag{4}$$

Here *m, r* and *n* are the initial, intermediate, and final molecular states involved in the scattering transition; μ_{rm} and μ_{nr} are the matrix elements of the electric dipole moment operator; and Γ_v is a damping constant and represents the lifetime of the intermediate state.

The intensity of Raman scattered light is about 10^{-6} to 10^{-8} times less than the intensity of incident laser light oscillating at $\omega_i << \omega_{rm}$. Advantage can be taken of the resonant Raman scattering mode, when ω_i approaches the value of the molecular vibrational frequency ω_{rm}. In such a case the first term in Eq. (4) becomes predominant and the intensity of the resonant Raman scattering departs from the fourth power $(\omega_{rm}$- $\omega_i)^4$ term in Eq. (3) and the intensity is increased by several others in magnitude.

Raman scattering from adsorbed molecules on an electrochemical interface was revitalized by Fleischmann et al. [85]. Soon after the first report, Van Duyne et al. [86] recognized that this kind of Raman scattering, surface-enhanced Raman scattering (SERS) or resonant surface-enhanced Raman scattering (RSERS), is about 10^5 to 10^7 times more intense than the ordinary Raman scattering.

The experimental data [87], have shown that a monolayer of molecules, including phthalocyanines [88] adsorbed on silver, gold, or copper electrodes, emit very intense Raman signals due to surface-enhanced Raman scattering. The origin of the enhancing mechanism has been a very controversial issue. The very high amplification factor on modified silver or gold substrates prompted the development of several theories that consider the enhancing mechanism or mechanisms involved in SERS. Research in the field of surface-enhanced Raman scattering has indicated that metal (silver, gold, copper, aluminum and a few alkali metals) requires a modification in order to obtain enhanced scattering. The modification of metal substrates usually consists of metal roughening or oxidation/dissolution procedures. A number of reports proposed the involvement of the image field effects [92], electromagnetic resonance [93], charge transfer [94], and a few others. Our experimental evidence [95,96] shows that the origin of surface enhancing is associated with metal clusters and a charge transfer involving a population inversion mechanism similar to that observed in lasers. An energy diagram that describes the enhancing mechanism and represents a surface laser model is schematically shown in Fig. 12. Detailed analysis will be given in section **C. vi.** This enhancing factor is one of the most important reasons for using phthalocyanine molecules in molecular electronic devices. The proposed method can provide a very intense optical signal and a simple and reliable method to read out molecular systems and devices.

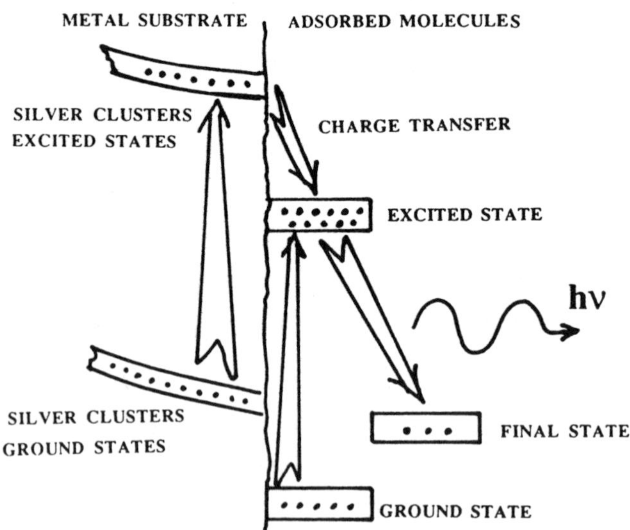

Figure 12. *Energy diagram showing charge transfers for an adsorbent-adsorbate system explaining surface-enhanced Raman scattering based on a surface laser model.*

v. Experimental Methods

The experimental setup for electro-optical studies of adsorbed phthalocyanine molecules on silver electrodes consists of a Raman spectrophotometer and a potentiostat for cyclic voltammetry, as shown in Fig. 13.

Phthalocyanine molecules are illuminated with a resonant He-Ne laser operating at 632.8 nm in the region of the *Q* bands and a nonresonant Ar^+ laser operating in the vicinity of about 500 nm. The scattered Raman light is collected at 90° and resolved with a standard spectrograph (Spex 1400 monochromator) which provides a number of resolved and very intense lines. Optical signals are detected with photomultipliers (ITT FW130) or by a photodiode array system (OMA Tracor Northern 1710) or, in the case of weak signals, by charge-coupled devices (CCD).

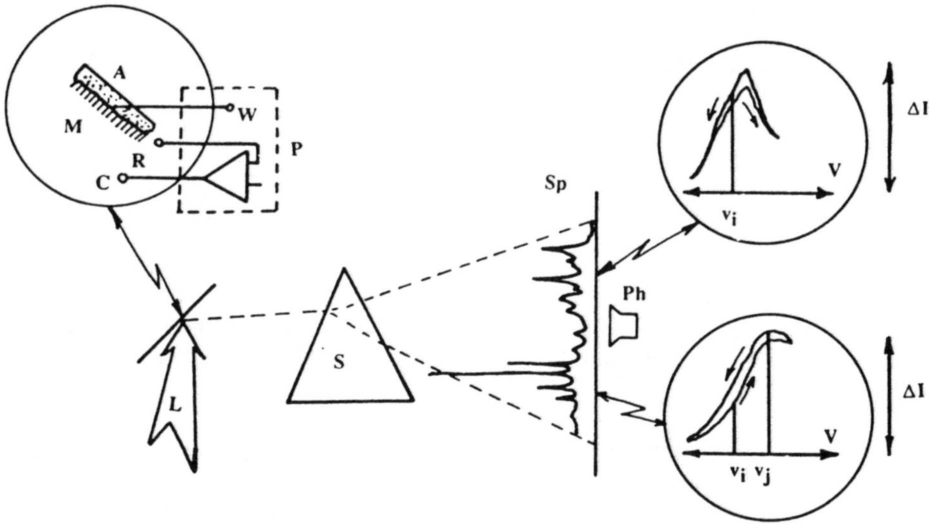

Figure 13. *Experimental setup for measurements of electro-optical properties of the adsorbed phthalocyanine monolayer M on the silver electrode A. W, R, and C are working, reference and counter electrodes, respectively. L, laser; S, spectrophotometer; Sp, optical signal; I-V, changes of the optical signals as a function of the electrode potential, Ph, photodetector; and P, potentiostat.*

Raman spectra obtained from Fe-TsPc phthalocyanine in the solution phase and adsorbed on a silver electrode are shown in Fig. 14. Four characteristic properties can be recognized:

1. Intensity of spectra.The polarized and depolarized spectra shown in Fig. 14(a) are obtained in the resonant scattering mode and indicate a low count rate (~300 counts/s). The spectra in Fig. 14(b) are two orders of magnitude stronger and indicate a net gain of about 10^6 [88].

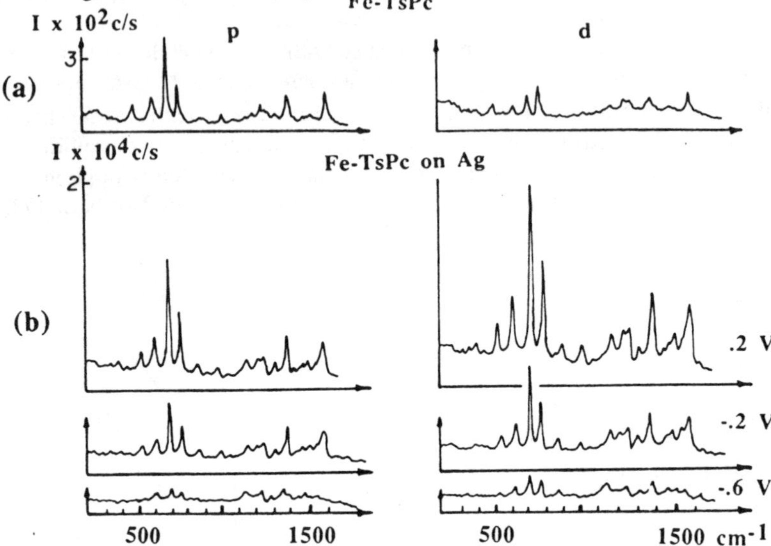

Figure 14. *Resonant Raman polarized (p) and depolarized (d) spectra obtained from (a) 10^{-5}* ***M*** *Fe-TsPc in aqueous media and (b) monolayer of Fe-TsPc adsorbed on silver electrode as a function of the electrode potential versus SCE. The supporting electrolyte is 0.05* ***M*** *Na_2SO_4 at pH ~ 7. Laser excitation at 632.8 with 20 mW output power.*

2. Vibrational Bands.The vibrational Raman bands are well resolved, and their assignment has been provided [88] and is summarized in Table 2. The similarity between the spectra obtained from the solution phase and the spectra obtained from adsorbed phthalocyanine species indicates a physisorption interaction without alteration of the Fe-TsPc structure.

3. Depolarization and molecular orientation. The depolarization ratio is normal (<3/4) for the spectra obtained from the solution phase, and (>3/4) for the adsorbed spectra, indicating a preferential orientationof the adsorbed molecules, with the molecular plane being normal to the electrode surface [88]. This is consistent with other results [69] and those obtained by scanning tunneling microscopy [97]. The orientation of the phthalocyanine adsorbed on an electrode interface is depicted in Fig. 15.

4. Modulation of optical signals. The intensity of the spectra in Fig. 14(b) is dependent on he applied voltage across the Ag/Fe-TsPc interface; that is, optical signals can be modulated by electrical means. This process is fully reversible.

TABLE 2. Vibrational Raman Bands.

	Co–TsPc		Co–Pr				Fe–TsPc		Fe–Pc				H_2–TsPc		H_2–Pc				pyrrole		assign-ment
	on Ag	in H_2O	in THF[a]	on Ag	solid[a]	IR[b]	on Ag	in H_2O	in THF[a]	on Ag	solid[a]	IR[b]	on Ag	in H_2O	THF[a]	Ag	solid[a]	IR[b]	*d*	*c*	
1	255						241			268											
2	344						332	327													
3	387			387			387	394					366	F	N	N					
3a						434						435						434			
4	519	518		498		520	517	523		493		518	506	L	O	O		492			χ_R
5	570			577		575	570					575	570					557			
6	610	612	593	598	593		612	616	587	598	596		624	U				620	586	585	χ_{HH}*
7	699	699	686	690	687	647	699	699	679	690	644	644	699				683		664	646	χ_R*
8	729	722				726	728	726	738			726	729	O			725	714	734	711	σ_R
																		736			
9			753	758	752	756				754	755	756	754					753			
10	763	765				772	763	767		785	780	771		R	D	D		766		768	•
11						780						780	799					778			
11a						804						804		E			800				
12	866	866	837	839	831	868	866	870	830	846	833	868			A	A		870	866	866	χ_{CH}
12a						877						877	891					880			
														S							
12b						915						910	911					914			
13	948			920		948	944			926		948		C	T	T		946			
14	981	983	962	968	960	956	977	983		964	951		968					958			
15									1007					E				1007			
16	1040			1014		1078	1040			1016		1072	1038		A	A	1030		1078	1072	σ_{CH}
16a						1092						1089		N				1094			
17	1114	1118	1114	1116	1108	1102	1112		1102	1118	1110	1121	1123				1085				
18	1136	1133	1141	1144	1140	1121	1136	1144	1156	1142	1147	1165		C			1142	1119	1138	1144	ν_R
18a						1165												1160			
19	1194	1204		1206		1174	1196	1208		1208	1200	1173	1185	E				1163			
20	1225	1229	1215	1230	1215		1223	1223		1230									1237	1237	δ_{CH}
21	1284	1289	1281	1284	1308	1291	1285	1285	1286	1283		1290	1283					1277			
21a																		1304			
21b										1312								1321			
22	1347	1352	1348	1347	1348	1334	1346	1352		1341	1343	1333	1341					1336	1392	1379	ν_RN—C
23	1408	1412		1406						1366										1418	
24	1437		1440	1436		1428	1433			1437	1422	1429	1429					1439			
25	1467	1476		1466	1465	1471	1462	1476	1445	1456	1450	1468						1461	1474	1467	ν_R
26	1488					1487	1489		1484	1476		1484						1478			
27				1521		1526	1515			1511	1520	1516	1521					1503			
28	1550	1551	1535	1546	1540		1543	1551		1535			1552				1544		1530	1532	C═C*
29				1592		1597	1595			1592		1592						1600			
30	1611					1612	1604					1609	1620					1617			

[a] Aleksandrov, I. V.; Bobovich, Ya. S.; Maslov, V. G.; Sidorov, A. N. *Opt. Spectrosc.* **1974**, *37*, 265–269. [b] Sidorov, A. N.; Kotlyar, I. P. *Opt. Spectrosc.* **1961**, *11*, 92–96. [c] Lord, R. C.; Miller, F. A. *J. Chem. Phys.* **1942**, *10*, 328–341. [d] Simic-Glavaski, B.; Zecevic, S.; Yeager, E. B. *J. Raman Spectrosc.* **1983**, *14*, 338–341. [e] χ out-of-plane bending mode, δ in-plane bending mode, and ν stretching mode. An asterisk indicates the principal strongest bands.

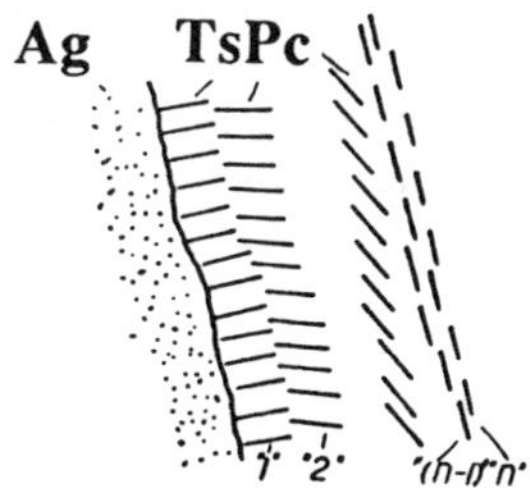

Figure 15. *A model of tthe adsorbed macrocycle molecules on a silver electrode.*

The structure of phthalocyanine molecules may have a profound influence on the electro-optical behavior of these molecular species. The monomeric and polymeric four-ring fused sheet type phthalocyanines shown in Fig. 16; have very similar electrochemical properties expressed in cyclic voltammograms.

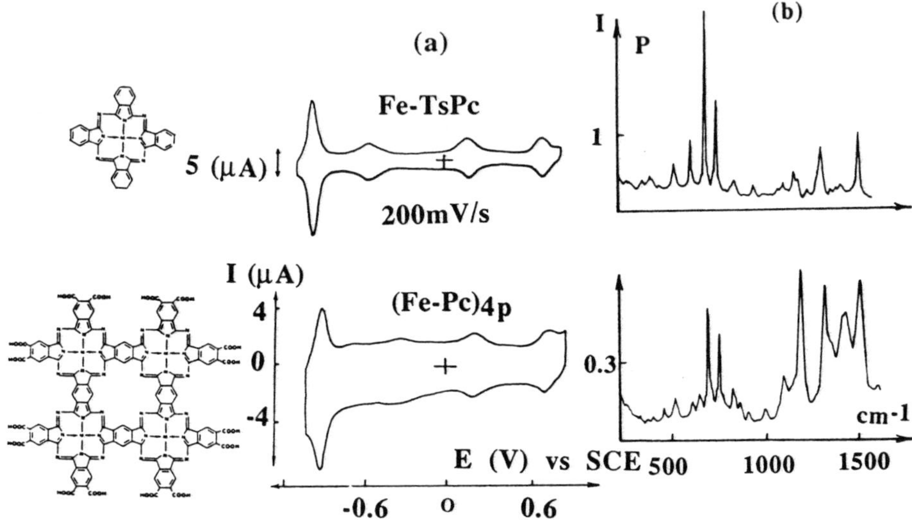

Figure 16. Electro-optical properties of Fe-TsPc monomer and polymeric sheet (Fe-Pc)$_{4p}$: (a) Cyclic voltammograms;(b) surface-enhanced resonant Raman spectra. Laser excitation at 632.8 nm with 20 mW output power.

This is interpreted [97] such that the redox states of the Fe atoms are unaffected by the process of polymerization. At the same time polymerization deforms the benzene and pyrrole rings, which alters the optical signals in the high-frequency region at about 1300 cm^{-1}. On the other hand, ring-stacked phthalocyanines shows substantial changes of its electrochemical properties, while optical Raman signals are less affected, in particular when one compares the electro-optical responses of monomeric iron phthalocyanine versus four-stacked-ring structures (see Fig.17). The multivalued redox states correspond to one-electron transfers between the rings [98].

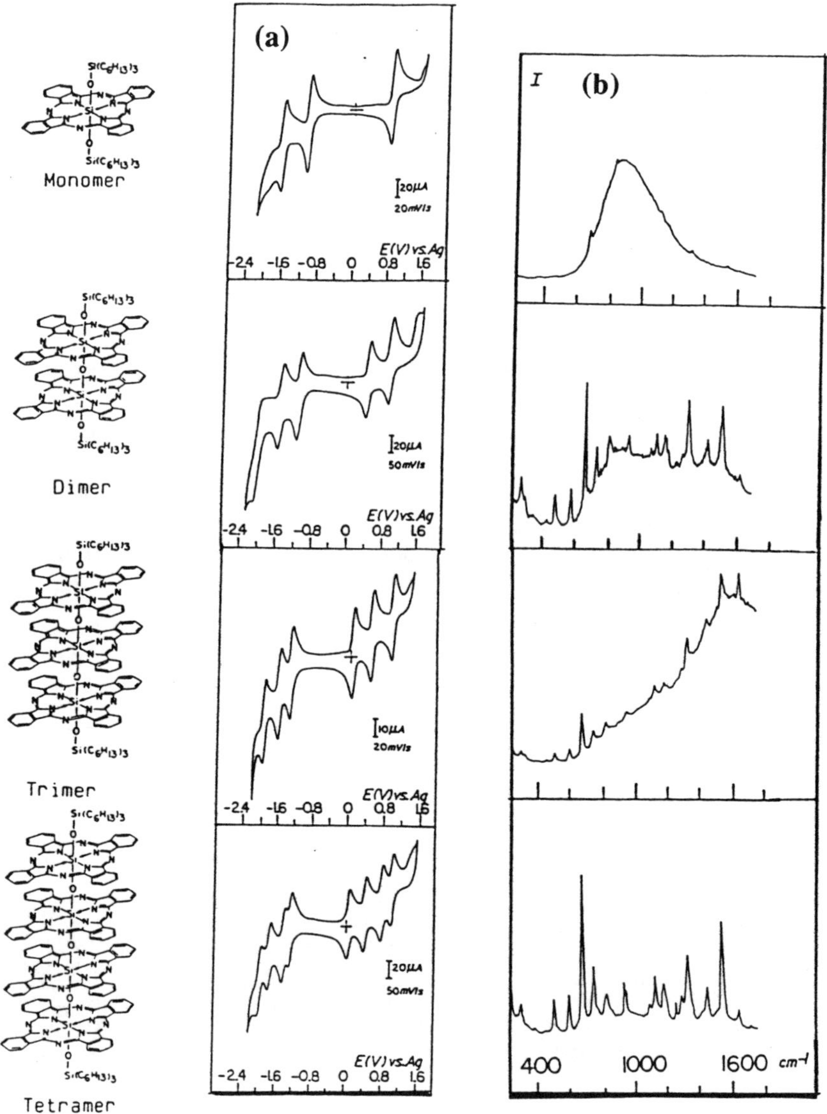

Figure 17. *Electro-optical properties of oxygen bridged (O-Si-Pc)$_n$ for n=1, 2, 3 and 4. (Midle) Cyclic voltammograms obtained from 10^{-3}**M** (O-Si-Pc)$_n$ in 0.1**M** tetra-n-butylammonium perchlorate in CH_2Cl_2 adsorbed on a platinum electrode and (Right) depolarized resonant surface-enhanced Raman spectra obtained from (O-Si-Pc)$_n$ adsorbed on a silver electrode at 0 V versus SCE. Laser excitation at 632.8 nm and 20 mW output power. The electrolyte is 0.05**M** Na_2SO_4 saturated with argon gas.*

vi. Mechanism of Surface Enhanced Raman Scattering

The high intensity level of surface-enhanced Raman scattering is a crucial element which is useful in molecular electronic devices, SERS can show very intense optical changes of the individual polarizabilities, thus avoiding the need for complex amplifying systems. Some fundamental elements concerning the issue of the mechanism involved in surface enhancement were briefly discussed in the previous section. The enhancement of Raman signal intensity is clearly demonstrated in Fig. 14.

The calculated gain factor for adsorbed phthalocyanines is between 10^5 and 10^7 [88]. In these Raman scattering measurements the laser beam is usually focused to a micrometer size; thus if the phthalocyanine coverage of the electrode is about 10^{-10} M/cm^2, then simple calculation indicates that only 10^4 to 10^5 molecules are illuminated to give the Raman signal. The measured intensity of the scattered light in a solid angle of $\pi/2$ shows that it is about 7% of the incident laser light, a value which agrees with the calculated one.

In addition, phthalocyanine molecules are very stable structures and can withstand very high temperatures. The surface-enhanced Raman scattering data shown in Fig. 18 demonstrate that the enhancement can be obtained at much higher temperatures than previously believed [99].

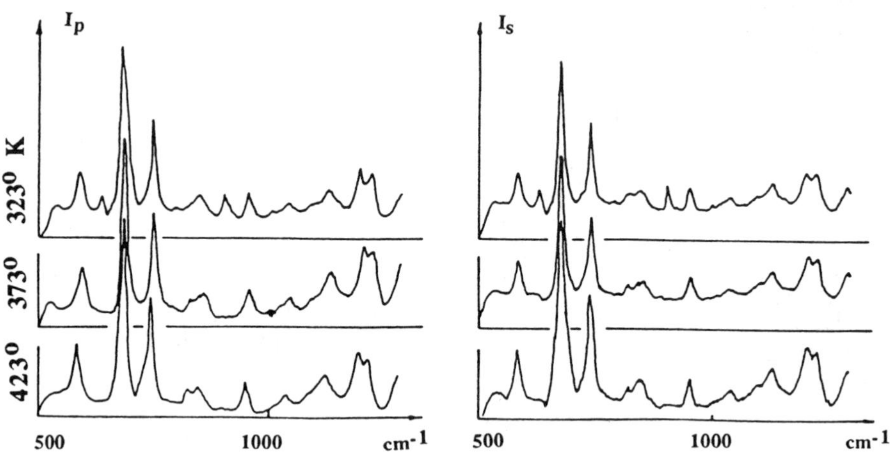

Figure 18. Resonant surface-enhanced Raman spectra obtained from Fe-TsPc adsorbed on a silver electrode as a function of temperature in K. The electrolyte is hydrated $AgNO_3$. Laser excitation at 632.8 nm with 20 mW output power.

The coincidence that metal substrates that provide surface enhancement are also photographically active raises questions: is there a correlation between the gain in photography which is of the order of 10^9, and the gain of the order of 10^7 associated with SERS? Could it be possible that both processes are governed and controlled by similar if not identical mechanisms ? These questions have been examined and analyzed in a report describing the surface-enhanced Raman scattering from Fe-TsPc adsorbed on chemically activated silver bromide crystals [95]. The AgBr-TsPc interface can be precisely analysed in terms of well-defined and documented positions of the electronic energy levels for both AgBr and Fe-TsPc molecules.

Semiconducting AgBr crystal plates are normally used as standard infrared windows. The plates were exposed to commercially available Kodak-76 developer with the intention of reducing silver ions to metallic silver clusters. Different time intervals (1, 2 and 3 min) were used to develop the AgBr plates, which were then covered with the appropriate amount of Fe-TsPc to form a monolayer. Partial surface Raman spectra were recorded from Fe-TsPc adsorbed on (a) nondeveloped substrate and (b) developed silver bromide and shown in Fig. 19.

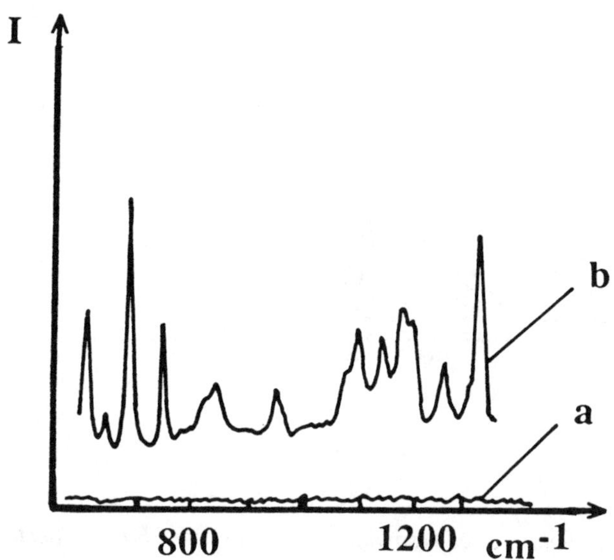

Figure 19. *Resonant Raman spectra obtained from Fe-TsPc adsorbed on (a) nondeveloped and (b) developed AgBr. Laser excitation at 632.8 nm with 20 mW output power.*

Surface-enhanced Raman spectra were also obtained from deposited Fe-TsPc on silver monocrystals, where photoactivation of the silver substrate had been performed without any electrochemical roughening procedure [96]. Silver monocrystals were polished in an aqueous solution of 30 vol % H_2O_2 and a solution of 21.5 g/l of NaCN mixed in a 1:1 ratio. The crystal was held for 5 sec in these solutions, with vigorous gas evolution occurring. It was taken out, held in air for 3 sec, and transferred into a solution of 37.5 g/l NaCN, where gas evolution ceased. This procedure was repeated

3-4 times until a highly reflective surface was obtained. After deposition of Fe-TsPc and formation of a monolayer with a coverage of 1.6 x 10^{-10} *M*/cm^2, the crystal was examined with LEED and the characteristic monocrystal pattern was obtained. The monocrystal with deposited Fe-TsPc was exposed to He-Ne laser radiation and spectrum a was recorded (see Fig. 20). Spectrum a is essentially background noise with a count rate of about 100/sec photons. The optical signal showed an intensity increase after about 5 min and continued to show further increments when recorded in time intervals of 10 minutes as shown in Fig. 20, spectra a-g. After about an hour of illumination the p-polarized spectra showed about 300,000 photons for the most intense Raman line at 699 cm^{-1}. At the same time the s-polarized spectra did not show any Raman line. The p and s polarized spectra in Fig. 20 are in complete agreement with the theoretical expectation that the Fe-TsPc molecular plane is normal to the surface of the monocrystal substrate.

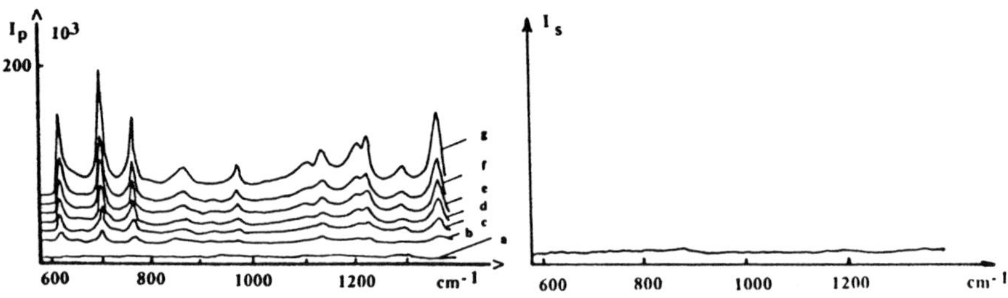

Figure 20. *Photoinduced resonant surface-enhanced Raman spectra a-g obtained as a function of illumination time, with recording time sequences of 10 minutes. p- and s-polarized spectra were recorded with laser excitation at 632.8 nm with 20 mW output power.*

Analysis of the electronic energy levels in semiconducting AgBr and the adsorbed Fe-TsPc provides a more detailed and deeper insight into the mechanism involved in both photography and surface-enhanced Raman scattering. The Fermi level for AgBr is around -5 eV [100]. The effect of the reducing developer on the AgBr plates is to form silver clusters and Ag_n and Ag_n^+ aggregates. The energy levels of the silver clusters' latent images are shown in Fig. 21. [95]. According to the theory of photography [100], prior [95], and more recent and comprehensive report [96], it appears that silver clusters are responsible for the photographic and surface-enhancing mechanisms.

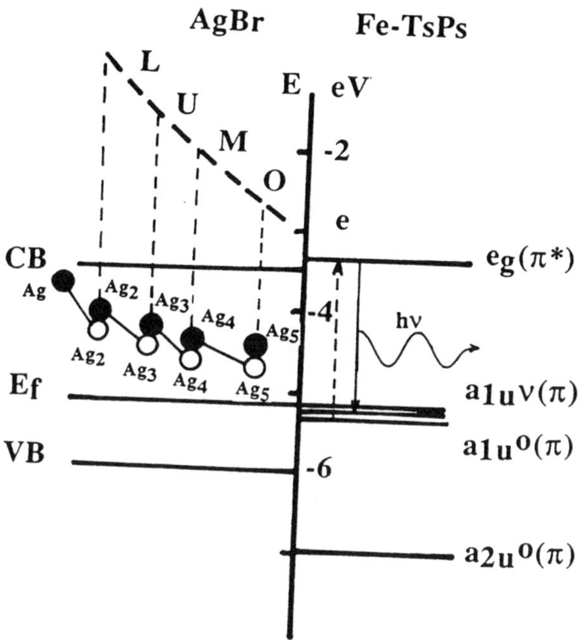

Figure 21. *A diagram of the energy levels for the AgBr/Fe-TsPc interface. VB, CB and E_f are valence, conduction, and Fermi levels,respectively. Ag_n are silver clusters.$a_{1u}(\pi)$ and $e_g(\pi^*)$ are occupied ground and excited phthalocyanine levels Raman radiation hv originates from the transition $e_g(\pi^*)$ to the vibrational level $a_{1u}(\pi)$.*

The common phenomenon in the activation of silver substrates involving electrochemical (anodic and cathodic), chemical, and photoactivation is the formation of silver Ag^+ and silver clusters Ag_n [101]. Neutral silver clusters Ag_n release an electron with an energy according to the scheme Ag_n + hν ---> Ag^+_{n+1} + e [100, 95, 96], this electron passes into the lowest unoccupied *s* molecular orbitals which are in the AgBr conduction band.

The optical activation energy of the electrons trapped by clusters or latent images is about 1.54 eV much lower than the ~3 eV of bulk silver [101]. In the subsequent analysis by Baetzold [102] of electron excitation from silver clusters, one finds that excitations of electrons from the highest occupied (HOMO) to the lowest unoccupied (LUMO) molecular orbitals to be between 1.8 and 2.4 eV for Ag_5 and Ag_4 clusters, respectively.

The He-Ne laser radiation at 632.8 nm or 1.96 eV pumps electrons from the silver clusters into higher energy levels or their LUMO states that are slightly higher than the $e_g(\pi^*)$ of the adsorbed Fe-TsPc. Electrons are transferred from excited LUMO states of silver clusters to the excited state of adsorbed Fe-TsPc and then to the final vibrational state of $a_{1u}(\pi)$ with emission of the appropriate Raman line as shown in Fig. 21. The charge transfer from LUMO -> $e_g(\pi^*)$ -> $a_{1u}(\pi)$ with a higher population of $e_g(\pi^*)$ than $a_{1u}(\pi)$, is similar if not identical to the population inversion mechanism involved in lasing [79]. The high gain in photography and surface-enhanced Raman scattering may be related and determined by the ratio of the population numbers of $Ne_g(\pi^*) / Na_{1u}(\pi)$.

In other words, the surface-enhanced Raman scattering is a type of surface lasing that is controlled with adsorbed molecules. This model is consistent with previously described processes involving dye sensitizing [100] and models involving four [103] and multilevel [104] charge transfers.

The model of a surface lasers for SERS or RSERS analyzed and obtained for the Fe-TsPc adsorbed on silver and other substrates involving silver clusters is also in agreement with a prior report that phthalocyanine molecules were used as lasing material in the first constructed dye laser [105]. The very high optical fluorescence conversion efficiency between 40 and 80% for Mg-Pc reported in earlier publications [106, 107], is in agreement with our measured value (in a solid angle of $\pi/2$) of 7% conversion efficiency for SERS.

vii.) Modulation of the Raman Optical Signals and Logic Elements

It was shown in the previous section that the intensity of the surface-enhanced Raman scattering is a function of the applied electrode potential. However, if a particular Raman band is selected, then its optical intensity can be modulated by varying the electrode potential within the limits of the oxidation-reduction cycle (ORC) [75]. The intensity variation ΔI of the optical Raman signal shown in Fig. 22 is an analog response to an analog change of potential. The cyclic voltammogram I-V curves for the silver substrate (dashed line) and the adsorbed H_2-TsPc on the same electrode are shown in Fig. 22. Two redox states are shown, and a monolayer of the adsorbed H_2-TsPc was calculated from the charge amount shown in the voltammogram.

Both the cyclic voltammogram and the Raman optical variation ΔI or Raman ORC are reversible and reproducible after numerous cycles. The optical signal changes are the most pronounced in the regions of the redox potentials, showing that the optical intensity is controlled by the redox states.

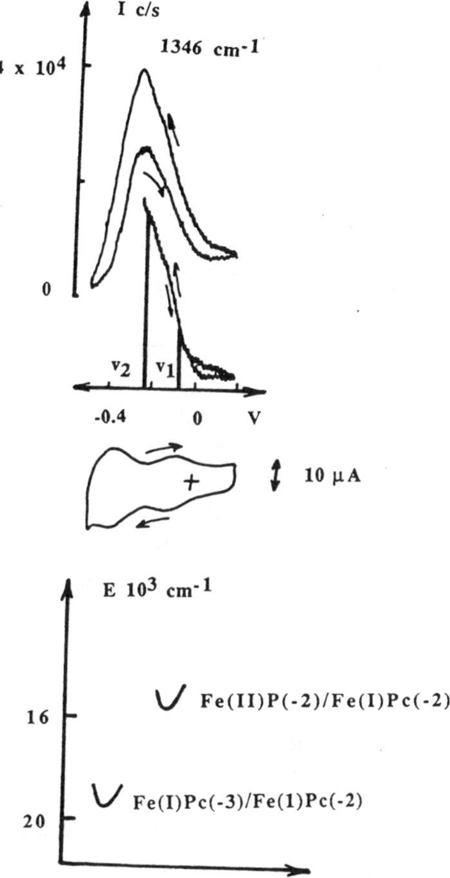

***Figure 22** Raman ORC for the 1346 cm^{-1} Raman band from an adsorbed monolayer of Fe-TsPc on a silver electrode. Supporting electrolyte 0.05* **M** *H_2SO_4 at pH 1. Optical signal changes I are a function of the biasing potential V. Argon ion laser excitation at 514.5 nm with 50 mW output power.*

The optical intensity changes are not caused by a Raman band frequency shift or adsorption -desorption processes. Figure 23 displays the optical signal changes for two Raman ORCs together with the cyclic voltammograms. The Raman ORCs are also characterized by a pronounced hysteresis that is a function of the *pH* of the supporting electrolyte. Optical signal changes without hysteresis can be also obtained, if the potential variation does not cross more than one redox state [88, 99]. The intensities and shapes of the Raman ORCs are a function of the specific bond and the central metal ion in the phthalocyanine molecule [88, 89]. The variation of the optical signal in Fig. 22 is in the range between 100 and about 48,000 photons, showing a contrast of about 480 for a potential change of only 0.7 V. This contrast is much greater than the recently reported record value >100 [49].

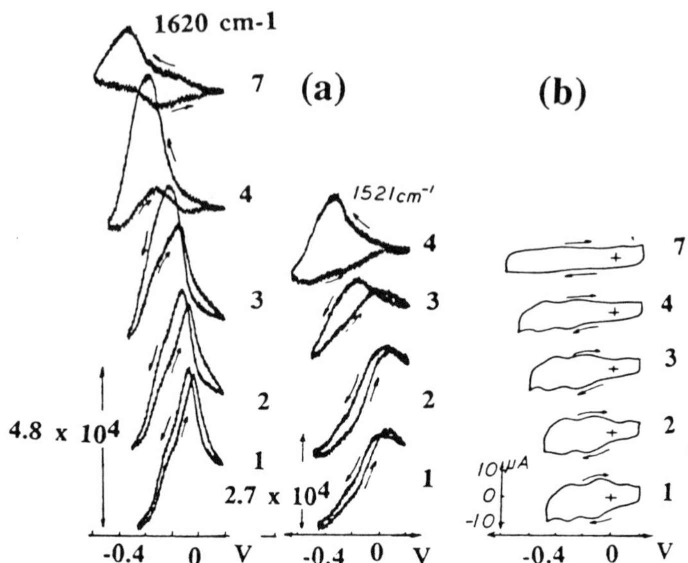

Figure 23. *(A) Raman ORCs for two Raman bands obtained from H_2-TsPc adsorbed on a silver electrode as a function of electrode potential V and pH. Vertical bar indicate 4.8 and 2.7 x 10^4 photons and (B) cyclic voltammograms. Argon ion laser excitation at 514.5 nm with 50 mW output power.*

The bistability of the optical Raman ORCs with a characteristic hysteresis loop is very similar to the optical bistability in nonlinear materials reported by Gibbs [46] and Smith [44].

Some phthalocyanines such as H_2-Pc, Mg-Pc, Zn-Pc and Si-Pc show very intense flourescence in solution phases and when deposited on electrode interfaces. The fluorescence radiation can also be modulated by changing the applied potential across the interface [89].

Optical signal modulation by electrical analog mode can be achieved using several methods [80]. If the electrode is biased at a potential V_j, and if the illumination is switched ***on*** and ***off***, then one obtains an optical pulse of intensity I_j. For other biasing potentials other values of I_j are obtained, but within the limits of the optical Raman ORCs. The optical switching time of I_j and its transition from a low 0 value to high I_j value is related to the lifetime τ of the particular Raman line and is measured and calculated from the linewidth of the Raman line. Typical measured values for τ are 2 to 5 x 10^{-13} s. By raising the temperature, switching times can be even shorter but cannot be shorter than the time required for the tunneling of electrons or excitation time of an electron, which is about 10^{-16} s.

In a digital or optical modulation mode one needs two lasers: A readout laser which operates in a nonadsorbing region at about 500 nm, and a writein laser whose incident frequency can be altered between the Soret and *Q* (~330 and ~660 nm) bands. These resonant laser excitations induce electronic transitions from two fundamental energy levels $a_{2u}(\pi)$ --> $e_g(\pi^*)$ and $a_{1u}(\pi)$ --> $e_g(\pi^*)$ and enhance high ($1000\text{-}1700\ cm^{-1}$) and low ($0\text{-}1000\ cm^{-1}$) frequency Raman lines, respectively. By altering the excitation frequency, rapid ($< 10^{-12}$ s) intramolecular charge transfer occurs which modulates the output optical signal from one intensity level to another. In other words, if one applies laser energies that correspond to the energies of the quantum wells (see Fig. 6), intramolecular charge transfers occur, and they can be used as a switch from one memory state to another.

A further characteristic of the Raman ORC is that they are bistable optical signals that depend on the redox state of the phthalocyanine molecule. Since the Raman ORCs are also a function of particular Raman vibrations, their redox state dependence can be used to form a logic gate operation. Two Raman ORCs from two different Raman lines are shown in Fig. 23. These ORCs originate from the same adsorbed phthalocyanine assembly, where about 10^4 to 10^5 molecules radiate in unison. The optical output from these two Raman ORCs function as **AND** and **NOT** logic gates (Fig. 24). If an electrical potential V and exciting laser I are considered as inputs, then the optical output of the Raman line functions are the output of a logic gate. The multiplicity of Raman lines of different intensities and with different optical responses also acts as a multiplexed output. This multiplexed output is nothing more than a state vector whose components I_j are the individual Raman lines.

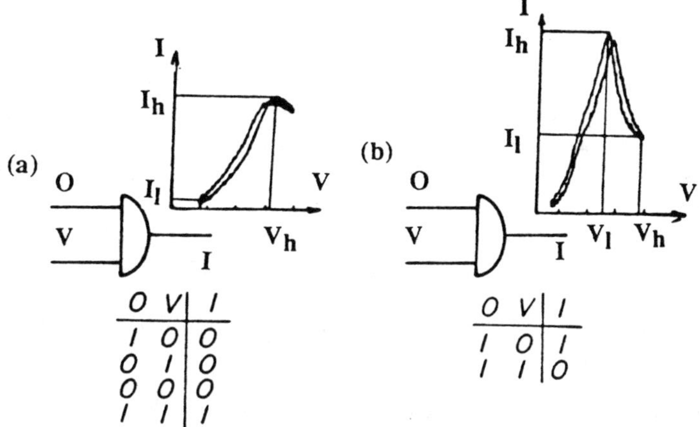

Figure 24. *(a)* ***AND*** *and (b)* ***NOT*** *logic gates obtained from the Raman ORCs in Fig. 23. V is electrical input, and I and O are optical input and output.*

The Raman line intensity changes depend on the electrode biasing potential (i.e., redox state), of the adsorbed Pc molecule or its memory state. In the case of phthalocyanines adsorbed on a silver electrode in an electrolyte, one observes two redox or memory states V_j and V_k. Optical ***off on*** excitation of phthalocyanine molecules biased at potentials V_j or V_k causes an optical signal change from 0 to I_j and I_k with a switching time of the order of 10^{-13} s. If the redox states are altered by optical means, then one eliminates interfacial capacitance effects and provides only rapid (<10^{-12} s) intramolecular charge transfers [108], which cause a rapid alteration of the state vectors.

$$\underset{\text{State j}}{\begin{bmatrix} 0 & I_{j1} \\ 0 & I_{j2} \\ -- & -- \\ -- & -- \\ 0 & I_{jn} \end{bmatrix}} \Longleftrightarrow \underset{\text{State k}}{\begin{bmatrix} 0 & I_{k1} \\ 0 & I_{k2} \\ -- & -- \\ -- & -- \\ O & I_{kn} \end{bmatrix}}$$

Where I_{jn} and I_{kn} are Raman line intensities at the redox potentials j and k.

viii. Modulation of Optical Signals from Phthalocyanines Adsorbed on Natural Signal Generators - Nerves.

a. Phthalocyanine and natural neurons

Some basic electro-optical properties of the phthalocyanine and similar porphyrin molecules have been described and discussed in earlier sections. Porphyrin molecules have a very important role in natural biosystems. Biocompatibility of these molecular species may be a very significant factor in the analysis and control of biosystems [80].

This section discusses the electro-optical behavior of adsorbed phthalocyanine molecules on more complex biological interfaces which obey the principal laws of bioelectrochemistry.

The electrical and functional properties of neural membranes have been studied extensively in order to elucidate the electrical signal of a neuron, which represents the basic unit of processing element in natural communication and information processing. Traditionally, microelectrodes have been the principal tool in studies of individual neurons [109] and their more complex assemblies, neural nets. The great disadvantage of this experimental methodology lies in its invasive aspect and, in the case of neural nets consisting of a few to billions of individual neurons, in the need for great number of individual microelectrodes.

The optical monitoring of the activity of a cell, particularly the electrical activity of neurons, has been an attractive alternative to the use of microelectrodes [110, 111]. In order to monitor the electrical state of a neuron reliably, both cell-intrinsic [110] and cell-extrinsic [111] photic probe molecules have been proposed and used. However, a description of the electro-optical interaction mechanism between the photic probe molecules and the nerve membrane substrates is still a challenging problem at the molecular level. Weak changes of the optical signal, nonspecific information, and toxicity are the main discouraging properties. Cohen [111] reported that changes in light scattering from giant squid axons are a function of the action potential. The changes of scattered light were about 10^{-6} times weaker than the incident light and required about 10,000 averaging sweeps. Ideally one would like to have large changes of the optical signal with high contrast in order to follow an action potential in a single sweep, avoiding the need for averaging. The analysis of the optical changes constituted a major problem in specifying the mechanism and origin of these changes. These findings have inspired efforts to measure other optical changes and to study changes in indicator molecules added to the axon. The optical signal response could be significantly improved using efficient fluorescent dyes; however, these added indicator molecules are usually not without a toxic effect.

The present discussion analyzes experimental data on the electro-optical behavior of adsorbed phthalocyanine molecules on nerve membrane interfaces. These data are correlated with the electro-optical data obtained from adsorbed phthalocyanine molecules in a controlled spectroelectrochemical environment. In this way, one obtains highly specific information at a submolecular level. Raman spectroscopy and electrochemical cyclic voltammetry yield very precise details on molecular structural-vibrational and redox properties. Combined techniques can provide and establish a relationship between electrical and optical properties of the molecule under investigation.

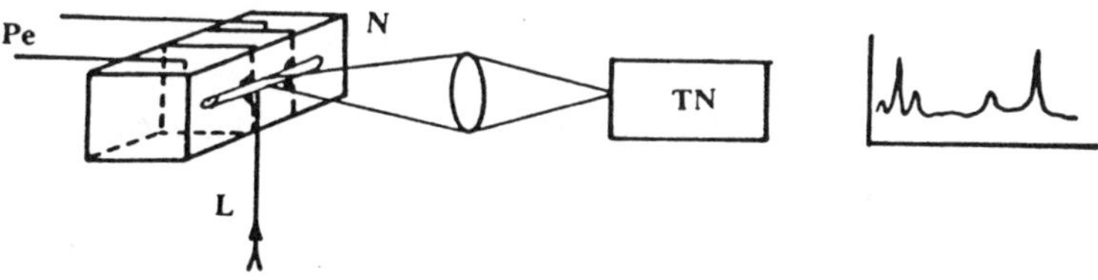

Figure 25. *The experimental setup for recording Raman spectra from adsorbed dyes on nerve membranes. N, neuron, Pe, polarization AgCl electrodes; L laser, TN, monochromator with a recording system. After [82].*

The extensive knowledge of the Fe-TsPc molecule is advantageous for its applications in more complex tasks.

The experimental setup used in the electro-optical studies of neurons is shown in Fig. 25. Sensory and motor neurons were extracted from the walking leg of a lobster and placed in an electro-optical scattering chamber that provided all the necessary conditions for physiological activity [80, 82]. The nerves in the central compartment were illuminated with a focused laser beam and the scattered light was collected and passed through a monochromator.

Raman spectra were recorded from light scattered from methyl orange and nerve bundles stained with optical indicator, as shown in Fig. 26.The Raman spectra in Fig. 26 are dissimilar, indicating a chemical interaction of the photic dye molecules with the neuron substrate.

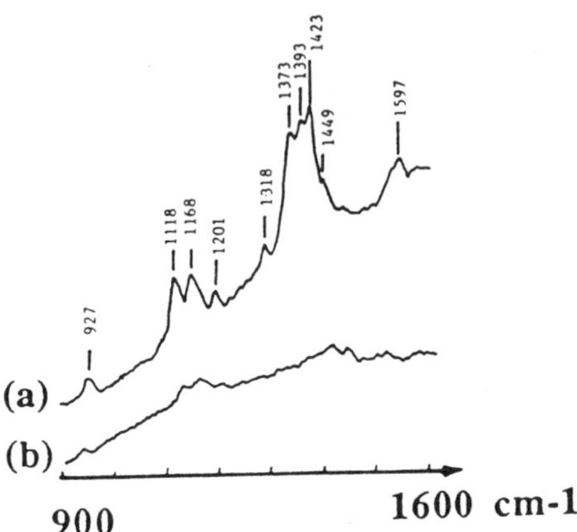

Figure 26. *(a) Resonant Raman spectrum obtained from 5×10^{-5}* ***M*** *methyl orange in sea water. (b) Raman spectrum from lobster nerve stained by 5×10^{-5}* ***M*** *methyl orange for 15 min. Nerve at resting potential, Argon ion laser excitation at 488.0 nm with 100 mW output power. After (115).*

On the other hand, the resonant Raman spectra obtained from Fe-TsPc in artificial sea water and Fe-TsPc adsorbed on nerve bundles shown in Fig. 27 are very similar, suggesting a physical interaction and the noninvasive properties of the phthalocyanine molecules. More important, the Raman spectra from the Fe-TsPc adsorbed on the nerve, membrane are a function of the electrical polarization state of the nerve, and the spectra change considerably during nerve depolarization. The change in the Raman band intensity in the range of 1200 cm^{-1} can be so specific that it can describe electron transfer and its delocalization within the pyrrole ring of a phthalocyanine molecule and the reduction of Fe(III) to Fe(II) [82].

Resonant Raman spectra obtained from Fe-TsPc adsorbed on a nerve membranes are also reversible as a function of depolarization and indicative of no adverse effects from the phthalocyanine molecules on biologically active molecules in the nerve bundles. The optical changes in the Raman spectra, as a function of nerve depolarization, are illustrative of bond changes in the molecules responsible for ionic transfer in nerve membranes. This information can explain the small ~ 10^{-6} changes of light scattered from nerve membranes previously reported by Cohen and coworkers [110] and is in agreement with the fundamental principle that the Raman scattered light is about 10^5 to 10^7 times less intense than the incident exciting light.

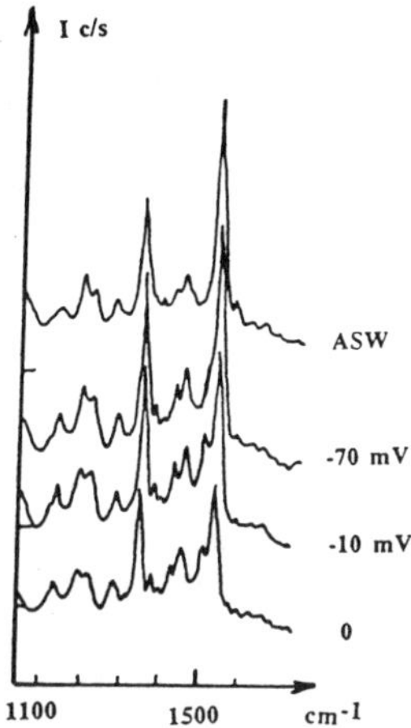

Figure 27. *Resonant Raman spectra obtained from Fe-TsPc in artificial sea water (ASW) and adsorbed on dissected lobster neuron. Electrode potentials are indicated in mV. Laser excitation at 632.8 nm with 20 mW output power.*

The other advantageous property is that Raman bands are spatially isolated from the intense incident light and, although changes in Raman scattered light are small, these changes have high contrast. This can be applied to optical studies of complex neural nets and may provide a better experimental methodology than microelectrode techniques. In this way, optical signals from phthalocyanine molecules adsorbed on individual neural cells can be observed in parallel from thousands of individual sources and thus provide parallel information processing on the behavior of complex systems.

b. Phthalocyanines as artificial neurons and in neural networks

The properties and functional versatility of phthalocyanine molecules are widely proposed for a variety of purposes, including artificial neurons and neural nets. Taomoto et al. [112] proposed to use the electrical properties of vacuum evaporated lead phthalocyanine films as one of the candidates for an artificial neural device. The current-voltage characteristics showed a rectifying phenomenon with field-induced reversible switching. The device is composed of three layers: The first layer has the function of transforming an input signal and memorizing the state caused by the signal; the second layer has a switching function; and the third layer a nonlinear processing function. It was reported that such devices are sensitive to gases with significant anisotropic conductivity. Simic-Glavaski [83] reported that different structural forms of phthalocyanine molecules can have interesting properties that are vitally important in modeling artificial neural networks. A sigmoid curve that simulates the pulse code firing rate of a neuron can be obtained from a single pixel of Fe-TsPc adsorbed on a silver electrode. This is demonstrated in Fig. 28, showing the variation of an optical signal as a function of applied potential. In order to obtain a similar electrical output, one needs approximately 20 CMOS elements occupying an area of about 40 x 40 μm^2.

In another mode of operation, based on the electrochemical properties of oxygen-bridged silicon phthalocyanine in the ring stacked configuration, one obtains the current-voltage characteristics shown in Fig. 17. The number of minima or redox states are related to the number of phthalocyanine rings involved in the stacked configuration. The redox peaks represent the addition or removal of electrons to the molecular orbitals, but in general the redox peaks correspond to one-ring electron transfer. The multiplicity of redox peaks is very similar to the Hopfield energy curve [113] with the multiplicity of minima representing stable states of a neural network. The optical output in the form of a Raman spectrum with a multiple Raman bands represents a multiplexed output and a vector with the same number of components. The values of the vector component can be zero or I_{rj} depending on the light illumination state of the phthalocyanine molecules. These vectors are determined for each redox state or stable states that can be easily altered and controlled.

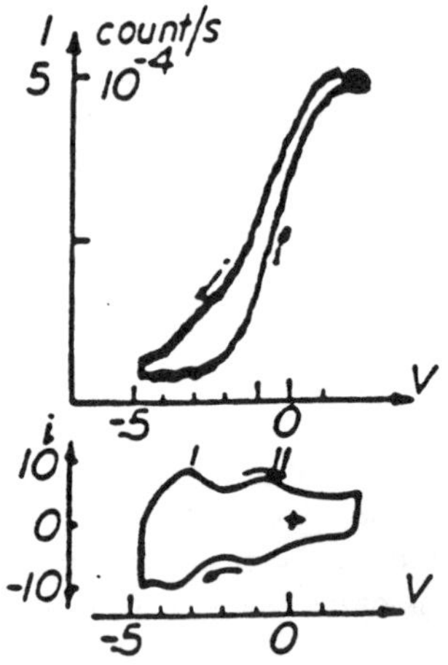

Figure 28. *A curve representing the pulse code firing rate of a neuron obtained from Fe-TsPc adsorbed on a silver electrode.*

Sakurai and Takano [114] consider stacked metallophthalocyanine rings as organic quantum devices with potential applications in cellular automata. The current-voltage characteristics have multiple states, similar to those shown in Fig. 17, which are considered as analogs to optical bistability and may perform as multiple-valued digital logic devices. These logic devices with multiple ***on*** and ***off*** states have been extensively considered in the information dynamics of cellular automata. The electron transfer between phthalocyanine rings is proposed to solve the interconnection dilemma in conventional $GaAs/Al_xGa_{1-x}As$ quantum devices.

D. COMMENTS AND CONCLUSIONS

Molecular electronic devices based on phthalocyanine molecules show a wide spectrum of interesting properties with flexible applications in various systems.

One of their most obvious potentials is for use in switching, memory, and logic processing elements within electro-optical and opto-optical computers with traditional and nontraditional functional responses. The fact that they can accept electrical, as well as optical, inputs allows them to be interfaced with systems based on traditional semiconductor technologies. The analog and digital responses of molecular devices based on phthalocyanine molecules provide additional attractiveness in their applications.

The high gain factor associated with surface states and adsorbed phthalocyanines demonstrated in surface enhanced Raman scattering allows these systems to be used for the amplification of optical signals. These surface microlasers may have wide applications in communication systems and also in integrated optical circuits.

The biocompatibility of phthalocyanine molecules is yet an additional and very important feature. Phthalocyanine molecules can be used in biosensors and hold promise for interfacing living with artificial systems. The property that phthalocyanine molecules can be placed on neural membranes without invasive effects can hardly be matched with toxic GaAs or similar materials. The possibility that phthalocyanine molecules can be intercalated between DNA bases is a very important property that may be effectively used in DNA sequencing and understanding the problems associated with the human genome project.

Phthalocyanine molecules in molecular electronic devices have the following characteristics:

1. Switching times less than 10^{-12} seconds;
2. Energy requirements of the order of $30kT$ per bit of information, allowing a construction of 10^6 processing elements /cm^2, dissipating less than 10 W operational at room temperature;
3. Operational up to about 300° C;
4. 2 x 2 μm^2 processing elements can be constructed using current technologies without complex systems or superclean room facilities;
5. Electrical and optical input/output properties make these devices compatible with semiconductor systems;
6. Optical output in the form of multiple Raman lines, make the devices attractive for analog and multisignal processing;
7. Cascadable;
8. High contrast, greater than 450;
9. Information processing capacity of about 10^{19} gate Hz/cm^2;
10. Use as dynamic memory elements;
11. Use as binary and multivalued logic gates;

12. Biocompatibility.

The demand for small, fast switching memory and processing elements, which have low energy requirements, and are inexpensive to fabricate, for use in optical computers and communication systems may be met by molecular electronic devices based on phthalocyanine molecules. The concept of controlled intramolecular charge transfer as a carrier of information has far-reaching consequences in information processing systems, both natural and artificial.

E. ACKNOWLEDGMENTS

The author wishes to express his thanks to his colleagues for valuable comments and discussions, in particular Drs. S. Zecevic, B. Nikolic, E. Yeager, A.B.P. Lever R. Adzic, T. Barrett and S. Gruber. Thanks are extended to the NAVYAIR grant MDA 903-85-K-0024 for partial support of this work.

F. REFERENCES

1. F. L. Carter, Ed., *Proceedings of the Molecular Electronic Devices Workshop*, NRL Memorandum Report 4662, Naval Research Laboratory, Washington, D.C.,1981.

2. F. L. Carter, Ed., *Molecular Electronic Devices II*, Marcel Dekker, Inc., New York, 1987.

3. F. L. Carter, R. E. Siatkowski and H. Wohltjen, Eds., *Molecular Electronic Devices*, North-Holland, Amsterdam, Ney York, Oxford Tokyo, 1987.

4. F. Hong, Ed., *Molecular Electronics: Biosensors and Biocomputers*, Plenum Publishing, New York, 1988.

5. A. Aviram, Ed., *Molecular Electronics-Science and Technology*, Engineering Foundation, New York, 1989.

6. M. Aizawa, Ed., *Second International Symposium on Bioelectronic and Molecular Electronic Devices,* Fujiyoshida, 1988; P. I. Lazarev, Ed., Second International Conference on Molecular Electronics and Biocomputers, Moscow, 1989.

7. L. Brillouin, *Science and Information Theory*, Academic Press, 1962.

8. E. Fredkin and T. Toffoli, *Int. J. Theor. Phys.*, 21 (1982) 219.

9. C. H. Bennett, *IBM J.Res. Dev.*, 17 (1973) 525.

10. C. H. Bennett, *Int. J. Theo.Phys.*, 21 (1982) 905.

11. R. P. Feyman, in *New Directions in Physics*, N. Metropolis, D. M. Kerr, G-C. Rota, Eds., 7, Academic Press, Boston, 1985.

12. K. E. Drexler, in [3], p. 39.

13. A. D. Fisher, L. C. Giles and J. N. Lee, *J. Optic. Soc. Am.*, 1 (1984) 1337.

14. D. Psaltis and N. Farhat, *Optics Lett.*, 10 (1985) 2.

15. D. Psaltis and Y. S. Abu-Mostafa, in *Computational Power of Parallelism in Optical Computers*, California Institute of Technology, 1985.

16. G. J. Lasher and A. B. Fowler, *IBM J. Res. Dev.*, 8 (1964)471.

17. P. J.van Heerden, *App. Optics* 2 (1963) 4.

18. T. K. Gaylord, in *Handbook of Optical Holography,* Academic Press, 1979.

19. A. Huang, in *Proc. Int. Optical Computing Conf.*, W. T. Rhodes Ed., S.P.I.E. 232 (1980).

20. H. H. Szu and H. J. Caulfield, *Proc. IEEE,* 72 (1984) 902.

21. D. Casasent, in *Optical and Hybrid Computing*, H. H. Szu and R. F. Potter Eds., S.P.I.E. 634 (1986) 439.

22. J. A. Neff, ibid. 24.

23. S. D. Smith, A. C. Walker, B. S. Wherrett and F. A. P. Tooley, ibid. 134.

24. J. W. Goodman, F. J. Leonberger, S-Y. Kung and R. A. Athale, *Proc. IEEE*, 72 (1984) 850.

25. R. W. Keyes, *Optica Acta*, 32 (1985) 525.

26. J. von Neumann, in *Theory of Self-Reproducing Automata,* A. W. Burks, Ed., University of Illinois Press, Urbana, 1966.

27. E. P. Wigner, "On time-energy uncertainty relation" in *Aspects of Quantum Theory*, A. Salam and E. P. Wigner, Eds., Cambridge University Press, Cambridge, 1982.

28. K. K. Likharev, *IEEE Trans.Magn.*, 23 (1987) 1142.

29. K. K. Likharev, *IBM J.Res.Devel.*, 32 (1988) 144.

30. K. K. Likharev, *Inter.J.Theor. Phys.*, 21 (1982) 311.

31. S. M. Sze, *Physics of Semiconductor Devices*, Wiley, New York, 1981.

32. F. H. Dill, A. S. Farber, and H. N. Yu, *IEEE J.Solid-State Circuits,* 3 (1968) 160.

33. R.van Tuyl and C. Liechti, *IEEE Spectrum*, 14 (1977) 41.

34. H. P. Zape, in *Advances in Superconductivity*, B. Deaver and J. Ruvalds, Eds., 51, Plenum, New York, 1983.

35. S. L. Hurst, *Threshold Logic,* Mills & Boon, London, 1971.

36. N. G. Basov, W. H. Culver, and B. Shah, in *Laser Handbook*, F. T. Arecchi and E. O. Schulz-Dubois Eds., North Holland, Amsterdam, The Netherlands; 1972.

37. R. A. Kingston, B. E. Burke, K. B. Nichols, and F. J. Leonberger, *App.Phys.Lett.*, 41 (1982) 413.

38. R. A. Boenning, V. B. Morris, and E. G. Vaerewyck, presented at the 29th Instrumentation Symp., Instrument Soc.America, Albuquerque, N.M. 1983.

39. R.V.Johnson, D.H.Hecht, R.A.Sprague, L.N.Flores, D.L.Steinmetz, and W.D.Turner, *Opt.Eng.*, 22 (1983) 665.

40. A. Korpel, *Proc.IEEE*, 69 (1981) 48.

41. A. F. Garito, K. D. Singer, K. Hayes, G. E. Lipscomb, S. J. Lalama, and K. N. Desai, *J. Opt. Soc. Am.*, 70 (1980) 1399.

42. P. W. Smith, E. H. Turner, and P. J. Maloney, *IEEE J. Quantum Electron.*, QE-14 (1978) 207.

43. R. C. Alferness, *IEEE J. Quantum Elec.*, QE-17 (1981) 946.

44. P. W. Smith and W. J. Tomlinson, *IEEE*, 18 (1981) 26.

45. S. D. Smith, *App. Optics*, 25 (1986) 1550.

46. J. L. Jewell, M. C. Rushford, and H. M. Gibbs, *App.Phys.Lett.*, 44 (1984) 172.

47. D. A. B. Miller, D. S. Chemla, T. C. Damen, A. C. Gossard, W. Weigmann, T. H. Wood, and C. A. Burrus, *App.Phys.Lett.*, 45 (1984) 13.

48. D. A. B.Miller, D. S. Chemla, T. C. Damen, T. H. Wood, C. A. Burrus, A. C. Gossard, and W. Weigmann, *IEEE J. Quantum Electronics,* QE-21 (1985) 1462.

49. M. Whitehead, A. Rivers, G. Parry, J. S. Roberts, and C. Button, *Electronics Lett.*, 35 (1989) 984.

50. A. R. Tanguay, *Opt.Eng.*, 24 (1985) 2.

51. D. S. Chemla and J. Zyss, *Nonlinear Optical Properties of Organic Molecules and Crystals*, Academic Press, New York, 1987.

52. R. W. Landauer, *IBM J. Res. Dev.*, 5 (1961) 183.

53. A. Aviram and P. Seiden, U.S. Patent, 3,833,894, 1974.

54. C. K. Chiang, M. A. Druy, S. C. Gau, A. J. Heeger, E. J. Louis, A. G. McDiarmid, Y. W. Park, and H. Shirakawa, *J. Am. Chem. Soc.*, 100 (1978) 1013.

55. K. M. Ulmer, *Science,* 219 (1983) 666.

56. H. Sasabe, T. Furuno, and K. M. Ulmer, [5], 285.

57. R. J. Butera, B. Simic-Glavaski, and J. Lando, *Macromolecules,* 23 (1990) 199 and 211.

58. R. R. Birge, A. F. Lawrence, and L. A. Findsen, in *Proceedings of the International Congress on Technology and Technology Exchange*, International Technology Institute, Pittsburg, Pennsylvania, 1986, 3.

59. J. J. Hopfield, J. N. Onuchic, and D. N. Beratan, *Science*, 241 (1988) 817.

60. B. Simic-Glavaski, in *Optical and Hybrid Computing,* H. H. Szu and R. F. Potter Eds., *SPIE*, 634 (1986) 195.

61. A. B. P. Lever, in *Advances in Inorganic Chemistry and Radiochemistry*, H. J. Emeleus and A.G.Sharpe, Eds., 7 (1965) 27.

62. T. M. Mezza, R. N. Armstrong, G. W. Riter, J. P. Iafelice, and M. E. Kenney, *J. Electroanal. Chem.*, 137 (1982) 227.

63. B. N. Achar, G. M. Fohlem, and J.A.Parker, J. Polym. Sci., 20 (1982) 1785.

64. F. L. Carter, in [1], 344.

65. T. J. Marks, C. W. Dirk, K. F. Scoch Jr., and J. W. Lyding, in [1], 146.

66. K. W. Wynne, T. J. Marks, and T. Inabe, U.S.Patent 4,622,170, 1986.

67. S. Baker, M. C. Petty, G. G. Roberts and M. V. Twigg, *Thin Solid Films*, 99 (1983) 53.

68. S. Baker, G. G. Roberts, and M. C. Petty, *IEE Proc.,* 130 (1983) 260.

69. Y. L. Hua, G. G. Roberts, M. M. Ahmad, M. C. Petty, M. Hanack, and M. Rein, *Philos. Mag. B*, 53 (1986) 105.

70. Y. L. Hua, M. C. Petty, G. G. Roberts, M. M. Ahmad, M. Hanack, and M. Rein, *Thin Solid Films*, 149 (1987) 163.

71. J. Batey, M. C. Petty, G. G. Roberts, and D. R. Wight, *Electr. Lett.*, 20 (1984) 491.

72.T. W. Barrett, H. Wohltjen, and A. Snow, *Nature,* 301 (1983) 694.

73. S. Volker, *Annu. Rev. Phys. Chem.*, 40 (1989) 499.

74. B. Simic-Glavaski, in [2], 537.

75. B. Simic-Glavaski, *Proc. First Int. IEEE VLSI Multilevel Interconnection Conference*, 1984, 167.

76. B. Simic-Glavaski, in [3], 203.

77. B. Simic-Glavaski, in Proc. 5th Int. *Congress on Applications of Lasers and Electro-Optics,* ICALEO'86, C. M. Penney and H. J. Caulfield, Eds., 233, Springer-Verlag, Berlin, Heidelberg, New York, London, Paris, Tokyo, 1986.

78. B. Simic-Glavaski, U.S. Patent 4,808,930, 1989.

79. B. Simic-Glavaski, U.S. Patent 4,855,243, 1989.

80. B. Simic-Glavaski, U.S. Patent 4,930,272, 1990.

81. M. Eckhauser, A. Bonamino, J. Perski, S. Kendrick, A. Imbembo, B. Simic-Glavaski, and K. Koehler, *Lasers in Med.Sci.*, 5 (1990) 21.

82. B. Simic-Glavaski, *Cell Biophys.*, 7 (1985) 205.

83. B. Simic-Glavaski, *Proc. Int. Joint Con. Neural Networks,* San Diego, 1990, II-809.

84. D. A. Long, *Raman Spectroscopy*, McGraw-Hill International Book Company, 1977.

85. M. Fleischmann, P. J. Hendra and A. McQuillan, *Chem. Phys. Lett.*, 26 (1974) 163.

86. D. I. Jearmaire, M. R. Suchanski, and R. P. Van Duyne, *J. Am. Chem. Soc.*, 97 (1974) 1699.

87. M. Moskovits, *Rev.Mod.Phys.*, 57 (1985) 783.

88. B. Simic-Glavaski, S. Zecevic, and E. Yeager, *J. Am. Chem. Soc.*, 107 (1985) 5625.

89. B. Simic-Glavaski, S. Zecevic and E. Yeager, *J. Phys. Chem.*, 87 (1983) 4555.

90. T. H. Weber and D. H. Busch, *Inorg.Chem*, 4 (1965) 469.

91. S. Zecevic, B. Simic-Glavaski, E. Yeager, A. B. P. Lever, and P.Minor, *J. Electroan. Chem.*,196 (1985) 339.

92. W. King, R. P. Van Duyne, and G. C. Schatz, *J. Chem. Phys.*, 69 (1978) 4472.

93. J. I. Gersten and A. Nitzan, *J. Chem. Phys.*, 73 (1980) 3023,

94. J. Billmann, G. Kovacs, and A. Otto, *Surf.Sci.* , 92 (1980) 153.

95. B. Simic-Glavaski, *J. Phys. Chem*, 90 (1986) 3863.

96. B. Simic-Glavaski, *J. Phys. Chem.*, submitted.

97. J. K. Gimzewski, J. H. Coombs, R. Moller, and R. R. Schlittler, in [5], 87.

98. B. Simic-Glavaski, A. A. Tanaka, M. E. Kenney, and E. Yeager, *J. Electroanal. Chem.*, 229 (1987) 285.

99. T. H. Wood, *Phys. Rev.,* B24 (1981) 2289.

100. W. Jaenicke, in *Advances in Electrochemistry and Electrochemical Engineering,* H. Gerischer and C. W. Tobias Eds., 10, 91, John Wiley and Sons, New York, 1977.

101. J. H. Webb, *J. Opt. Soc. Am.*, 40 (1950) 3.

102. R. C. Baetzold, *J. Chem. Phys.*, 55 (1971) 4363.

103. D. A. Weitz, S. Garnoff, J. I. Gersten, and A. Nitzan, *J. Chem. Phys.*, 78 (1983) 5324.

104. S. Efrima, *J. Phys. Chem.*, 89 (1985) 2843.

105. P. P. Sorokin, J. R. Lankard, E. C. Hammond, and V. L. Maruzzi, *IBM J. Res. Dev.*, 11 (1967) 130.

106. P. P. Sorokin and J. R. Lankard, *IBM J. Res. Dev.*, 10 (1966) 162.

107. P. P. Sorokin, J. J. Luzzi, J. R. Lankard, and G. D. Pettit, *IBM J. Res. Dev.*, 8 (1964) 182.

108. A. Harriman, G. Porter, and A. Wilowska, *J. Chem. Soc. Faraday Trans.*, 80 (1984) 191.

109. A. L. Hodgkin and A. F. Huxley, *J. Physiol. (London)*, 117 (1952) 500.

110. L. Cohen, R. D. Keyes, and E. Hille, *Nature*, 218 (1968) 433.

111. I. Tasaki, A. Watanabe, R. Sadlin, and L. Carnay, *Proc. Nat. Acad. Sci. USA*, 61 (1968) 883.

112. A. Taomoto, K. Waragai, K. Nichogi, Y. Saitoh, Y. Machida, and S. Asakawa, 2nd International Symp. on Bioelctronics and Molecular Electronic Devices, R7D Association for Future Electron Devices, 129, 1988, Fujiyoshida, Japan.

113. J. J. Hopfield, *Proc. Nat. Acad. Sci. USA*, 79 (1982) 2554.

114. K. Sakurai and S. Takano, *Proc. 12th Ann. Int. Conf. IEEE Eng. Med. Biol. Soc.*, 12 (4) (1990) 1756.

115. B. Simic-Glavaski, in *Nonlinear Electrodynamics in Biological Systems*, W. R. Adey and A. F. Lawrence, Eds., 519, Plenum Press, New York and London, 1984.

116. R.S. Potember, T. O. Poehler, and R. C. Bensen, *Appl. Phys. Lett.*, 41 (1982) 548.

4

Raman Spectra of Phthalocyanines

W. E. Smith and
B. N. Rospendowski

A. INTRODUCTION

Raman scattering provides detailed electronic and structural information on phthalocyanines. The method has not enjoyed widespread popularity because of the sophistication of the equipment required and because of the complexity involved in the interpretation of the data. However, recent work has demonstrated that these difficulties may be overcome. Practical applications of Raman scattering include the characterisation of phthalocyanines in situ in pigments, in solution, in solid state devices, in biological matrices and at metal surfaces. In addition, it is now possible to use the experimental versatility and theoretical methodology of Raman scattering to probe the vibronic properties and electron–nuclear dynamics of phthalocyanines. The result is that Raman scattering could become a key technique for the solution of problems with the use of phthalocyanines such as the effect of phonon coupling on conduction [1], or the molecular mechanism of conduction in solid–state electron transfer [2,3]. The ready accessibility and ease of use of modern lasers as excitation sources is also a powerful fillip in the development and acceptance of the technique. The purpose of this article is to review the progress made in technique development, to establish a system for the assignment of individual bands, and to illustrate the potential of the method.

Raman spectroscopy is used mainly to provide vibrational information although other forms of Raman spectroscopy such as rotational Raman spectroscopy [4] are becoming more prevalent. However, most reported phthalocyanine spectra comprise either vibrational resonance Raman scattering or surface enhanced resonance Raman scattering (SERRS). These techniques are much more powerful than normal Raman scattering. The main advantages are:

1. They provide electronic information [5]. This can enable quite detailed assessment of the nature of an excited state, can define the oxidation state of the central metal ion [6], and can define the nature of charge transfer interactions [7].

2. They are sensitive and selective. Small amounts of material can be detected and characterised. In addition, phthalocyanines can be detected in the presence of other

coloured materials in situ. Therefore studies can be made of phthalocyanines absorbed on a metal surface [8], or mixed with other dyestuffs [9].

3. The Raman scattering from water is weak. Therefore, small amounts of material can be in detected in aqueous solution or from electrode surfaces in aqueous electrolytes.

B. THEORY OF RAMAN AND RESONANCE RAMAN SCATTERING FROM PHTHALOCYANINES.

When a beam of light is scattered by a molecule or particle most of the light scatters with the same frequency as the incident light (Rayleigh or Mie scattering). However, a weak component scatters one vibrational unit different in energy from that of the incident beam. This weak component is Raman scattering [10]. When the excitation is tuned to the same frequency as a real electronic transition of the molecule, resonance Raman scattering is obtained [5,11,12,13]. Surface enhanced Raman scattering (SERS) is obtained from molecules adsorbed on specifically designed dielectrics. These are usually roughened noble metal and especially silver surfaces.

Surface-enhanced resonance Raman scattering is obtained from adsorbed molecules on SERS active surfaces that are chromophoric at the wavelength of excitation used. Resonance Raman scattering [11-13] and SERS/SERRS [14, 15] spectroscopies have been well reviewed. In this chapter, only the essential features are dealt with, with a view to providing the information required for a practical interpretation of phthalocyanine spectra. It is possible to make quite selective use of the spectroscopy and the analysis given has a direct implication for the practical examples described later. The basic theory is written specifically for phthalocyanines, and consequently it is necessary to begin with a synopsis of the main results from studies of their electronic structure (see Chapter 3 of Volume 1 of this series [16]).

The largest orbital contribution to the electronic spectrum of phthalocyanine dianions in the visible region are filled a_{2u} and a_{1u} orbitals and an unfilled e_g orbital. They give rise to two electronic transitions as in Figure 1. The molecular energy level diagram at the top right-hand side of the figure depicts the difference in energy of the HOMOs for porphyrin (P^{-2}) dianions as compared with phthalocyanine dianions (PC^{-2}) [17].

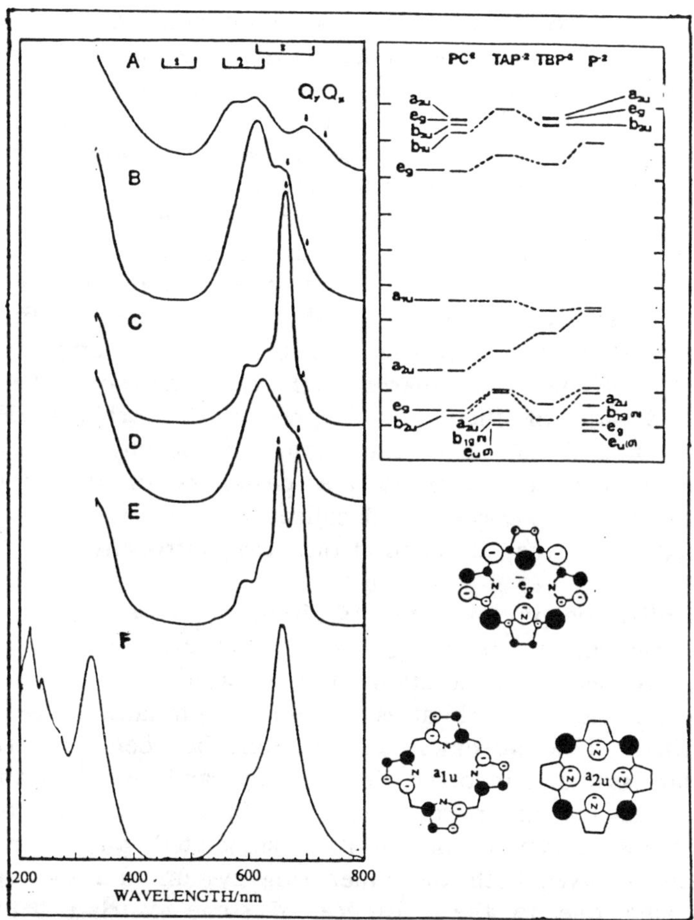

Figure 1 Electronic Spectra of Phthalocyanines in Different Environments:

The spectra given are, (A), solid copper phthalocyanine; (B), sulfonated copper phthalocyanine in water; (C), sulfonated copper phthalocyanine and tetraethylammonium bromide in ethanol; (D), sulfonated metal free phthalocyanine and tetra–ethylammonium chloride in water; (E), sulfonated metal free phthalocyanine and tetraethylammonium chloride in ethanol [18] and (F), vapor–phase copper phthalocyanine [19]. (Right–hand side reprinted with permission from J. Phys. Chem., Vol. 86. Copyright 1982 American Chemical Society (see text for details). Spectrum F reprinted with permission from J. Mol. Spec. Copyright 1970 Academic Press Inc.

The electronic structure is analogous to that in the porphyrin case, and the same labels are used. Thus, the lower-energy transition gives rise to an absorption band between 500 and 700 nm, which is called the Q band, and the higher-energy transition gives rise to an absorption band between 450 and 380 nm, which is called the B band.

The stablization of the a_{2u} orbital, which is the origin for the Soret or B-band transition, can be seen as a benzene ring is added to the periphery of the ring system in the case of tetra-benzoprophyrin (TBP^{-2}) and by the replacement of the methine bridge by aza linkages in the case of tetraazaporphyrin (TAP^{-2}). The molecular-orbital diagrams on the bottom right-hand side of the figure represent the orbitals responsible for the B and Q band transitions of porphyrins [6]. The addition of benzene rings to the periphery of the ring system enables more extensive delocalization of the π electrons associated with the B band from the nitrogens to the outer carbons.

Unlike porphyrins, the transition $a_{1u} \longrightarrow e_g$ is allowed [17] and consequently both Q and B-bands are intense in phthalocyanines. In addition, the separation between the a_{2u} orbital and the a_{1u} orbital is greater in phthalocyanines than in porphyrins. The movement of electrons between the overlapping molecular orbitals involved in the Q-band $\pi-\pi^*$ transition is restricted to the inner ring system. In contrast, the movement of electrons between the orbitals associated with the B-band transition involves both the inner ring system and the outer part of the ring due to the existence of some overlap between the a_{2u} and e_g orbitals on the phenyl ring.

A consequence of the double degeneracy of the e_g orbital is that a Jahn-Teller distortion can occur [20]. This leads to a splitting of the Q and B bands into two components. These components are readily identified in solution (Fig. 1 and [18]), where phthalocyanine spectra tend to be relatively sharp. For the Q-band transition, vibrations of the correct symmetry and with displacements that assist the movement of electron density in the inner ring system are the most likely to produce intense scattering. The occurrence of the Jahn-Teller effect will influence the scattering [5, 18, 21, 22] and this is left for discussion later.

Essentially, the intensity and polarization of Raman scattered light is a function of vibrational symmetry and of the derivative of the molecular polarizability with respect to the normal mode and the frequency of the incident light. The intensity of the scattered light is proportional to $\alpha^2\omega^4$ [10] α is the polarizability of the molecule, and ω is the frequency of excitation of the incident light. The Kramers–Heisenberg–Dirac (KHD) expression for light scattering is [5, 11, 12, 13]

$$\alpha_{xy} = \sum_i \frac{\langle f|\mu_{xy}|i\rangle \langle i|\mu_{xy}|g\rangle}{\omega_{g\rightarrow i} - \omega_L - i\Gamma_i} + \frac{\langle f|\mu_{xy}|i\rangle \langle i|\mu_{xy}|g\rangle}{\omega_{g\rightarrow i} + \omega_L - i\Gamma_i} \tag{1}$$

This expression describes the polarizability of the molecule and Σ is a sum taken over all the intermediate excited states $|i\rangle$.

The expression is in the frequency domain and there is no explicit time variable. $|f\rangle$ is the final vibronic state of the ground electronic state of the molecule, and $|g\rangle$ is the ground vibrational state of the ground electronic state. The numerator of each of the two terms consists of two integrals. They represent an upward transition from the ground state to the virtual state and a downwards transition from the virtual state to the final state. The meaning of the integrals is illustrated in Fig. 2 by the upward and downward vertical arrows.

In Raman scattering, these processes are linked so that the polarization of the scattered light is related to the polarisation of the incident light.

In the denominator [Eq. 1], the term Γ_i represents the homogeneous broadening of the intermediate state and is small. In the first term, when the incident light is set to give resonance, $\omega_L=\omega_{g\rightarrow i}$ for the one excited state that has the same energy as the laser ω_L. Scattering from this term becomes very large and domantes the summation. Consequently the scattering from other states can be ignored. Since the second term has a denominator in which ωL is added to $\omega_{g\rightarrow i}$, it can also be neglected. This means that the Raman scattering is dependent on the nature of one excited state, and we can expect electronic as well as vibrational information to be obtained. The effect is illustrated here for α–copper phthalocyanine (Figure 3).

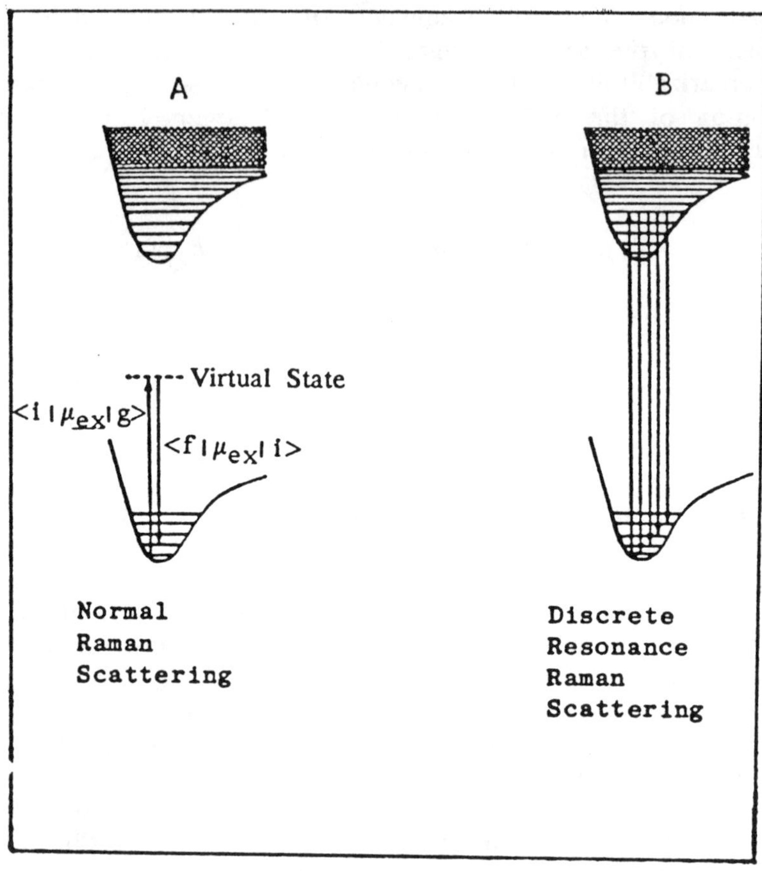

Figure 2 Types of Raman scattering, depending on the proximity of the exciting light energy to the molecular transition energies. (A) Normal Raman scattering with excitation far from any real molecular transitions. (B) Resonance Raman scattering with excitation energies in the region of discrete rotational – vibrational levels of a single electronic intermediate state.

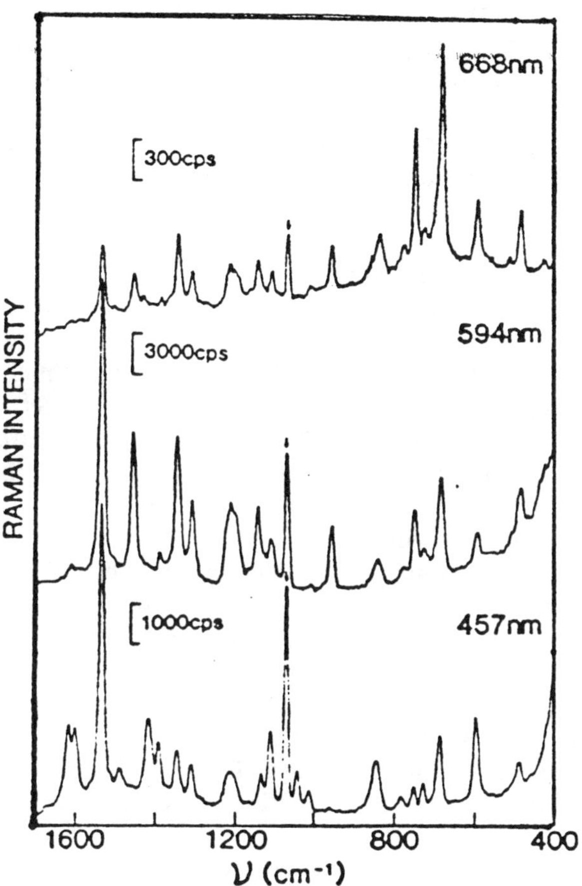

Figure 3 Raman spectra of αCuPc at room temperature taken with different excitation frequencies. The spectrum with 457 nm excitation exhibits preresonance from the B band whereas those with 594 and 668 nm excitation exhibit resonance with the Q band. The nitrate standard is arrowed.

The changes in relative intensity for different vibrations are due to changes in scattering efficiency as different vibronic states are brought into resonance. With visible laser radiation, phthalocyanines usually give resonant or preresonant spectra. As well as Raman scattering other processes such as absorption and fluorescence can occur. The main difference between scattering and stepwise absorption – emission is in the lifetime of the excited state and in the temporal coherence (i.e., phasing). For normal (i.e., off resonance) Raman scattering the excited state lifetime is short (10^{-15} s), with the consequence that the incident and scattered beams are coherently linked, as demonstrated by the relationship between the polarization direction of the two beams. Absorption – emission involves loss of coherence or phase due to other physical or chemical processes occurring in the molecule while it resides in the excited state. These processes [23] may include photophysical processes such as intersystem crossing, internal conversion and vibrational/rotational energy transfer, and photochemical processes such as bond dissociation and formation. At resonance, a relatively long (10^{-8} to 10^{-13}s) lifetime often obtains. In the case of phthalocyanines, the lifetime is reported as about 3 ps for the Q state [24]. Therefore the probability of a photophysical or a photochemical event occurring that perturbs the phase of the transition is higher than in the case of off resonance Raman scattering.

Clearly, in resonance, it would be possible for fluorescence processes with very short excited–state lifetimes gradually to approach those involved in Raman scattering [23b]. Some processes of this type have been identified and are usually called resonance fluorescence. Normally, the bands are somewhat broader than in Raman scattering. In the more common fluorescence processes that require longer lifetimes, the excited state usually relaxes, and emission occurs from the lowest vibrational level. Thus, fluorescence is a function of the electronic separation of the excited state from the ground state and therefore should not vary in absolute energy with variation in laser frequency. Thus, it is usually quite simple to differentiate Raman spectroscopy from other emission processes. From a practical point of view, if fluorescence emission obscures a Raman emission, it may be possible to change the laser excitation wavelength in order that the Raman emission occurs at higher or lower energies than the fluorescence maximum, and thus prevent the fluorescence interference. Time resolved methods may also be used to separate the processes.

The relative efficiency of scattering, emission and sample decomposition is mainly a property of the molecule. In the case of phthalocyanines, the molecules are fairly robust, and the natural partition of emission processes provides a significant Raman scattering contribution. Sample presentation is important and can be used to maximize one or other process, and methods of measurement of Raman scattering for phthalocyanines have been developed that enable good signals to be obtained. However, in some cases fluorescence is sufficiently strong that the Raman spectrum is completely obscured. Figure 4 illustrates the spectrum of an intermediate case where both Raman scattering and fluorescence are observed in zinc phthalocyanine [25].

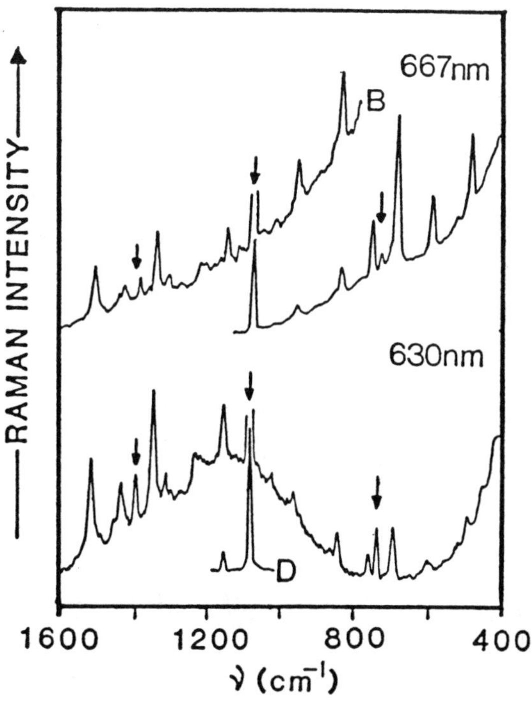

Figure 4 Raman spectra of zinc phthalocyanine at room temperature recorded with 630 and 667 nm radiation. The nitrate standard is arrowed.

There are important differences between resonant scattering and basic Raman scattering that require a brief resume of some points in the theory. The time for the Raman excitation and emission processes to occur is short compared to that for one complete vibration of the nuclei although the nuclei may execute several vibrations in the excited state before the eventual Raman emission [26a]. Using the adiabatic Born – Oppenheimer approximation [Eq. (2)], combined with a Taylor series expansion of the electronic wave function about the equilibrium internuclear distance [Eq. (3)], it is possible to obtain the dependence of the electronic wave functions with respect to vibrational movement.

$$|i\rangle = |\theta\Gamma elec(r,R)\rangle|\Phi\Gamma vib(R)\rangle \qquad (2)$$

$$\theta\Gamma elec(r,R) = \theta\Gamma elec(r,R{=}0) + R_{\Gamma vib}\left[\frac{\delta\theta\Gamma elec(r,R)}{\delta R_{\Gamma vib}}\right]_{R=0} + \ldots \qquad (3)$$

By substitution of [Eq. (3)] into the KHD [Eq. (1)], the following expression, is obtained:

$$\alpha_{xy}^{Q} = \left\{\frac{(\langle\theta_{A1g}^{Q}|\mu_{xy}^{Q}|\theta_{Eu}^{Qo}\rangle)^{2}\langle 1_{\Gamma vib}|0'\rangle\langle 0'|0\rangle}{\omega_{o-o'}-\omega_{L}-i\Gamma_{Eu}^{Qo}}\right\}$$

$$*$$

$$\left\{\langle\theta_{A1g}^{B}|\mu_{xy}^{B}|\theta_{Eu}^{B}\rangle\langle\theta_{Eu}^{Q}|\mu_{xy}^{Q}|\theta_{A1g}^{Qo}\rangle\frac{\langle\theta_{Eu}^{Q}|\dfrac{\delta H}{\delta R_{\Gamma vib}^{HT}}|\theta_{Eu}^{B}\rangle}{\omega_{Eu}^{B1}-\omega_{Eu}^{Qo'}}\,\frac{\langle 1_{\Gamma vib}|R_{\Gamma vib}|0'\rangle\langle 0'|0\rangle}{\omega_{o-o'}-\omega_{L}-i\Gamma_{Eu}^{Qo'}}\right\}$$

$$**$$

$$\left\{(\langle\theta_{A1g}^{Q}|\mu_{xy}^{Q}|\theta_{Eu}^{Qo}\rangle)^{2}\langle\theta_{Eu}^{Qo'}|\frac{\delta V}{\delta R_{\Gamma vib}^{JT}}|\theta_{Eu}^{Qo'}\rangle\frac{\langle 1_{\Gamma vib}|R_{\Gamma vib}^{JT}|0'\rangle\langle 0'|0\rangle}{\omega_{o-o'}-\omega_{L}-i\Gamma_{Eu}^{Qo'}}\right\}$$

$$***$$

Equation (4) has been written in terms of the electronic states of the phthalocyanines that are responsible for the Q and B bands. It is written for a resonant condition and consequently the sum over states of Eq. (1) has been dropped. Generalisation to situations with more than two contributing excited states is possible. The first (A, after Albrecht) term [* in Eq. (4)] does not contribute to the Raman emission, for normal Raman scattering. The closure theorem, ($\Sigma|k><k|=1$), applied to the sum in Eq. (1) leads to the collapse of the Franck - Condon overlap integrals, and only Rayleigh scattering is produced by this term [11]. For normal Raman scattering, the second (B, or Hertzberg - Teller) term [** in Eq. (4)] should be modified by a sum that is to be taken over all the excited vibrational states $|k>$, rather than the two k=0 vibronic states explicitly used here. Normal Raman scattering arises from the B term. In this term, there are matrix elements that involve a coordinate operator ($R_{\Gamma vib}$). The difference in vibrational quantum number between the states for there to be a finite value for matrix elements involving $R_{\Gamma vib}$ is one. Thus, normal Raman scattering from overtones is very weak.

At resonance with the 0 - 0 transition in the Q band, the inelastic scattering arises from both the A and B terms. A-term scattering arises from totally symmetric modes, and long overtone progressions are usually observed for small molecules such as iodine [12]. The absolute scattering intensity is related to the strength of the electronic transition. The relative intensity between the fundamental and overtone bands depends on the nuclear displacement in the excited state. A-term scattering allows scattering from the 0–0, 0–1, 0–2, and higher vibronic states. Because large molecules such as phthalocyanines do not undergo large overall displacements in the excited state, it may be thought that A-term scattering would not occur. However, a larger number of small nuclear displacements can produce some intensity in A-term scattering from large molecules [26b, c]. Phthalocyanines do show A-term enhancement.

A simple illustration of this is their ability to scatter strongly with totally symmetric modes [18, 27a]. B-term scattering dominates with partly forbidden transitions involving coupling with an allowed transition. This is the case with the Q-state transition of porphyrins, which, in contrast to phthalocyanines, is forbidden. Only scattering from the 0–0 and 0–1 transitions is possible with B term. Scattering from non-totally-symmetric modes (unless they are Jahn-Teller active) arises mainly from B-term scattering.

As a consequence of the double degeneracy of the e_g

excited state of phthalocyanines, a Jahn–Teller distortion may occur. This distortion is shown in Fig. 5.

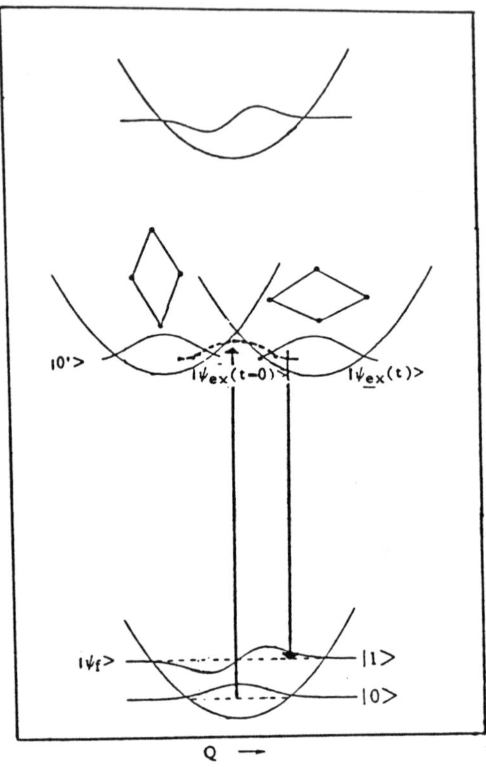

Figure 5 Representation of the Jahn Teller effect under D_{4h} symmetry. The potential wells of the doubly degenerate e_g orbital are shown as distorted upon excitation of the 0–0 transition of the Q–band. The potential well lying above the Jahn–Teller distorted Q state represents the B state. It is shown in a non Jahn–Teller distorted form and is displaced, for clarity, from the position it would occupy (i.e., at the same energy as the Jahn–Teller distorted orbital, the difference in energy being due to the ground state). Coupling of the $|0\rangle$ and $|1\rangle$ states of the Q and B excited states is responsible for the Hertzberg–Teller contribution to the scattering. Also shown is the time-dependent [see Eq. (5)] nature of the scattering. $|\psi_{ex}(t=0)\rangle$ is the propagation of the initial wave packet into the excited state. $|\psi_{ex}(t)\rangle$ is the moving wave packet propagated along the Jahn–Teller active coordinate (Q). $|\psi_f\rangle$ is the final vibronic state.

To account for the Jahn–Teller effect, a third term must be added to Eq. (4). The scattering produced can be regarded as essentially an additional contribution to the A–term scattering.

The existence of a Jahn–Teller contribution to the A–term scattering is revealed by the relative magnitude of scattering from totally symmetric modes, and by the symmetry of the non totally symmetric modes, which provide the greatest scattering. The former scatter as a consequence of the Jahn–Teller induced nuclear displacement in the excited state whereas the latter scatter by virtue of relaxation of the symmetry restriction for A–term scattering. A_{2g} modes cannot contribute to the Jahn–Teller effect and hence scattering from such modes cannot be enhanced by this mechanism, unlike A_{1g} modes and Jahn–Teller active B_{1g} and B_{2g} modes.

Therefore, for Q–band excitation of the phthalocyanine, A–term, B–term, and Jahn Teller distortion induced contributions to the scattering mechanisms are possible. [Note that the Jahn–Teller effect may influence the $a_{1u} \rightarrow e_g$ and $a_{2u} \rightarrow e_g$ transitions to different degrees. The electronic states in Eq. (4) are identified with the Q band].

To obtain an understanding of the electronic properties of the molecule, it is often useful to plot the intensity of the scattering against the frequency of the laser light used. Thus, by varying laser excitation, it is possible to obtain a spectrum that is analogous to that of an electronic spectrum although the selection rules are different and therefore the shapes are different. Examples (Fig. 6) of such plots, known as resonance excitation profiles (REPs), are shown for α copper phthalocyanine [18] (for other profiles see refs 27b and 28).

The profiles clearly indicate the nature of the different types of vibronic coupling that can occur. They highlight the different behavior of a low–frequency macrocycle mode as compared with a higher–frequency mode, which is essentially due to the isoindole ring. Note that the electronic origins for copper phthalocyanine which are difficult to see in the broad UV–visible spectra, are clearly identified by resonance Raman spectroscopy. This method has also been used to identify and assign other transitions that may occur in the visible region in certain cases such as a ligand to metal charge–transfer transition identified in hydroxophthalocyaninato manganese(III) [28].

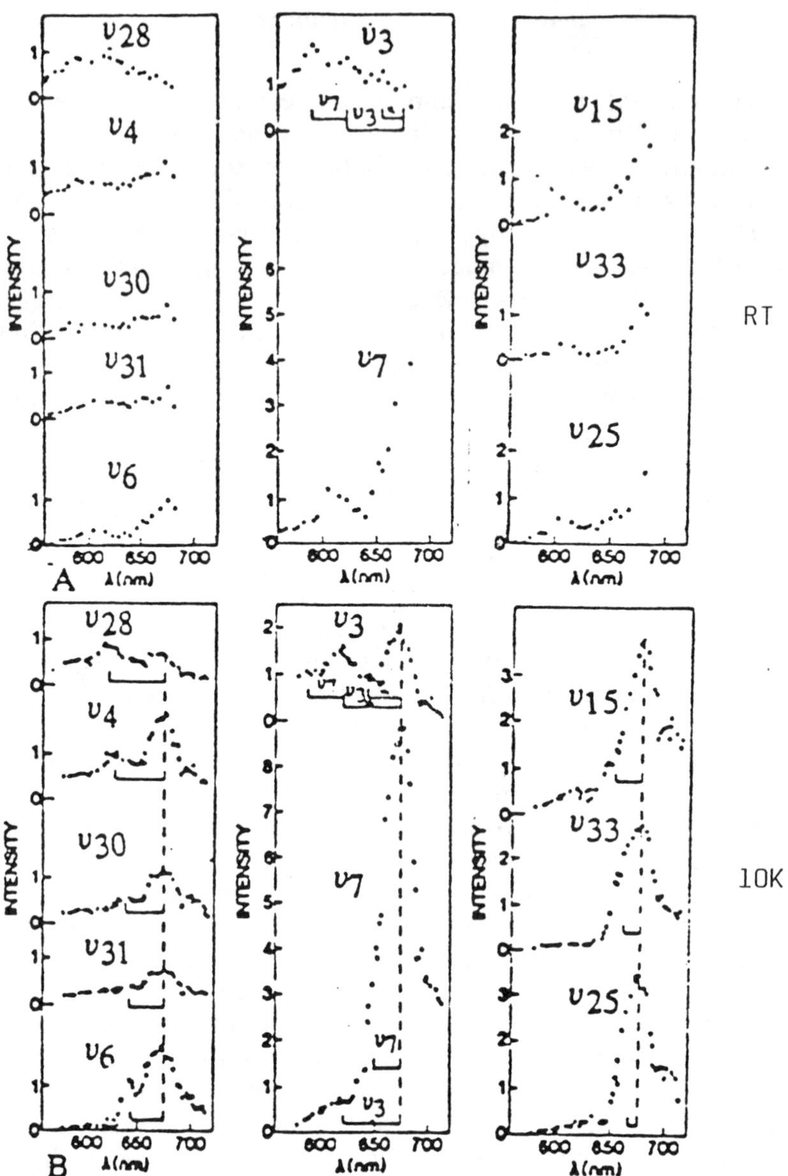

Figure 6 Low temperature (10K) resonance profiles of αCuPc. ν_3 and ν_7 presented in the middle diagram illustrate the type of profile found at high and low frequencies, with the transition between them being illustrated by the high–frequency vibrations on the left–hand side [18]. These profiles show improved resolution as compared with those at room temperature [18].

Time–dependent theory of Raman scattering [26a, 29, 30], has the advantage that it can give details of the wave packet dynamics in the excited state. Therefore, while the frequency–domain approach enables the study of the specific involvement of vibrations, the states involved in vibronic coupling, and the consequence of excited state distortions, only a time–dependent approach can lead to a picture of the evolution of an excited state distortion or the dynamics of vibronic coupling. The appropriate expression for the polarizability is:

$$\alpha_{xy}^{Q} = C\frac{i}{h}\int_{o}^{\infty} dt\, e^{[i(\omega_{g\rightarrow i}-\omega_L)t \;-\; \Gamma_i t/\hbar]} \langle 1|\mu_{ex}^{Q}|\psi_{ex}(t)\rangle \tag{5}$$

At t = 0,

$$|\psi_{ex}(t=0)\rangle = \mu_{ex}^{Q}|0\rangle \tag{6}$$

describes the propagation of the initial state, $|0\rangle$, into the excited–state potential surface.

$$|\psi_{ex}(t)\rangle = e^{-iH_{ex}t/\hbar}\,\mu_{ex}^{Q}|0\rangle \tag{7}$$

is the vibrational wave packet moving on the excited potential energy surface.

Extensions of the time–dependent theory have been made to wave packet movement that corresponds to non–totally–symmetric modes [31]. These propagate via the existence of two close–lying excited states such as the Q and B system of tetrapyrrolic macrocycles. However, further theoretical refinement is required in order for a satisfactory time–dependent theory to be readily applied to the multimode system of phthalocyanines. Once this is accomplished, the frequency domain and time dependent approaches will be complementary. The former approach possibly has the advantage that it can describe the Raman spectrum as experimentally determined, that is, it predicts <u>where</u> in the spectrum bands should occur. The latter approach describes <u>when</u> in the spectrum bands should occur.

C. EXPERIMENTAL DETERMINATION AND INTERPRETATION OF PHTHALOCYANINE RAMAN SPECTRA.

For some phthalocyanines, the emission process that dominates is fluorescence, and the best method of obtaining Raman scattering is to use excitation frequencies outside the absorption range. In this case the advantage of resonance will be lost and only Raman spectra will be obtained. The technique usually used for this purpose is Fourier transform Raman spectroscopy. In order to discriminate against fluorescence, and thus bias towards detection of Raman scattering, off-resonance, near-IR excitation is used [32a]. Since Raman scattering depends on the fourth power of the frequency, it is inherently weaker for near-infrared as compared with visible excitation, but the Fourier transform system efficiently collects the light and compensates for this. As well as measuring basic Raman spectra, this system may also be of value for the measurement of resonance Raman spectra from phthalocyanines designed to absorb in the infrared, although the fluorescence may also occur. Reports of resonance Raman scattering from near-IR absorbing dyes [32b] and from phthalocyanines on gold island filsm have appeared [32c]. More recently, the range of conventional laser spectrometers has been extended towards the near infrared, and dye lasers and detectors sensitive to wavelengths less than 1 μm can be used.

Raman spectra can be obtained from solids, liquids or gases. Commonly, solid powders or compressed discs are used. The advantage of a disc or powder pressed into a holder is that the sample can be spun and so the laser beam remains in contact with any one part of the sample for a limited time. This helps prevent sample decomposition. However, for phthalocyanines the signal-to-noise ratio of powder or disc spectra is poor due to self-absorption. To obtain better spectra, a fine dispersion of the particles or a thin film is more effective. The simplest method is to mix the phthalocyanine with silver particles in a ratio of about 99% silver to 1% phthalocyanine [33a,b]. The phthalocyanine should be finely ground. Usually with pigment grade copper phthalocyanine the particle size is less than 1 μm whereas the silver particles used are about 50 μm in diameter. This mixture of powders is compressed into a silver disc. The disc consists of the larger silver particles sintered together to form a surface with many roughened interstices. The phthalocyanine particles are contained within these interstices and are also attached to the silver. This surface provides a very effective scattering medium, and the improvement in Raman scattering is clear in Fig. 7.

It is desirable to obtain Raman scattering from electronically unperturbed phthalocyanines. However, it has been noted that with the silver disc method fluorescence from certain phthalocyanines is either quenched or shifted in frequency [25]. Fluorescence quenching may arise from non-radiative interactions of the phthalocyanine excited state with the electron-hole pairs in the metal. Hence the excited-state lifetime is shortened, and there is energy transfer between the excited phthalocyanine molecule and the silver surface. Comparison of the Raman spectra of phthalocyanines as presented in KBr and silver discs indicate that there is little or no difference in the phthalocyanine spectra. Thus, the silver disc method appears to be effective for the measurement of the Raman spectra of inert phthalocyanines. It should not be used with phthalocyanines with functional groups that may react with a silver/silver oxide surface. Low-temperature spectroscopy can also be performed [34].

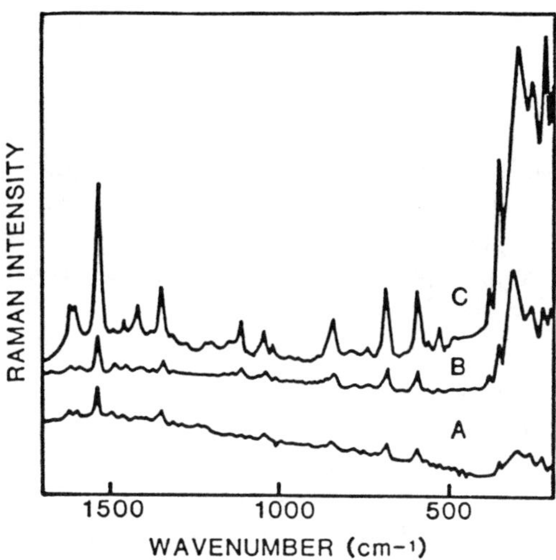

Figure 7 Raman spectra (A) from a powder of αCuPc, (B) from a pressed disc of αCuPc in KBr, and (C) from a pressed disc of αCuPc in silver. The excitation wavelength was 488.0 nm.

There is a wider range of studies where phthalocyanines presented as thin films have been used to observe Raman scattering (for references, see Section F). The incorporation of phthalocyanines in devices often requires such films. Again, the main advantage is that a thin film enables scattering because of reduced self-absorption. The main problem with the method is that the thickness of the film requires tuning to obtain the maximum scattering efficiency. Spectra can be obtained from glass, silicon, etc. (Fig. 8) [35].

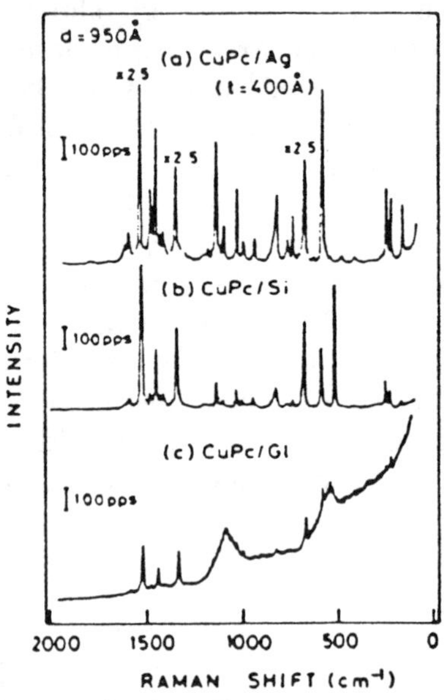

Figure 8 Comparison of the Raman spectra of 950 Å thick copper phthalocyanine films deposited on (a) silver (t=400 Å), (b) silicon, and (c) glass substrates. (Reprinted with permission from Surf. Sci., Vol. 137. Copyright 1984 Elsevier Science Publishers).

In solution, the phthalocyanine concentration and the frequency of the laser beam can be adjusted, in order to overcome self-absorption and so enhance the scattering efficiency. The laser beam should be set as near as possible to the surface of the cell at which the scattering is to be collected. Phthalocyanines can be studied on electrode surfaces. Resonance Raman spectra have been obtained on copper [8], gold [8, 36], polypyrrole [37], gold-coated glassy carbon electrodes [36], and on oxidized alumina [38]. In most cases the electrodes were not SERS active (see Section F), and hence enable the interaction of the phthalocyanine with the metal to be studied using a molecular property (i.e., resonance) rather than one that depends on surface roughness.

In the experimental determination of the Raman spectra of phthalocyanines, it is often necessary to change the frequency of the laser in order to compare scattering intensities from different vibrations. Raman spectrometers are usually single–beam machines, and scattering intensity can vary with time. With a double–beam spectrometer, sample positioning would be extremely critical since the directional laser beam can change slightly with changing frequency and so a slightly different point of the sample would be irradiated. In addition, the angle of the cone of Raman scattering will vary depending on changes in the angle of the incident light. Thus, a comparative standard in the sample is essential for good–quality Raman excitation profiles (REPs). It is usual to use a non–resonant standard such as sulfate or nitrate, and this has been effectively demonstrated for copper phthalocyanine [18]. An improvement in this method is to place a thinned crystal of nitrate or sulfate on the surface and to observe the Raman scattering through the thin crystal [39]. The advantage of this method is that the standard is not in contact with the phthalocyanine layer. In the more conventional method of adding sodium sulfate powder along with the phthalocyanine, more sulfate than phthalocyanine is required, since it produces a non–resonant signal, and consequently the bulk can mask the phthalocyanine signal, reducing the signal–to–noise ratio.

Obtaining standards in SERRS experiments is extremely difficult. Electrolytes containing phosphate and sulfate can be used, where the signal might be strong enough to measure, but at normal concentrations of electrolyte, the signals are either not observed or are weak, and consequently there is no simple way of obtaining a standard at the surface. In this case more complex methods are used in which either ratios of vibrations of the species under study are compared or a second resonant standard is added. This may be an improvement with respect to the case where there is no standard but only large clearly defined features should be interpreted in this manner.

D. SCHEME TO DESCRIBE VIBRATIONS

The number and symmetries of the irreducible representations of the Raman active in–plane vibrations of metal phthalocyanines is given by:–

$$\Gamma_{VIB} = 14A_{1g} + 14B_{1g} + 14B_{2g} + 13A_{2g}$$

Identification and rationalization of the vibrations that give rise to Raman scattering is a sine qua non for the practical exploitation of Raman spectra. Normal coordinate analyses have been performed for phthalocyanine [40, 41a, b]. At this stage of development, it is still possible for all these approaches to be rationalized, and one numbering system is proposed here.

There are strong arguments for the use of a vibrational numbering system that is related to that which is standard for porphyrins [18]. The porphyrin normal-mode descriptions have been well refined [42,43–46,47], and there are strong similarities in the Raman spectroscopy of porphyrins and phthalocyanines. Thus, when displacement vectors for the vibrations of the porphyrins are transcribed to the phthalocyanines, considerable headway is made in the interpretation of the phthalocyanine spectra. The main difference between the two molecules is that phthalocyanines have isoindole groups that are bridged by aza nitrogens whereas porphyrins have pyrrole groups which are bridged by methine carbons. A spectrum of copper phthalocyanine obtained with an excitation wavelength of 642 nm [48a] is shown with assignments in Fig. 9.

The symmetries of the vibrations and the region of the electronic spectrum where the vibrations scatter strongly are given in Table 1.

Table 1:

Frequency/cm^{-1}	Symmetry	Assignment
1613	B_{1g}	ν_{10}
1594	A_{2g}	ν_2
1530	A_{1g}	ν_3
1450	A_{1g}	ν_{28}
1428	B_{2g}	ν_{29}
1380	B_{2g}	ν_{20}
1341	A_{2g}	ν_4
1305	A_{1g}	ν_{12}
1213	B_{1g}	ν_{13}
1190	B_{1g}	ν_{21}
1170	-	-
1142	B_{2g}	ν_{30}
1131	A_{2g}	ν_{22}
1108	B_{1g}	ν_{14}
1040	A_{1g}	ν_5
1010	A_{2g}	ν_{23}
955	B_{2g}	ν_{31}
837	A_{1g}	ν_6
775	B_{2g}	ν_{32}
747	B_{1g}	ν_{15}
738	B_{1g}	ν_{16}
682	A_{1g}	ν_7
590	A_{2g}	ν_{25}
484	B_{2g}	ν_{33}

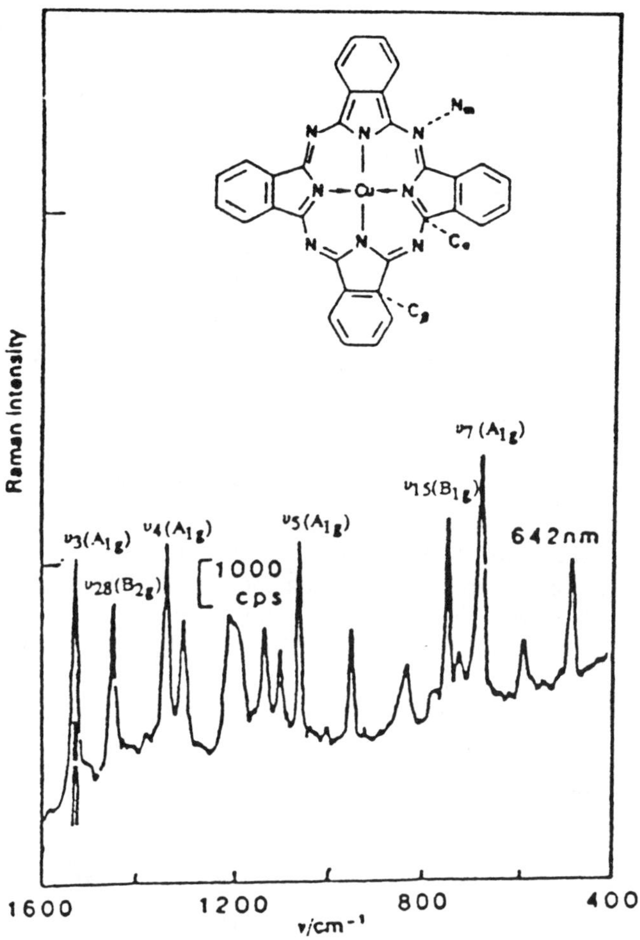

Figure 9 Raman spectra of copper phthalocyanine using excitation at 642 nm [48a]. The bands have been labeled as in [42–47].

Resonance with the Q–band transition causes an electron density shift from the pyrrole ring to the aza nitrogens. Hence, A–term scattering from vibrations that assist this movement of charge with the largest displacement in the excited state should be selectively enhanced. As this movement involves mainly the inner ring system of the macrocycle (i.e., that comprising the eight nitrogens, scattering from vibrations that affect this ring tend to be enhanced. In contrast, the transition that is responsible for the B band involves more extensive electron–density delocalization (see Section E). Therefore the ring that includes the outer pyrrole carbons and the six–membered rings must be considered.

The most intense bands arise from A_{1g}, B_{1g} and B_{2g} modes. The intense band at 482 cm^{-1} is tentatively assigned to a B_{2g} mode [49a]. The form of the displacements on each of the six major bands is shown in Fig. 10.

The major displacements demonstrated here are consistent with the normal coordinate analysis of [40–47]. The exact amount of displacement is an approximation, but they do help to show the nature of the displacements that produce the strongest scattering.

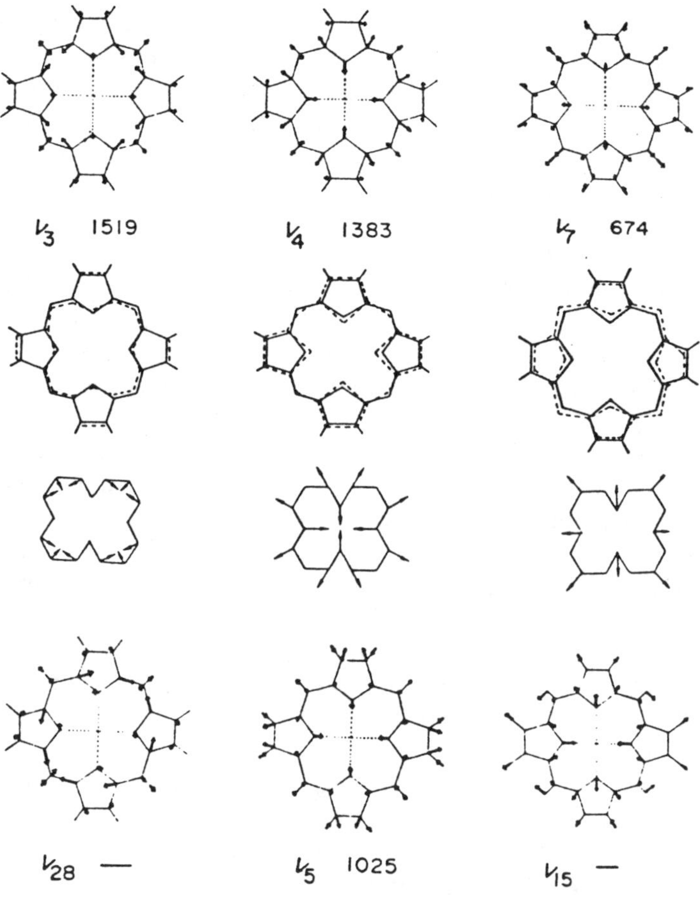

Figure 10 An illustration of the displacements for six of the bands labelled ν_3, ν_4, ν_5, ν_7, ν_{15}, and ν_{28}. For three of the bands, ν_3, ν_4, and ν_7, three different methods of depicting the vibrational motion are illustrated. The different approaches have value in a conceptual understanding of the changes taking place. (Top and foot two lines reprinted with permission from J. Chem. Phys., Vol. 69. Copyright 1978 American Institute of Physics).

E. INTERPRETATION OF SPECTRA

Vibrations with A_{1g}, B_{1g}, B_{2g} and A_{2g} symmetry in D_4h have all been shown to produce strong Raman scattering. Since totally symmetric A_{1g} modes are strongly enhanced as well as B_{1g} and B_{2g} modes, it would appear that both A and B term enhancement is obtained. Contributions to the A-term enhancement are expected both from the inherent strength of the transition dipole associated with the Q band transition, as represented by the first term of Eq. 3, and second from the additional displacement induced by the Jahn–Teller distortion in the e_g excited state [as represented by the third term of Eq. (4)].

The consequences of this Jahn–Teller distortion have been considered in detail for porphyrins by Shellnut [21]. Additional enhancement of scattering from A_{1g} modes is predicted. Jahn–Teller distortion induced A-term enhancement is also predicted for scattering from Jahn–Teller active B_{1g} and B_{2g} modes.

Vibronic coupling between the Q and B states will produce B-term enhancement for scattering from non-totally symmetric modes. Thus, under D_4h symmetry B_{1g}, B_{2g} and A_{2g} modes can scatter via B-term effects [see the second term of Eq. (4)]. Significantly, A_{2g} modes cannot scatter via a Jahn–Teller distortion, and therefore the presence of bands due to A_{2g} modes can be a measure of the amount of the contributions from B-term versus Jahn–Teller distortion to the scattering from non-totally-symmetric modes.

This is reflected in the spectrum shown in Fig. 9. The vibrations are predominantly of A_{1g}, B_{1g}, and B_{2g} symmetry, as predicted for a Jahn–Teller distorted D_4h system. The failure of A_{2g} modes to scatter strongly, unlike the situation in porphyrins, tends to suggest that B term enhancement for scattering from phthalocyanines is a less important feature. A caveat concerning the resonance Raman scattering from A_{2g} modes arises from considerations of depolarisation dispersion effects with phthalocyanines [49a, b]. In particular, a B_{1g} electronic and vibronic peturbation is identified through the depolarisation dispersion of scattering from this B_{2g} mode [49a] in platinum phthalocyanine. Contributions due to inhomogeneous effects on line width are also observed [49b].

High-frequency modes can be regarded as essentially isoindole ring vibrations. Because of the constrained nature of the ring, the degree of displacement of any one carbon or nitrogen atom is likely to be relatively small compared to macrocyclic modes. Thus, greater overlap with higher vibronic modes is possible, and high-frequency modes such as ν_3 do show more intense scattering upon resonance with higher vibronic states.

Q band excitation appears to favor scattering from lower-frequency modes such as macrocyclic vibrations. Therefore, a vibration such as ν_7, which is largely a macrocyclic breathing mode and involves large coordinate displacements of the bridging nitrogens, is strongly enhanced with Q band excitation, but not with B band excitation. In contrast, the intensity of the ν_7 band decreases as compared with the intensity of the ν_3 band as resonance with the B band is approached.

A comparison of the high-frequency vibration ν_3 and the lower-frequency vibration ν_7 is exemplified by the REP taken at 10 K, [Fig. 6(a)]. ν_7 scatters only from the lowest vibronic level (i.e., 0–0) whereas ν_3 scatters from the lowest vibronic level and from higher vibronic levels (i.e., 0–1 and 0–2) involves coupling to specific vibrations. The main coupling appears to be with symmetric vibrations, ν_3 itself and ν_7. In the ν_7 profile, there is a shoulder on the low-energy side. In fact, the e_g excited state if Jahn–Teller distorted will be split, and consequently two bands are expected. It is believed that the shoulder is the other electronic origin. Compared to electronic spectroscopy, the ability of resonance Raman spectroscopy to pick out the no-phonon position is quite striking. The two components of the Q band are small shoulders in the broad UV-visible spectrum.

The vibronic interactions in the case of ν_3 are much more clearly identifiable and indicate a simpler coupling scheme. Therefore, simple A-term enhancement [term 1 in Eq. (4)] appears to predominate for the high-frequency vibrations (e.g. ν_3) whereas Jahn–Teller induced A-term enhancement [term 3 of Eq. (4)] appears to predominate for the low-frequency vibrations (e.g. ν_7). This is consistent with the predicted behavior from a Jahn–Teller distorted system [21].

From the foregoing it may be concluded that the Jahn–Teller active vibration for the Q-band transition is ν_{15} and that low-frequency vibrations such as ν_7 can couple more effectively with this vibration than high frequency vibrations such as ν_3. The Jahn–Teller distortion argument can be extrapolated to the $a_{2u} \longrightarrow e_g$ transition, which is responsible for the B band. However, because of the different nature of the electronic transition responsible for the B band, a different Jahn–Teller active mode may be implicated. As the electron density shift associated with the B band transition appears to be more delocalized over the outer, rather than inner, macrocycle with less involvement of the aza nitrogens, a higher-frequency vibration may be the Jahn–Teller active mode. A Jahn–Teller distortion propagated by this mode may be expected to selectively enhance scattering from higher-frequency vibrations. Indeed, this appears to be the case, as ν_3 and ν_{10} increase in relative intensity as B band resonance is approached.

Studies of pigment particles and dyes on metal surfaces indicate that changes can occur that are due to alteration in the electron density in the molecular orbitals particularly in the excited states [48]. Electrodes of different potentials produce different intensities for the values of ν_3 and ν_7. This can be done both with molecules (see Section E) and with particles [Fig. 11(a)] and consequently is additional to the effect due to surface enhancement (see Section F). In the latter case molecular orientation is a determining factor in the relative intensity of scattering from different modes. This change appears to be due to a polarization of the electrons in the excited states, possibly due to the electric field at the metal–particle interface. These effects can be probed by constructing REP's. A profile on an electrode indicated a shift of 50 nm towards the red compared to an REP of the phthalocyanine alone, indicating that the surface-to-particle interaction had reduced the $\pi-\pi^*$ separation in the molecules [48b] [Fig. 11(a) – see also Fig. 17].

REPs have been used to assign a ligand-to-metal charge transfer transition in hydroxophthalocynaninatomanganese(III) [28] (Fig.12).

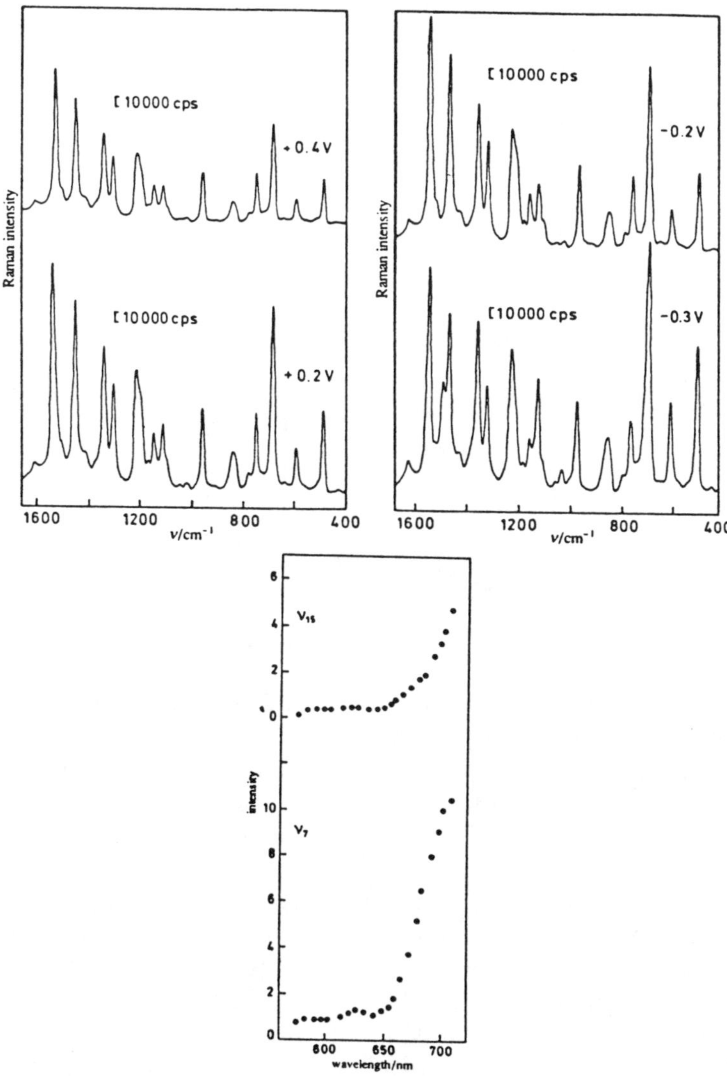

Figure 11 (A) Raman spectra from a compacted electrode before cycling and at selected voltages after cycling at 200 mVs^{-1} for 5 min between 0 and +0.3 V. Excitation wavelength equals 622 nm. (B) Excitation profiles of ν_7 and ν_{15} for a compacted electrode cycled at 200 mVs^{-1} between 0 and 0.3 V for 5 min and finally held at 0 V. ν_4, scattering varies only slightly in intensity and is used as reference [48b].

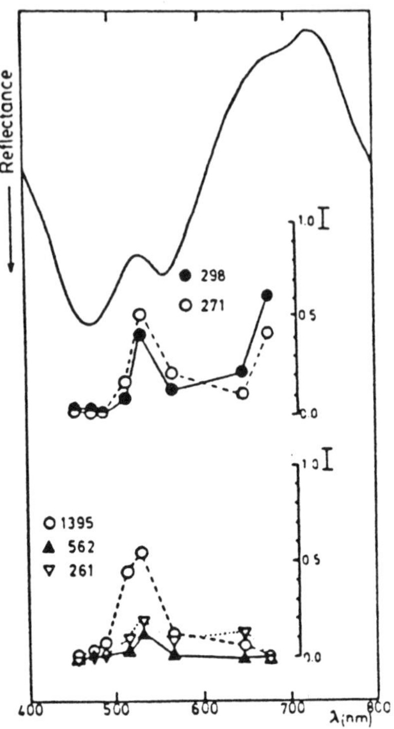

Figure 12 Excitation profiles and diffuse reflection spectra of solid hydroxophthalocyanatomanganese(III)[28]. The profiles depict scattering from vibrations which are enhanced in the region of the n–π^* transition. (Reprinted with permission from J. Raman Spectrosc., Vol. 20. Copyright 1989 John Wiley and Sons Ltd).

The transition is particularly characterized by a high-frequency (1395 cm^{-1}) mode. From the foregoing analysis, this mode may well be ν_{29} (B_{2g}) or ν_{20} (A_{2g}). These correspond respectively, to B_2 or A_2 modes under the C_{4V} symmetry, which should obtain for this phthalocyanine. Since an excited-state electronic displacement is required for strong resonant scattering with a charge-transfer transition, particular modes strongly influence this process. Modes with large metal-to-ligand displacements are strong candidates for such transitions. If ν_{29}/ν_{20} does correctly describe the 1395 cm^{-1} vibration, then the distortion of the macrocycle in the ligand-metal charge-transfer excited state may be in the direction of the major displacements of these modes. In addition, multimode [22] and Dushinsky effects [5] may contribute to the unique selective enhancement of scattering from this high-frequency mode.

Finally, vibronic coupling can be studied in this way. Q band excitation of copper phthalocyanine dispersed in a silver powder compacted electrode enables scattering from overtones to be detected [50]. These bands are mainly based on symmetric vibrations (ν_3, ν_4, ν_7 and ν_{15} – the first three are A_{1g} modes in D_4h symmetry). The pattern is consistent with the much greater part played by symmetric vibrations in phthalocyanines compared to the B_{1g} and B_{2g} vibrations which tend to dominate the porphyrins Q–band excited resonance Raman spectra. The broader nature of the overtones, may be due to breakthrough of emission processes such as resonance fluorescence. This overtone scattering suggests higher harmonics are influenced more by the excited state lifetime than the fundamentals (Fig. 13).

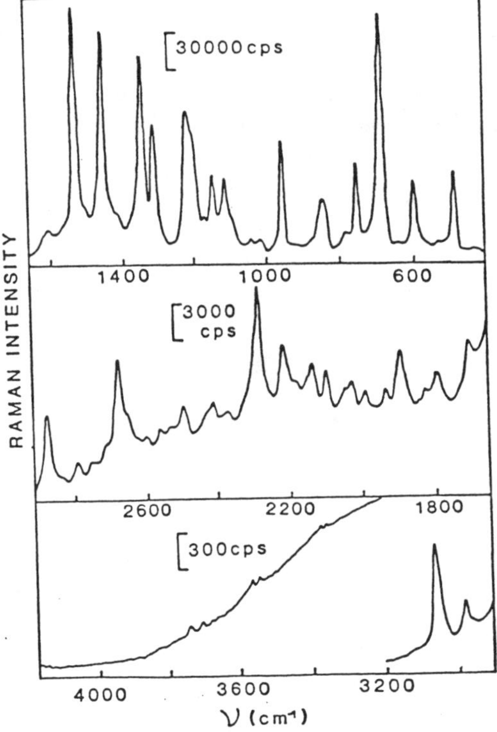

Figure 13 Raman spectra from a 10/1 Ag/αCuPc compacted electrode that has been electrochemically treated in 0.05 M H_2SO_4 held at 0 V versus SCE using 600 nm excitation [50]. First and very weak second overtones are observed. The relationship to the fundamental spectrum is discussed briefly in the text.

F. SURFACE ENHANCED RESONANCE RAMAN SCATTERING

In practice, many phthalocyanines are absorbed as thin films on metal surfaces. With specifically prepared metal surfaces, surface–enhanced Raman scattering (SERS), is observed [14, 15]. The theory of SERS is not as satisfactorily developed as that for resonance Raman scattering. Experimentally, it was discovered that Raman scattering from a monolayer of pyridine on a roughened silver electrode was readily observed [51]. The enhancement of the signal over basic Raman scattering was of the order of 10^6. Several theoretical explanations have been advanced. It is believed that a form of electromagnetic field enhancement is experienced by molecules on the metal surface due to coupling with surface plasmons. On a smooth surface plasmons cannot assist in reradiating incident light. However, on a rough surface the plasmons are localized and radiative and can assist in the coupling of the incident light field with the metal/adsorbate system and in the re–emission processes. Elastic and inelastic scattering from the adsorbate is transmitted back to the surface, from which an amplified signal is produced for detection. The theory used to rationalise the phenomenum for the enhancement [14, 15] identifies two main contributing terms.

1. Electromagnetic, distance–dependent enhancement.

2. Charge–transfer enhancement from the surface layer. This involves bond formation and/or photon–induced electron–transfer between the adsorbate and the metal surface.

Both terms can be appreciable, but the relative importance varies depending on the molecule. The extent of enhancement decreases the further a molecule is spaced from the surface. However, interference effects between the scattering from different layers of adsorbates have been observed from multilayer assemblies on SERS active substrates [52]. This can cause adsorbate–layer thickness controlled periodicity in the SERS intensity.

In SERS the intensity of scattering from specific vibrations is different from that in Raman or resonance Raman scattering since specific surface selection rules apply [53]. Essentially, vibrations which are associated with polarizability changes perpendicular to the surface will be enhanced. There is also a dependence on the symmetry of the vibrations involved. Voltage-dependent signals also arise from potential-controlled metal Fermi-level-to-adsorbate photon-induced charge transfer.

1. Specific dielectric properties of the metal are required for activation of surface plasmon resonances in the visible region. The most effective metals - silver, copper, and gold - are well characterized, and there are claims for a few others.

2. The technique requires a specifically roughened surface that is often annealed during chemical reactions leading to a loss of signal.

3. Sensitivities vary enormously between different molecular species and impurities are a problem, especially since the spectra can be quite different from unenhanced Raman spectra of the same molecules.

4. The theory requires to be improved. A common approach is to use a clean roughened metal surface but from a chemists point of view it is clear that the electrodes may be coated with other compounds. For example, chloride electrolytes are often used with silver electrodes [51]. In this case the surface of the electrodes is coated with tightly adhering silver chloride. Thus, the scattering observed is from a coated surface.

Since visible lasers are usually used, resonance and SERS are often combined to give surface-enhanced resonance Raman scattering
(SERRS). Very large enhancements can be obtained when resonance reinforces the SERS effect.

There is considerable advantage in the SERRS method, which goes a long way to overcoming the disadvantages of SERS. First, by combining resonance and SERS, enhancements of up to 10^{10} can be obtained, enabling very small quantities of material to be sensed. Second, by constructing experiments so that both smooth and rough surfaces are considered and spectra are measured on and off resonance, it is possible to disentangle the effect of resonance and SERS, leading to conclusions with regard to surface orientation and electronic structure, including the nature of the interaction between the surface and the dye [54, 55].

Van Duyne has studied cobalt phthalocyanine using SERRS [56]. He has shown that the scattering from a monolayer and a multilayer are quite different, indicating that both charge transfer and electromagnetic terms are operating.

There are many examples of SERRS, one of the more extensive of which is a series on phthalocyanines on silver electrodes [57–60]. The orientation of the phthalocyanine can be obtained, and the effect of electrochemical alteration of the surface can be studied.

In addition to electrode surfaces, other forms of roughened silver can be used. For example, infrared [61] and Raman scattering [62, 63] from thin films of phthalocyanine are enhanced by depositing layers of silver in the form of silver island films. Other methods that have proved suitable for obtaining SERRS active surfaces include microlithographically deposited silicon posts coated with silver [64] or compacted silver discs [65]. In addition, colloidal suspensions provide effective methods of measuring phthalocyanine spectra [66, 67].

Suspensions of colloidal particles are usually made by citrate or borohydride reduction of silver nitrate [66, 68a]. Phthalocyanine–coated silver particles can be prepared from these citrate–reduced silver colloids and give rise to very strong SERRS. It is possible to replace the citrate layer with other species and to obtain sols in nonaqueous environments [68b], where some phthalocyanines that are not soluble in water can be made soluble and therefore easily added to the sol surface.

For SERRS, a standard three–electrode cell can be constructed which will fit into a 1 cm fluorescence cell [69]. This provides a convenient way of obtaining Raman scattering. The main requirements are that the electrodes are firmly supported and that the distance between the scattering window and the working electrode is not too large. This latter requirement depends on the strength of the signal and the amount of self–absorption occurring. It is not nearly as critical as in infrared spectroscopy, and this enables electrodes to be designed so that good electrochemistry can be carried out.

SERRS provides additional information because surface selection rules apply. A simple illustration of this is for soluble phthalocyanines such as sulfonated phthalocyanines (Fig. 14) [57–60, 70, 71]. The intensity of Raman scattering from molecules absorbed on the surface depends on the selection rules of the molecule and on the surface selection rules. Essentially bands that have a polarizability component perpendicular to the surface are the most enhanced. Thus for any one molecule, as the molecule changes orientation, certain bands should be seen to change in intensity. This effect has been seen on an electrode. Using sulfonated phthalocyanine, both a parallel and perpendicular orientation of the molecule to the surface can be identified. This conclusion was made on the basis of the Raman spectra [70].

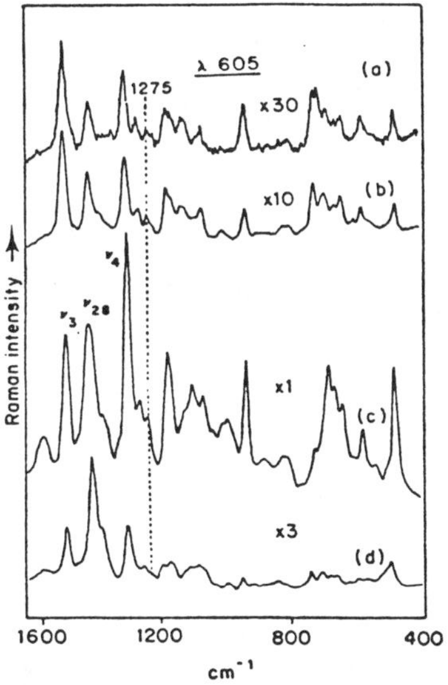

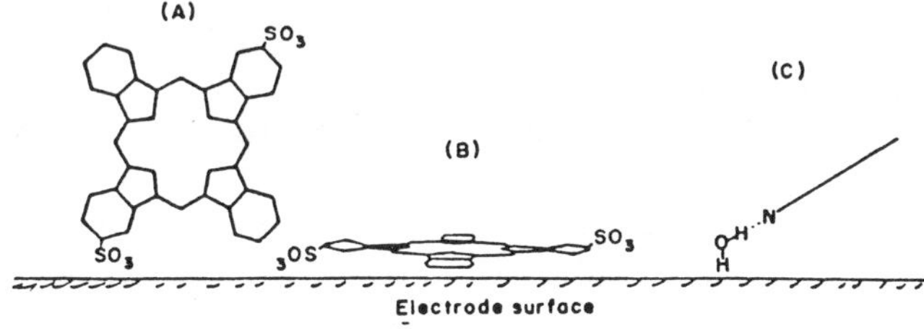

Figure 14 Top: Spectra of sulfonated copper phthalocyanines absorbed on a silver wire electrode surface which is held at (a) +0.6V, (b) +0.2 V, (c) −0.2 V, (d) −0.6 V versus SCE. Appropriate scale expansion factors are shown for each.
Bottom: Postulated molecular orientation with respect to the electrode surface as deduced from top spectra in for (a) −0.2 V, (b) +0.2 V and +0.6 V and (c) −0.6 V versus SCE. [70]

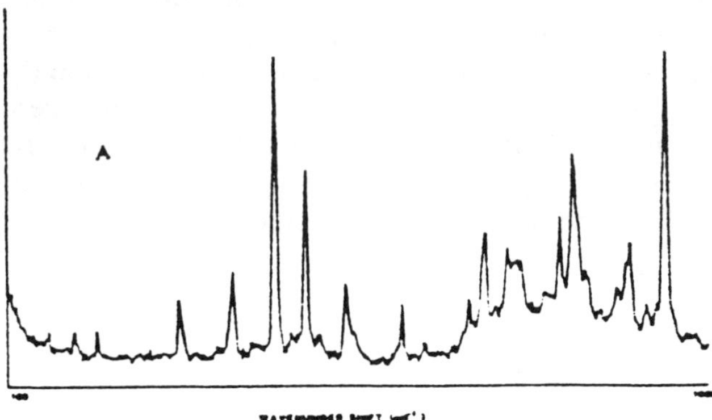

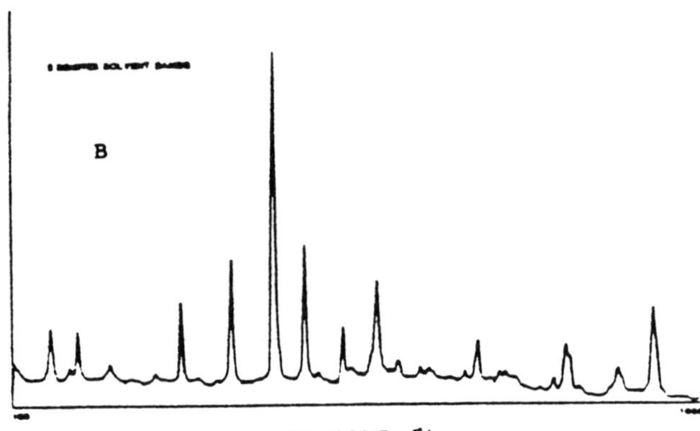

Figure 15 (A) Resonance Raman spectrum obtained from cobalt(II) phthalocyanine powder (0.5% in KBr) recorded using λ=647.1 nm. (B) Resonance Raman spectra obtained from 10^{-5} M cobalt(II) phthalocyanine in THF recorded using λ=647.1 nm [85]. (Reprinted with permission from J. Raman Specrosc., Vol. 18. Copyright 1987 John Wiley and Sons).

Finally, the effect of substituents on the ring can be observed. In the Q band region this effect is likely to be smaller than in the B band region since substitution is on the six-membered ring system. However, the relative intensity of the ν_3 and ν_7 bands is affected markedly [86]. Scattering from vibrations such as ν_2 can increase in intensity (Fig. 16).

G. APPLICATIONS

The primary application of phthalocyanines is in their use as pigments and dyes. Technological developments in electrophotography [73], optical data storage [74], solar energy conversion [75] and chemical sensors [76,77] have led to the creation of new markets where the demand is for robust phthalocyanines with novel and controllable properties. Future requirements will be for (non-linear optical [78a]) materials for optical computing and molecular electronics [78b]. The previous volume of this series described some of the applications of phthalocyanines and polymeric phthalocyanines in chemical sensors and photobiology. Their properties have been characterised using many different physical techniques including infrared and UV-visible spectroscopy [79], X-ray absorption [80] and ellipsometry [81]. In this section a review is presented of the unique ability of Raman scattering to assist in the characterisation and elucidation of the structure, mechanism and function of phthalocyanines in applications and devices [82].

i. Solid-State Resonance Raman Scattering

The application of the technique depends largely on the nature of the system to be studied. As a consequence of self-absorption by phthalocyanine single crystals and polycrystalline phthalocyanine particles, Raman spectra of bulk materials are often obtained from dispersed particles in matrices, thin films and suspensions. The main purpose of this work, as already indicated, is to characterize structural and electronic properties.

There have been a number of studies directly targetted to specific applications. For example, it should be possible to correlate electrochromic effects [83a] intrinsic molecular semiconductivity [83b] and laser-induced intramolecular charge transfer [83c] of diphthalocyanines with the resonance Raman spectra [83d]. Changes in potential affect the protonation of the octacyanophthalocyanine rings [83e, f] and the redox state of the metal particularly in the Q band region. As these changes occur, the ring geometry and π structure change and attendant changes in the Raman and UV-visible spectra are observed [25, 83e, f].

Raman spectra in a KBr matrix have been obtained from the bulk molecular metal species formed by doping nickel phthalocyanine with iodine [84]. The spectra show long overtone progressions of I_3^- with 514.5 nm excitation. Only very weak phthalocyanine peaks were observed. From the downshift in the I_3^- fundamental frequency, it was concluded that partial back donation from the I_3^- to the phthalocyanine macrocycle was occurring. The authors do not report whether the phthalocyanine vibrational frequencies or Raman scattering intensities are affected by the partial donation of 1/3 of an electron from the phthalocyanine macrocycle to the tri–iodine or by back donation from the tri–iodine to the metallo–phthalocyanine

Electrochemical reduction of CO_2 at a cobalt(II) phthalocyanine–impregnated PTFE bonded carbon gas diffusion electrode has been studied using resonance Raman scattering [85]. In metalloporphyrins, specific vibrations are particularly sensitive to spin and oxidation states of the central metal ion. The marker for oxidation state most commonly used is ν_4. Similarly, the oxidation state of the cobalt in the spectrum of the phthalocyanine is sensitive to the oxidation state. Upon reduction of cobalt(II) to cobalt(I) phthalocyanine at −1.8 V the spectrum excited with a 647.1 nm laser line collapses. However, with excitation with a 676 nm laser line, the spectrum is regenerated. Therefore, the electronic origin of the phthalocyanine appears to shift upon metal reduction. Resonance Raman spectra are reported for cobalt(II) phthalocyanine in a KBr matrix and in THF solution (Fig. 15). The relative intensities of the 682 (ν_7) and 1527 cm^{-1} (ν_3) modes change markedly when the solid–state and solution phase spectra are compared. Whether this effect is due to axial ligation of the cobalt by the THF, or to some solid–state versus solution state electronic and/or Jahn–Teller effect remains to be resolved, although a similar effect on an electrode surface has been observed.

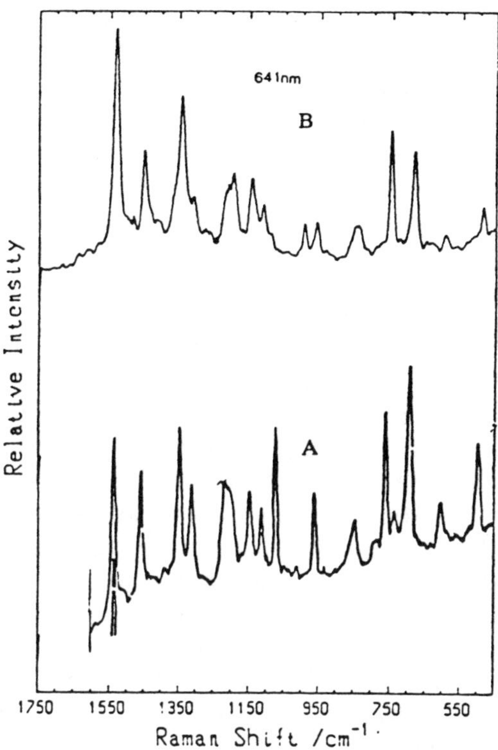

Figure 16 The effect of substitution on the phenol ring. (A) copper phthalocyanine in a silver disc (1:100; CuPc:Ag) and (B) p–nitrobenzoic acid amide derivative of copper phthalocyanine [CuPc(PnBc)] in a silver disc [1:100:140; CuPc (PnBc):Ag:Na_2SO_4)]. The band at 994 cm^{-1} is due to $SO_4{}^{2-}$ λ=641 nm [86].

ii. Thin Films

Phthalocyanines which have been studied as thin films include uranium superphthalocyanines [87], aluminum–oxo phthalocyanine [88], chloro–aluminum phthalocyanines [89], lutetium and ytterbium diphthalocyanines [83d], chloro–indium phthalocyanine [90] and dysprosium and holmium diphthalocyanines [91]. In the last study [91] the effect of NO_2 gas chemisorption was monitored by Raman spectroscopy. It would appear that the scattering from ν_2 and C–H bending is enhanced.

Iron phthalocyanine thin films have been studied on copper

and gold electrodes [8]. Potential dependent behavior is observed for many of the bands from the adsorbed iron phthalocyanine. In particular ν_3 is affected as the potential is swept negative [8]. Double-layer electric field distortion of the phthalocyanine is proposed. This should affect the Jahn-Teller distorted excited state and hence scattering from modes that gain intensity from Jahn-Teller induced A term would alter in relative intensity. The interaction of the iron phthalocyanine with gold electrodes has implications for the use of phthalocyanines for solar energy conversion, since ohmic contact of phthalocyanine with gold is required in photovoltaic cells [74].

Conducting polypyrrole-copper phthalocyanine tetrasulfonate thin films and their potential dependent behavior have been studied with Raman and resonance Raman scattering [37]. Raman scattering from platinum phthalocyanine on gold is insensitive to the electroreduction of oxygen in the potential region examined [36]. This indicates that the activity of the platinum phthalocyanine catalyst is dependent on film structure and not simply on molecular considerations.

Phthalocyanines form effective Langmuir-Blodgett films and Raman spectra of such films can give good signal-to-noise ratios. A Langmuir-Blodgett film of a substituted phthalocyanine on a silicon surface gave good spectra from which the orientation of the molecule with respect to the n-doped single crystal silicon surface was predicted [92a]. REPs were constructed for ν_3, ν_7, ν_{15} and ν_{28} (Fig. 17). Striking differences were revealed for the REP peak positions between the copper phthalocyanine tris-$CH_2HNC_3H_7$ film on silicon and the same material dispersed in a silver matrix. A silicon-induced shift in the energy of the π-π^* transition is proposed. In another study [92b, c] of Langmuir-Blodgett films from substituted phthalocyinato-polysiloxanes on silicon, polarized resonance Raman scattering reveals the anionational order of the films.

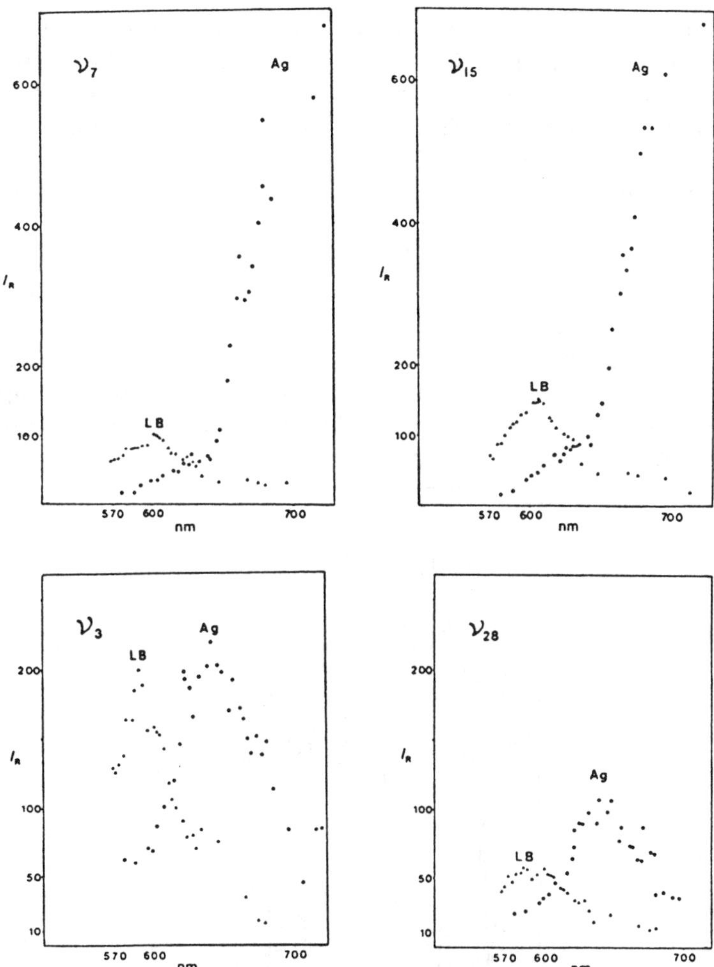

Figure 17 Resonance Raman excitation profiles for scattering from molecules of CuPc <u>tris</u>-$CH_2NHC_3H_7$ modified phthalocyanine dispersed in silver (·) and as a Langmuir-Blodgett film (▵) on <u>n</u>-type silicon [92].

iii. Tunnel Junctions and Raman Scattering

Many of the devices that incorporate the phthalocyanines involve an interface between a metal/metal oxide layer and the molecular film. The voltage tunneling current characteristics of this interface contain information on the inelastic electron tunneling due to interaction of the tunneling electron with the electronic and vibrational levels of the species in the molecular film [93]. Copper phthalocyanines have been studied by inelastic tunneling spectroscopy (IETS) and by Raman spectroscopy in aluminum/aluminum oxide/phthalocyanine/metal devices (Fig. 18) [94–96]. The phthalocyanines were either nickel or copper, and the metals were silver, lead, or thalium. Tunneling phenomena are important in existing applications that involve metal/metal oxides/phthalocyanine contacts. For example, these contacts are used in photovoltaic cells, metallized toner particles in photoelectrophotography, optical data laser storage systems and chemical sensors. Interestingly, bias-dependent changes are observed in the IETS of the tunnel diodes, but these changes do not appear to be reflected by resonance Raman scattering or SERRS [96]. However, the combination of tunneling spectroscopy, Raman spectroscopy, and scanning tunneling microscopy [97] offers a powerful method for the control and characterization of solid-state devices.

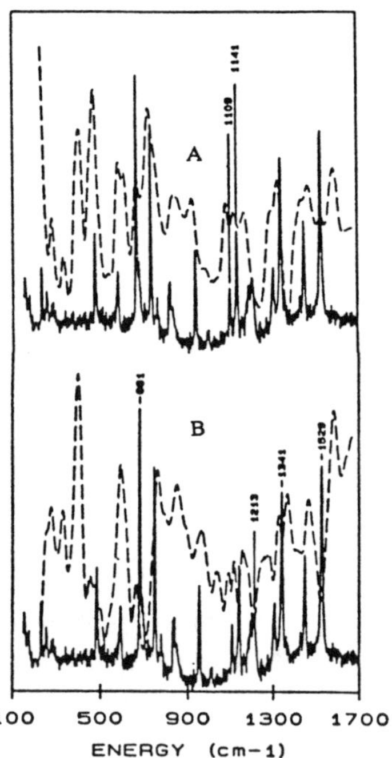

Figure 18 (A) Comparison of Raman and tunneling spectra obtained from Al–AlO_X–CuPc (1 nm)–Pb tunnel diodes. The inelastic electron tunneling spectrum (broken line) was obtained from the aluminum electrode biased positive. (B) The same spectra as in (A), but the inelastic electron tunneling spectrum (broken line) was obtained with the aluminum electrode biased negative [95]. (Reprinted with permission from Langmuir, Vol. 7. Copyright 1991 American Chemical Society).

iv. Surface-Enhanced Resonance Raman Spectroscopy

SERRS produces strong, environmentally specific signals and offers the prospect of molecular-level photoelectrochemical control, recognition, and signal processing. SERRS investigations of phthalocyanines can be assessed for the incorporation of phthalocyanines into devices whose optoelectronic characteristics may be determined by surface-enhanced optical properties [72].

Potential-dependent SERRS signals from sulfonated phthalocyanines that are adsorbed on electrochemcally roughened silver wire electrodes can be used to determine the orientation of the macrocycle with respect to the electrode surface [60a, 70]. This information is important in device applications should the orientation of the phthalocyanine determine the sensing or optoelectronic function.

Fe, Co, and metal-free sulfonated phthalocyanines (CoTSPc, FeTSPc, H_2TSPc) have been studied across a wide pH range and at different potentials on silver [58]. Large changes in the relative intensities of SERRS bands of CoTSPc are observed as the potential is swept to −0.6 V. Sulfonate involvement in adsorption of the TSPCs to the surface is suggested from the enhancement of scattering from a band assigned to SO_3^{2-} at 1275 cm^{-1}. The different proposed adsorption schemes are shown in Figure 19.

More detailed experimentation indicates potential-dependent behavior for the intensity, and occasionally for the frequency, of the SERRS bands assigned to specific modes. In particular, bands at 612, 699, 1346, and 1550 cm^{-1} are affected. The latter three bands can be assigned as ν_7, ν_4 and ν_3. Surprisingly, the frequency of the band at 1346 cm^{-1} is not affected by potential, although this band can be correlated with ν_4, which is the oxidation state marker in porphyrins. Orientation information for multilayer assemblies on the electrode surface can be deduced from polarization measurements. A gradual change in orientation is proposed [58] from adsorption-induced perpendicular orientation to a stacked phthalocyanine morphology.

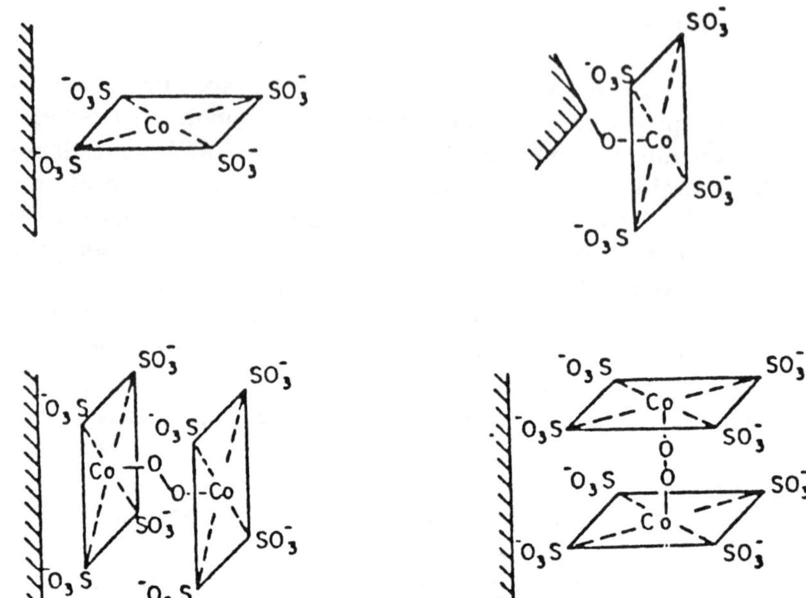

Figure 19 Possible configurations for tetrasulfonated cobalt phthalocyanine absorbed on an electrode surface [57]. (Reprinted with permission from J. Electroanal. Chem. Interfac. Phenom., Vol. 113. Copyright 1980 Elsevier Sequoia, S.A.).

Evidence for formation of an O_2 adduct with the phthalocyanine macrocycle is obtained from the substantial spectral changes in the SERRS of sulfonated phthalocyanines adsorbed on silver upon introduction of O_2 [58]. Thus it is concluded that the phthalocyanine is adsorbed onto the electrode surface by the sulfonate group and that little perturbation to the macrocycle arises upon adsorption.

Different oligomeric forms of silicon phthalocyanines of the type $R_3Si-O-(SiPcO)n-SiR_3$ have been studied [60b]. While fluorescence is quenched for the tetrameric form on the electrode surface, it dominates the spectrum for the adsorbed monomeric form. It is possible that a flat orientation of the ring with respect to the electrode surface is responsible for the strong fluorescent emission whereas a perpendicular orientation is preferred for the tetrameric form upon adsorption, and hence fluorescence quenching. Bulky R = C_6H_{13} groups that cap the central silicon ion may act to space the macrocycle from the metal surface, reducing the influence of nonradiative decay channels as a means of fluorescence quenching for the monomeric form. Steric considerations may result in adsorption of the tetrameric species with the macrocycle plane perpendicular to the metal surface.

Disulfonated copper phthalocyanine has been studied on a silver wire electrode [70]. Remarkable spectral changes are observed as the potential of the electrode is changed from +0.6 to −0.6 V versus the standard calomel electrode (Figure 11a). These changes resemble some of those reported in [60a]. In particular, the enhancement of a band at 1450 cm^{-1} (1495 cm^{-1} in [60a]) is observed for negative potentials in both studies. This band is assigned to ν_{28} (B_{2g}) [18]. The behavior of bands that involve movement on different parts of the macrocycle is analyzed as a function of potential, and the potential-dependent molecular orientations of the macrocycle with respect to the metal surface are deduced [Figure 14(b)].

With this substantial library of potential-dependent SERRS spectra it should now be possible to predict the behavior of phthalocyanines in potential-controlled devices. In particular the mode of operation of photovoltaic cells for solar energy conversion may be rationalized with this approach. Many of the existing and proposed devices incorporating phthalocyanines are fabricated by thin film deposition techniques. SERRS from thin films of phthalocyanines on "SERS-active" semiconductors [98, 99] and metals has been obtained.

The dependence of adsorbate coverage for phthalocyanine on 'SERS-active' silver and on non-SERS-active chromium has been studied [56]. A strong dependence on the relative intensity of the SERRS active bands is observed for submonolayer versus multilayer coverage on silver, whereas no such dependence is observed for chromium [56]. Quenching of the surface plasmon by the phthalocyanine overcoat is held to be resonsible for the dimunition in overall intensity as the film thickness was increased. However, no interpretation of the spectra was conducted on a mode-by-mode basis. The relative intensity of scattering from ν_7 (683 cm^{-1}) and ν_3 (1533 cm^{-1}) is most intense for submonolayer coverage (Fig. 20). The question of whether this mode-specific effect is due to phthalocyanine surface-coverage-dependent shifting of the 0–0 π–π^* transition and/or surface-induced distortion along the Jahn-Teller active coordinate of the absorbed phthalocyanine molecules remains to be resolved.

SERRS studies on silver island films include a comparison of silicon versus silver and glass as substrates [35]. For thick phthalocyanine films (approximately 80 nm) interference effects between the layers of the phthalocyanine films on silver can cause an enhancement of the signal as compared with that obtained from glass, whereas for silicon de–enhancement can occur. Electromagnetic coupling of a monomolecular phthalocyanine film with the silver metal surface plasmons produced a 10^4 signal amplification [35]. With silver–coated silicon [100] and gallium phosphide [98] particles, activation of the surface plasmons does not occur, but electric field concentration at the surface can still give rise to enhancement of the signals from the copper phthalocyanine layer. The signal enhancement increases as the gallium phosphide particle size increases. This is contrary to the behavior observed for silver, where the silver enhancement decreases as the particle size increases for particles greater than 10 nm in diameter.

Copper phthalocyanine has been studied in the Kretschman attenuated total reflection (ATR) configuration [101–103]. The angular dependence of the scattering is pronounced with this arrangement, and maximum intensity is found for $\theta_i = \theta_c$. θ_i is the angle of incidence of the laser light, and θ_c is the critical angle for the ATR. This indicates that electromagnetic enhancement rather than chemical term enhancement is responsible for the amplified signal. From ATR measurements, a negative value of the real part of the dielectric constant of a copper phthalocyanine layer on silver was found [103]. This suggests that the copper phthalocyanine layer has a metallic nature. Controversy surrounds the determination of the orientation of the copper phthalocyanines using ATR. From Raman measurements a flat orientation is indicated, whereas polarization modulation electronic absorption measurements indicate a perpendicular orientation [104]. ATR is used in communication systems, and hence the study of phthalocyanines in such devices may be applicable to fields such as optical computing and communication.

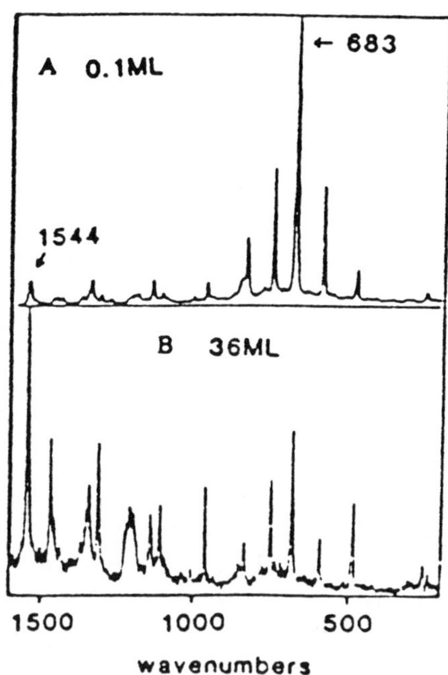

Figure 20 SERS spectra of cobalt phthalocyanine on a CaF_2 roughened silver substrate (A) with coverage of 0.1 monolayers, (B) with coverage of 36 monolayers. λ=647.1 nm [56]. (Reprinted with permission from J. Chem. Phys., Vol. 87. Copyright 1987 American Institute of Physics).

Langmuir–Blodgett films of mixed phthalocyanine/arachidic acid monolayers deposited on silver–coated tin spheres [105] and silver [106, 107], gold [108], and indium [103] island films have been studied. Both the distance– [102] and coverage–dependent [103] SERRS signals of tetra–tert–butyl substituted phthalocyanines have been obtained. Maximum intensity is obtained for submonolayer amounts of the phthalocyanine, which is spaced from the metal surface by no more than one monolayer of arachidic acid. Interestingly, the monolayer coverage dependence for the phthalocyanine Langmuir–Blodgett layer deposited on gold films is different from that on silver. Some form of metal–dependent metal/dye interaction is proposed. Only physisorption is believed to occur, as evidenced by the similarities in frequency of the bands of the phthalocyanine on glass and on the metal films. Activation of the surface plasmon resonance for gold requires red light with λ>600 nm. For silver, the surface plasmons may be activated with light λ>350 nm. Therefore, the different coverage–dependent behavior for silver versus gold may be rationalized by consideration of the coupling of the electronic transitions of phthalocyanine and the different surface plasmon λ_{max} for each metal.

The coverage-dependent changes in the relative intensities of scattering from ν_3 versus ν_7 [56] is not observed [107, 108]. This may be a consequence of physisorption rather than chemisorption from the surface by the tert-butyl group and other substituents of the phthalocyanine molecules. These studies indicate how spacial spectroscopic tuning of dye-metal interfaces may be accomplished. This has been achieved with Langmuir-Blodgett monolayers of tetra-tert-butyl phthalocyanines and with N,octyl-substituted perylene dyes deposited on silver-coated tin spheres [9, 109]. Langmuir-Blodgett monolayers of tetra-sulfonated phthalocyanines have also been studied on gold island films [71]. These studies are relevant to the incorportion and mode of operation of phthalocyanines in P-N junctions for solar cells and electrophotography.

The interaction of the reversible NO_2/N_2O_4 gas adsorption with a langmuir-Blodgett monolayer of metal-free tetra-*tert*-butyl phthalocyanine (H_2TTPC), (CuTTPc) and ytterbium bisphthalocyanines has been monitored with SERRS [110]. Interestingly, the bands which appear to be most affected by exposure to the film to NO_2/N_2O_4 are that due to the postulated Jahn-Teller active mode, ν_{15}, at approximately 750 cm^{-1} and those due to ν_7 at approximately 680 cm^{-1} and ν_3 at approximately 1530 cm^{-1}. This suggests that the adsorption of the gas may perturb the Jahn-Teller distortion in the excited state for the phthalocyanine by either a shift of the 0-0 transition and/or by influencing the Jahn-Teller active coordinate. SERRS monitoring of the interaction of NO_2/N_2O_4 with Langmuir-Blodgett layers of lutetium and ytterbium diphthalocyanines [111] and praseodymium bisphthalocyanine has also been reported [112].

Using an electrode of compacted silver particles rather than a silver wire and by cycling in sulfuric acid, very strong phthalocyanine signals are obtained [48(b)]. The strong signals are a consequence of roughening the electrode in the acid and of the existence of an extensive hydrogen bonding network bonding the phthalocyanine particle to the silver surface. A new band at 1475 cm^{-1} is observed when the potential of the compacted electrode is swept below −0.2 V. This is attributed to a SERRS active mode [Fig. 11(a)]. In this way, organization of the orientation of molecules spaced from the surface could be obtained. Cationic surfactants are reported to assist in binding of tetrasulfonated nickel (II) phthalocyanine to silver colloidal surfaces, and hence enable detection of the negatively charged macrocycle by SERRS [113]. Studies of supermolecular interactions between phthalocyanines have been reported for thin films of trivalent and tetravalent metal phthalocyanines [114].

A patent has been issued for a laser that uses silver or silver halide as a conducting substrate covered with a complex material [115]. Iron–tetrasulfonated phthalocyanine is deposited on a silver bromide crystal from which SERS is obtained. Applications for SERRS from phthalocyanines have also been proposed for molecular electronic devices [78(a)] with fast switching times, which could be used in the coupling of artificial systems with neurobiological systems [116]. SERRS for phthalocyanines has also been proposed for use in optical data storage [117]. In other uses, the relationship between photography and SERRS of phthalocyanines has been pointed out. Essentially, the process of energy transfer between the metal and the molecule is very similar to that proposed in color photography using dye sensitisation. Thus, SERRS is an apposite probe for changes that occur in photography [59, 118].

H. CONCLUSIONS

It is clear that the method is potentially a powerful one, and at present few experiments make use of its full power. It is capable of using very small samples either as thin films or as single crystals down to less than 1 μm in size. It can provide both electronic and vibrational information. In addition it is possible to study such processes as phonon coupling in electron transfer that are at the forefront of modern research. The latest developments in laser Raman technology involve pulsed lasers capable of studying femtosecond processes [119]. With such equipment it is possible to study bond making, breaking, and twisting, adding yet more power to what is already a technique with considerable unused potential.

REFERENCES

1. M. Cardona, in Light Scattering in Solids II, Vol. 50 of Topics in Applied Physics, M. Cardona and G. Guntherodt, Eds., Springer-Verlag, Berlin, 1982, p.19.
2. S.H. Lin, J. Chem. Phys., 90 (1989) 7103.
3. K.V. Mikkelsen and M.A. Ratner, Chem. Rev., 87 (1987) 113.
4. L.D. Zeigler, Y.C. Chung, P. Wang and Y.P. Zhang, in Time Resolved Spectroscopy, Vol. 18 of Advances in Spectroscopy, R.J.H. Clark and R.E. Hester, Eds., John Wiley & Sons, Chichester, 1989, p.55.
5. W. Siebrand and M.Z. Zgierski, in Excited States, Vol. 4, E.C. Lim, Ed., Academic Press, London, 1979, p.1.
6. T.G. Spiro and T.C. Strekas, J. Am. Chem. Soc., 96 (1974) 338.
7. L.K. Orman, Y.J. Chang, D.R. Anderson, T. Yabe, X. Xu, S.-C. Yu and J.B. Hopkins, J. Chem. Phys., 90 (1989) 1469.
8. C.A. Melendres, C.B. Rios, X. Feng and R. McMasters, J. Phys. Chem., 87 (1983) 3526.
9. R. Aroca and U. Guhathakurta-Ghosh, J. Am. Chem. Soc., 111 (1989) 7681.
10. D.A. Long, Raman Spectroscopy, McGraw-Hill, New York, 1977.
11. R.J.H. Clark and B. Stewart, Struct. Bonding, 38 (1979) 1.
12. D.L. Rousseau, J.M. Friedman and P.F. Williams, in Raman Spectroscopy of Gases and Liquids, Vol. 11 of Topics in Current Physics, A. Weber, Ed., Springer Verlag, 1979, 203.
13. (a) R.J.H. Clark and T.J. Dines, Angew. Chem. Int. Ed. Engl., 25 (1986) 131.
    (b) A.B. Myers and R.A. Mathies, in Biological Applications of Raman Spectroscopy, Vol. 2, T.G. Spiro, Ed., John Wiley and Sons, Chichester, 1987, p.1.
14. M. Moskovits, Rev. Mod. Phys., 57 (1985) 783.
15. (a) A. Otto, in Light Scattering in Solids IV, Vol. 54 of Topics in Applied Physics, Springer-Verlag, Berlin, 1984, p. 289.
    (b) R.L. Birke and J.R. Lombardi, in Spectroelectrochemistry: Theory and Practice, R.J. Cale, Ed., Plenum Press, New York, 1988, p. 263.
16. M.J. Stillman and T. Nyokong, in Phthalocyanines: Properties and Applications, C.C. Leznoff and A.B.P. Lever, Eds., VCH Publishers, New York, 1989, p. 133.
17. L.K. Lee, N.H. Sabelli and P.R. LeBreton, J. Phys. Chem., 86, (1982) 3926.
18. A.J. Bovill, A.A. McConnell, J.A. Nimmo and W.E. Smith, J. Phys. Chem., 90 (1986), 569.

19. L. Edwards and M. Gouterman, J. Molec. Spectrosc., 33 (1970) 292.
20. I.B. Bersuker, The Jahn-Teller Effect and Vibronic Interactions in Modern Chemistry, Plenum, New York, 1984.
21. J.A. Shelnutt, L.D. Cheung, R.C.C. Chang, N.-T. Yu and R.H. Felton, J. Chem. Phys., 66 (1977) 3387.
22. V. Hizhnyakov, I. Tehver and G. Zavt, J. Raman Spectrosc., 21 (1990) 231.
23. (a) A. Gilbert and J. Baggott, Essentials of Molecular Photochemistry, Blackwell, London, 1991
    (b) H. Kono, Y. Nomura, Y. Fujimura, Adv. Chem. Phys., 80 (1991) 403.
24. R.R. Millard and B.I. Greene, J. Phys. Chem., 89 (1985) 2976.
25. C.R. Bartholomew, "A Raman and Electrochemical Study of First Row Transition Metal Phthalocyanines and Rare Earth Diphthalocyanines." Ph.D. awarded 1987 by University of Strathclyde, U.K.

26. (a) S.O. Williams and D.G. Imre, J. Phys. Chem., 92 (1988) 3363.
    (b) R.J.H. Clark and T.J. Dines, Molec. Phys., 42 (1981) 19.
    (c) R.J.H. Clark and T.J. Dines, Molec. Phys., 45 (1982) 115.
27. (a) I.V. Aleksandrov, Ya.S. Bobovich, V.G. Maslov and A.N. Sidorov, Opt. Spectrosc., 37 (1974) 265.
    (b) T.-H. Huang, K.E. Rieckhoff and E.-V. Voight, Can. J. Chem., 56 (1978) 976.
28. O.V. Quinzani, A.H. Jubert and P.J. Aymonino, J. Raman Spectrosc., 20 (1989) 141.
29. (a) S.-Y. Lee and E.J. Heller, J. Chem. Phys., 71 (1979) 4777.
    (b) E.J. Heller, Acc. Chem. Res., 14 (1981) 368.
    (c) E.J. Heller, R.L. Sundberg and D. Tannor, J. Phys. Chem., 86 (1982) 1822.
30. K.-S. Shin, R.J.H. Clark and J.I. Zink, J. Am. Chem. Soc., 111 (1989) 4244.
31. (a) S.-Y. Lee, J. Chem. Phys., 92 (1990) 6376.
    (b) K.-S. Shin, R.J.H. Clark and J.I. Zink, J. Am. Chem. Soc., 112, (1990) 3754.
32. (a) I.W. Levin and E.N. Lewis, Anal. Chem., 62 (1990) 1101A.
    (b) S.K. Doorn, J.T. Hupp, D.R. Porterfield, A. Campion and D.B. Chase, J. Am. Chem. Soc., 112 (1990) 4999.
    (c) C.A. Jennings, G.J. Kovacs and R. Aroca, J. Phys. Chem., 96, (1992), 1340.
33. (a) J.A. Nimmo, A.J. Bovill, A.A. McConnell and W.E. Smith, J. Raman Spectrosc., 18 (1985) 245.
    (b) J.A. Nimmo, "The Electronic Properties of Metal Phthalocyanines and the Effect of Metal Interactions." Ph.D. awarded 1983 by University of Strathclyde, U.K.
34. C.R. Bartholomew, A.A. McConnell and W.E. Smith, J. Raman

Spectrosc., 20 (1989) 595.

35. S. Hayashi and M. Samejima, Surf. Sci., 137 (1984) 442.
36. C. Paliteiro, A. Hamnett and J.B. Goodenough, J. Electroanal. Chem., 249 (1988) 167.
37. C.S. Choi and H. Tachikawa, J. Am. Chem. Soc., 112 (1990) 1757.
38. J. Dowdy, J.J. Hoagland and K.W. Hipps, J. Phys. Chem., 95 (1991) 3751.
39. C.R. Bartholomew, A.A. McConnell and W.E. Smith, J. Raman Spectrosc., 18 (1987) 277.
40. C.A. Melendres and V.A. Maroni, J. Raman Spectrosc., 15 (1984) 319.
41. (a) R. Aroca, Z.Q. Zeng and J. Mink, J. Phys. Chem. Solids, 51 (1990) 135.
    (b) D. Christen, V. Hoffman, A. Rager and W. Gopel, Thin Solid Films, Vol. 208 (1992) 284.
42. T. Kitagawa, M. Abe and H. Ogoshi, J. Phys. Chem., 69 (1978) 4516.
43. M. Abe, T. Kitagawa and Y. Kyogoku, J. Phys. Chem., 69 (1978) 4526.
44. T. Kitagawa and Y. Ozaki, Struct. Bonding, 64 (1987) 71.
45. X.-Y. Li, R.S. Czernuszewicz, J.R. Kincaid, Y.O. Su and T.G. Spiro, J. Phys. Chem., 94 (1990) 31.
46. X.-Y. Li, R.S. Czernuszewicz, J.R. Kincaid, P. Stein and T.G. Spiro, J. Phys. Chem., 94 (1990) 47.
47. X.-Y. Li, R.S. Czernuszewicz, J.R. Kincaid and T.G. Spiro, J. Am. Chem. Soc., 111, (1989) 7012.
48. (a) A.J. Bovill, A.A. McConnell and W.E. Smith, J. Chem. Soc. Faraday Trans., 86 (1990) 4065.
    (b) A.J. Bovill, A.A. McConnell and W.E. Smith, J. Chem. Soc. Faraday Trans. 1, 85 (1989) 3695.
49. (a) M.Z. Zgierski and M. Pawlikowski, Chem. Phys., 65 (1982) 335.
    (b) M.Z. Zgierski, Chem. Phys., 65 (1982) 369.
50. A.J. Bovill, A.A. McConnell, B.N. Rospendowski and W.E. Smith, J. Chem. Soc. Faraday Trans., 88 (1992) 455.
51. M. Fleischmann, P.J. Hendra and A.J. McQuillan, J. Chem. Soc. Commun., (1973) 80.
52. D. Blue, K. Helwig, M. Moskovits and R. Wolkow, J. Chem. Phys., 92 (1990) 4600.
53. J.A. Creighton, in Spectroscopy of Surfaces, Vol. 16 of Advances in Spectroscopy, R.J.H. Clark and R.E. Hester, Eds., John Wiley & Sons, Chichester, 1988, p. 37.
54. Rospendowski, B. and W.E. Smith, Inorg. Chem., 27 (1988) 4509.
55. U. Guhathakurta-Ghosh and R. Aroca, J. Phys. Chem., 93 (1989) 6125.
56. E.J. Zeman, K.T. Carron, G.C. Schatz and R.P. Van Duyne, J. Chem. Phys., 87 (1987) 4189.
57. R. Kotz and E. Yeater, J. Electroanal. Chem., 113 (1980) 5625.
58. B. Simic-Glavaski, S. Zecevic and E. Yeager, J. Electroanal. Chem., 150

(1983) 469.
59. B. Simic-Glavaski, J. Phys. Chem., 90 (1986) 3863.
60. (a) B. Simic-Glavaski, S. Zecevic and E. Yeager, J. Am. Chem. Soc., 107 (1985) 5625.
(b) B. Simic-Glavaski, A.A. Tanaka, M.E. Kenny, J. Electroanal. Chem., 229 (1987) 285.
61. Y. Nakao and H. Yamada, J. Electron Spectrosc. Related Phenom., 45 (1987) 189.
62. A. Brotman and E. Burstein, Physica Scripta, 32 (1985) 385.
63. R. Aroca and R.O. Loutfy, J. Raman Spectrosc., 12 (1982) 262.
64. P.F. Liao, J.G. Bergman, D.S. Chemla, A. Wokaun, J. Melngailis, A.M. Hawryluk and N.P. Economou, Chem. Phys. Lett., 82 (1981) 355.
65. C. Campbell, J. Clarkson, K.P.J. Williams and W.E. Smith, Proceedings of the Twelfth International Conference on Raman Spectroscopy, J.R. Durig and J.F. Sullivan, Eds., John Wiley & Sons, Chichester, 1990, p. 252.
66. P.C. Lee and D. Meisel, J. Phys. Chem., 86 (1982) 3391.
67. I.K. Stewart, B.N. Rospendowski and W.E. Smith, to be published.
68. (a) J.A. Creighton, C.G. Blatchford and M.G. Albrecht, J. Chem. Soc. Faraday Trans. 2, 75 (1979) 790.
(b) J. Clarkson, C. Campbell, B.N. Rospendowski and W.E. Smith, J. Raman Spectrosc., 22 (1991) 771.
69. A.J. Bovill, A.A. McConnell and W.E. Smith, Surf. Sci., 158 (1985) 333.
70. A.A. McConnell, J.A. Nimmo and W.E. Smith, J. Raman Spectrosc., 20 (1989) 375.
71. Y. Bai, Y. Zhao, L. Zhang, K. Tian, X. Tang and T. Li, Thin Solid Films, 180 (1989) 249.
72. R. Aroca and G.J. Kovacs, in Vibrational Spectra and Structure, Vol. 19, J.R. Durig, Ed., Elsevier, Oxford 1991, 55.
73. K.-Y. Law, J. Phys. Chem., 92 (1988) 4226.
74. M. Emmelius, G. Pawlowski and H.W. Vollman, Angew. Chem. Int. Ed. Engl., 28 (1989) 1445.
75. P. Leempoel, F.-R. F. Fan and A.J. Bard, J. Am. Chem. Soc., 87 (1983) 2948.
76. (a) M.S. Nieuwenhuizen, A.J. Nederlof and A.W. Barendsz, Anal. Chem., 60 (1988) 230.
(b) D.G. Zhu, M.C. Petty and M. Harris, Sensors and Actuators B, 2 (1990) 265.
(c) C.L. Honeybourne and J. O'Donnell, Anal. Proc., 28 (1991) 333.
(d) S. Kurosawa and N. Kamo, Langmuir, 8, (1992), 254.
77. A.W. Snow and W.R. Barger, in Phthalocyanines: Properties and Applications, C.C. Leznoff and A.B.P. Lever, Eds., VCH Publishers, New York, 1990, p. 341.
78. (a) D. Li, M.A. Ratner and T.J. Marks, J. Am. Chem. Soc., 110 (1988)

1707.
(b) B. Simic-Glavaski, Proc. SPIE-Int. Soc. Opt. Eng., 634 (Adv. Opt. Hybrid Comput.) (1987) 195.
79. D. Wohrle, V. Schmidt, B. Schumann, A. Yamada and K. Shigehara, Ber. Bunsenges. Phys. Chem., 91 (1987) 975.
80. G. Tourillan, R. Cote, D. Guay and J.P. Dodelet, J. Electrochem. Soc., 136 (1989) 2931.
81. J. Martensson and H. Arwin, Thin Solid Films, 188 (1990) 179.
82. Phthalocyanines: Properties and Applications, C.C. Leznoff and A.B.P. Lever, Eds, VCH Publishers, New York, 1990.
83. (a) L.G. Tomilova, N.A. Ovchinnikova and E.A. Luk'yanets, Zh. Obshch. Khim., 57 (1987) 2100.
(b) G. Guilland, M. Al Sadoun, M. Maitrot, J. Simon and M. Bouret, Chem. Phys. Lett., 167 (1990) 503.
(c) T.-H. Tran-Thi, D. Markovitsi, R. Even and J. Simon, Chem. Phys. Lett., 139 (1987) 207.
(d) A. Aroca, R.E. Clavijo, C.A. Jennings, G.J. Kovacs, J.M. Duff and R.O. Loutfy, Spectrochim. Acta, 45A (1989) 957.
(e) K. Takeshita, Y. Aoyama and M. Ashida, Bull Chem. Soc. Jpn., Vol. 64 (1991) 1167.
(f) K. Takeshita and M. Ashida, J. Electrochem. Soc., Vol. 138 (1991) 2617.
84. C.J. Schramm, R.P. Scaringe, D.R. Stojakovic, B.M. Hoffman, J.A. Ibers and T.J. Marks, J. Am. Chem. Soc., 102 (1980) 6702.
85. D. Masheder and K.P.J. Williams, J. Raman Spectrosc., 18 (1987) 391.
86. I.K. Stewart, "Synthesis and Raman Spectroscopy of Novel Copper Phthalocyanine Derivatives." Ph.D. awarded 1991 by University of Strathclyde, U.K.
87. R.E. Clavijo, R. Aroca, G.J. Kovacs, C.A. Jennings, J. Duff and R.O. Loutfy, J. Raman Spectrosc., 20 (1989) 461.
88. Z.Q. Zeng, R. Aroca, A.-M. Hor and R.O. Loutfy, J. Raman Spectrosc., 20 (1989) 467.
89. R. Aroca, C. Jennings, R.O. Loutfy and A.-M. Hor, Spectrochim. Acta, 43A (1987) 725.
90. R. Aroca, C. Jennings, R.O. Loutfy and A.-M. Hor, J. Phys. Chem., 90 (1986) 5255.
91. J. Souto, R. Aroca, J.A. DeSaja, J. Raman Spectrosc., Vol. 22 (1991) 349.
92. A.A. McConnell and W.E. Smith, J. Raman Spectrosc., 20 (1989) 31.
93. N.M.D. Brown, in Spectroscopy of Surfaces, Vol. 16 of Advances in Spectroscopy, R.J.H. Clark and R.E. Hester, Eds., John Wiley & Sons, Chichester, 1988, p.215.
94. (a) K.W. Hipps, J. Phys. Chem., 93 (1989) 5958.
(b) K.W. Hipps and J.J. Hoagland, Langmuir, 7 (1991) 2180.

95. K.W. Hipps, J. Dowdy and J.J. Hoagland, Langmuir, 7 (1991) 5.
96. J.J. Hoagland, J. Dowdy and K.W. Hipps, J. Phys. Chem., 95 (1991) 2246.
97. J.K. Gimzewski, E. Stoll and R.R. Schlitter, Surf. Sci., 181 (1987) 267.
98. S. Hayashi, R. Koh, Y. Ichiyama and K. Yamamoto, Phys. Rev. Lett., 60 (1988) 1085.
99. R. Aroca, C. Jennings, G.J. Kovacs, R.O. Loutfy and P.S. Vincett, J. Phys. Chem., 89 (1985) 4051.
100. R. Koh, S. Hayashi and K. Yamamoto, Solid State Commun., 64 (1987) 375.
101. M. Osawa and W. Suetaka, Surf. Sci., 186 (1987) 583.
102. R. Aroca, D. Battisti, G.J. Kovacs and R.O. Loutfy, J. Electrochem. Soc., 136 (1989) 2902.
103. D. Battisti, R. Aroca and R.O. Loutfy, Chem. Mater., 1 (1989) 124.
104. M. Osawa, S. Yamamoto and W. Suetaka, Appl. Surf. Sci., 33-34 (1987) 890.
105. S. Ushioda, J. Electron Spectrosc. Related Phenom., 54/55 (1990) 881.
106. A. Hatta, S. Suzuki and W. Suetaka, Appl. Surf. Sci., 40 (1989) 9.
107. R. Aroca and D. Battisti, Langmuir, 6 (1990) 250.
108. G.J. Kovacs, R.O. Loutfy, P.S. Vincett, C. Jennings and R. Aroca, Langmuir, 2 (1986) 689.
109. D. Battisti and R. Aroca, J. Molec. Struct., 218 (1990) 351.
110. D. Battisti and R. Aroca, J. Am. Chem. Soc., 114 (1992) 1201.
111. R.E. Clavijo, D. Battisti, R. Aroca, G.J. Kovacs and C.A. Jennings, Langmuir, Vol. 8 (1992) 113.
112. J. Souto, L. Tomilova, R. Aroca and J.A. DeSaja, Langmuir, 8 (1992) 942.
113. J-S. Ha, M. Yoon, M. Lee, D-J. Jang and D. Kim, J. Raman Spectrosc., Vol. 22 (1991) 349.
114. T.D. Sims, J.E. Pemberton, P. Lee and N.R. Armstrong, Chem. Mater., 1 (1989) 26.
115. B. Simic-Glavaski, Eur. Pat. Appl. EP 216, 551
116. B. Simic-Glavaski, PCT Int. Appl. WO 87 01, 807.
117. M. Kamiyama, T. Hashimoto, O. Oka, M. Tsuchida, Jpn. Kokai Tokkyo Koho JP 01 50,251 [89 50,251].
118. K. Kneipp, J. Molec. Struct., 218 (1990) 357.
119. Z.Z. Ho and N. Peyghambarian, Chem. Phys. Lett., 148 (1988) 107.

KEY TO SYMBOLS USED IN EQUATIONS

α_{xy} is the polarisability in the (xy) plane of the phthalocyanine molecule.

$\alpha_{xy}{}^{Q}$ is the polarisability in the (xy) plane of the phthalocyanine molecule for excitation of the Q–band 0–0 transition.

$\sum_i$ is a sum over the intermediate vibronic excited states, $|i\rangle$.

$|g\rangle$ and $|f\rangle$ are the initial, ground vibrational and final, excited vibrational states (for Raman scattering), of the ground electronic state.

$|\theta_{A1g}{}^{Q}\rangle$ and $|\theta_{Eu}{}^{Q}\rangle$ are the Born Oppenheimer electronic wavefunctions for the ground and excited Q–state of the phthalocyanine dianion molecule.

$|\theta_{A1g}{}^{B}\rangle$ and $|\theta_{Eu}{}^{B}\rangle$ are the Born Oppenheimer electronic wavefunctions for the ground and excited B–state of the phthalocyanine dianion molecule.

$|0\rangle$ and $|0'\rangle$ are the ground vibrational states of the ground electronic state and Q excited state of the phthalocyanine molecule, respectively.

$|1_{\Gamma vib}\rangle$ is the wavefunction, of symmetry Γvib, for the first vibrationally excited state of the ground electronic state.

$\omega_{g\to i}$ and ω_L are the energies (frequencies) of the electronic transition from $|g\rangle$ to $|i\rangle$ and of the incident laser light, respectively.

$\omega_{o-o'}$ is the energy (frequency) of the 0–0 Q–state electronic transition.

$\omega_{Eu}{}^{B1} - \omega_{Eu}{}^{Qo'}$ is the difference in energy (frequency) between transitions to the first vibrational excited state of the B–state (B1) and to the ground vibrational state of the

Q–state (Qo')

Γ_i is a damping term which reflects the homogeneous width of |i>.

$i\Gamma_{Eu}{}^{Qo'}$ is the damping term for the |0'> state of the Q excited state

μ_{xy} is the transition dipole moment operator in the plane of the phthalocyanine molecule.

$\mu_{xy}{}^{Q}$ is the transition dipole moment operator for the Q–state electronic transition in the plane of the phthalocyanine molecule.

$\mu_{xy}{}^{B}$ is the transition dipole moment operator for the B–state electronic transition in the plane of the phthalocyanine molecule.

$|\theta_{\Gamma elec}(r,R)\rangle$ is the Born Oppenheimer electronic wavefunction (r is the electron coordinates), with symmetry, Γelec, which is parametrically dependent on the nuclear coordinates, R. R=0 represents the nuclear position at the equilibrium bond length.

$|\Phi_{\Gamma vib}(R)\rangle$ is the Born Oppenheimer vibrational wavefunction, with symmetry, Γvib.

$R_{\Gamma vib}$ is the vibrational coordiate, or coordinate operator, with symmetry, Γvib.

For the time dependent equations, 5 to 7, the terms and symbols are as described above or in the text plus:

$\frac{i}{\hbar}$ is the $\sqrt{-1}$ divided by Plancks Constant/2π

C is a constant

t is time

H_{ex} is the excited state Hamiltonian.

5

Absorption and Magnetic Circular Dichroism Spectral Properties of Phthalocyanines

Martin J. Stillman

A. INTRODUCTION

While the color of the phthalocyanine ring was the attraction for much of the early work on the chemical and spectral properties of metal complexes of the phthalocyanine dianion ring [1], the diversity of species formed following oxidation and reduction of phthalocyanine complexes has led to a very active redox literature today. Several areas of research have particularly stimulated the measurement of spectroelectrochemical data for cationic and anionic complexes. Among these are (1) studies on photochemically initiated electron transfer reactions involving the phthalocyanine ring [2-16], (2) the use of phthalocyanines as photosensitizers in photodynamic therapy of tumors [17], (3) the one-dimensional metal [18-21] and semiconductor [22] properties, and (4) the development of color displays [23-25]. The possibilities for new chemistry based on the redox states of the essentially planar metallophthalocyanine are exciting (see as an example the x-ray structure reported by Scheidt and Dow [26] for ZnPc as an example), especially when metallophthalocyanine complexes in both solution and solid-state phases exhibit multiple, reversible redox states that can be studied.

Advances in the redox chemistry of the phthalocyanines and porphyrins have followed developments in the use of spectroelectrochemical techniques that enable spectral data to be measured directly from the electrochemical cell, for example, see [27]. Recently in our laboratory a cell that can be used inside the superconducting magnet used in magnetic circular dichroism (MCD) measurements has been built and tested [28, 122, 123]. In addition, spectroscopic techniques like MCD and resonance Raman spectroscopy, now provide even more information about the electronic configurations involved in the optical transitions of the ring-oxidized and ring-reduced complexes.

It is useful to consider briefly the development of one of the most thoroughly studied areas of the redox chemistry of the phthalocyanines and porphyrins: the modeling of the heme group in horseradish peroxidase compound I. The reported redox chemistry and spectral properties of cation and anion radical complexes of the parent porphyrins, particularly the catalase, peroxidase, and cytochrome P-450 classes of heme enzyme, the chlorophylls, and the tetraphenylporphyrin (TPP) and octaethylporphyrin (OEP) derivatives, are far more extensive than those of the phthalocyanines, for example, [29-40].

The magnetic circular dichroism spectrum of porphyrin π cation radical complexes exhibits an overlapping series of bands between from 300 to 700 nm [29,36-45]. Because these complexes tend to dimerize readily in solution, especially at low temperatures, the expected g=2.0025 EPR signal [39] is often quenched [46]. Analysis of the origins of the unusual optical spectra of the compound I derivatives of horseradish peroxidase and catalase led to the hypothesis that these species involved two-electron oxidation of the ferric heme group, resulting in both metal and ring oxidation to give $[Fe(IV)\text{-heme}(-1)]^{\cdot 2+}$ [37]. Because of the effects of the proximity of the paramagnetic Fe(IV) and π radical cation, the magnetic properties of the protein species were not easy to interpret, so that a range of spectroscopic techniques have been applied to the assignment problem [31, 36,37, 39, 47-50]. The ring oxidation assignment was based results of on model compound studies, in particular the correspondence between the optical spectra of $[Co(III)OEP(-1)]Br_2$ and $[Co(III)OEP(-1)](ClO_4)_2$ [39, 40]. Magnetic circular dichroism spectra of the protein and model compound species clearly supported the assignment of a π cation ring in these compound I species [29, 36, 49, 51, 52].

While the MCD spectra of porphyrin π cation radical complexes offer a clear diagnostic tool with which to identify π ring oxidation [36], as we will discuss in more detail in the following, the a_{2u} highest occupied molecular orbital in the porphyrins is "accidentally" degenerate with the a_{1u} orbital. The order of these two orbitals in ring-oxidized porphyrin complexes is currently the subject of considerable interest [36,47,50].

The involvement of a magnesium chlorin π cation in early stages of photosynthesis has also prompted much development work on the electronic and magnetic properties of a range of magnesium porphyrin π cation radical species.

Much less has been reported for porphyrin and phthalocyanine π anion complexes compared with the literature for π cations, although the interest in the role of bacteriochlorophyll as an electron acceptor has stimulated studies of the redox chemistry of bacteriochlorophylls, for example, reduction to the anion radical [32] and the one-electron reduction of Ni chlorin [53]. The papers by Dodd and Hush in 1964 [54] and Clack and Yandle in 1972 [55] still remain excellent sources for spectral data of reduced porphyrins and phthalocyanines.

i. Spectral Properties of the Phthalocyanine Dianion

It is convenient to use Gouterman's four-orbital model in discussions of the origins of the spectral intensity in neutral phthalocyanines [56]. We note that several more recent calculations reproduce this order of molecular orbitals in the region of the top filled and lowest unoccupied orbitals. We have used the order reported by Ohno et al. [57] and Orti et al. [58] to construct the diagram shown in Fig. 1. In this theoretical model, transitions to the lowest unoccupied molecular orbital ($6e_g$) give rise to a band near 670 nm, the *Q* band, and a band near 330 nm, the *B* or Soret band. (The relationship between the absorption and MCD spectra and reported theoretical calculations of neutral phthalocyanines has been discussed in Volume 1 of this series [1]). The four-orbital model, proposed first by Gouterman [56], has been supported in the main by the more recent calculations of Sundbom and co-workers [59, 60], Orti et al. [58], and Ohno et al. [57]. However, while these authors predict a similar stack of orbitals, they use more complicated mixing to generate the excited states.

Our absorption and MCD data of main group phthalocyanines, for example, of MgPc(-2) [61] and ZnPc(-2) [62], confirm the degenerate nature of the lowest four excited states from the measurement of MCD *A* terms. However, the MCD data do suggest that the *B* band region is much more complicated than the analogous band in the porphyrins (the Soret band), with two transitions closely overlapped. Fig. 2 shows absorption and MCD data (the dashed lines in the figure) for the neutral $(im)_2$MgPc(-2), which acts as a reference point for the spectral data of the ring-oxidized and ring-reduced phthalocyanines. The intense absorption and MCD *A*-term signal near 670 nm, followed by a "window" region at 500 nm, leading to a series of much broader, overlapping bands near 350 nm, is characteristic of MPc(-2) spectra in the absence of charge transfer between the metal and the ring [1]. It is worth noting that the significant differences between the optical spectra of neutral phthalocyanines and the comparable spectral data for porphyrins (using tetraphenylporphyrin and octaethylporphyrin as examples [38, 63, 64]) are also found in the spectral data of oxidized and reduced phthalocyanines [36, 45, 54].

ii. Formation of Ring-oxidized and Ring-Reduced Complexes

Phthalocyanine π cation radical complexes can be formed chemically with oxidizing agents such as Br_2 and HNO_3 [2, 65], electrochemically in

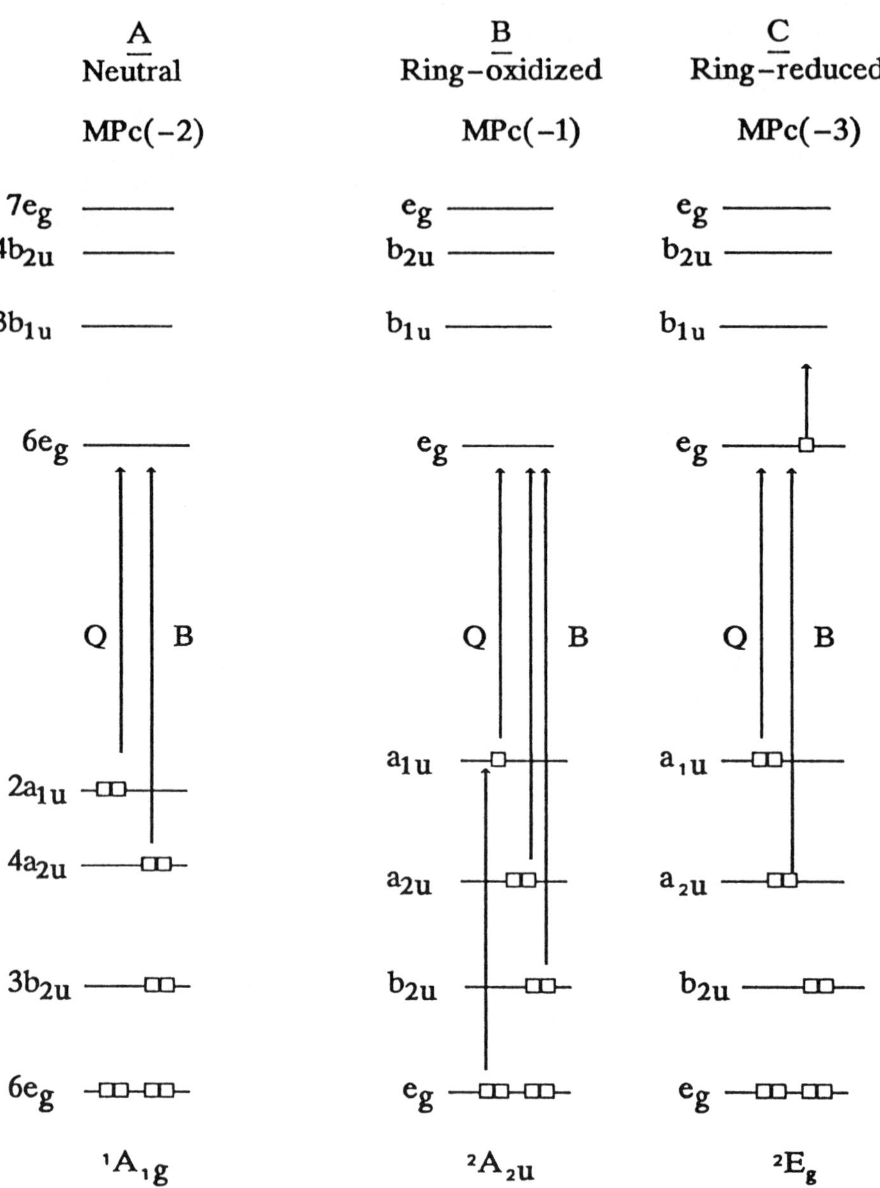

Figure 1. Molecular orbitals involved in the major absorption transitions with energies between 10,000 and 50,000 cm^{-1}. The order of the orbitals for the oxidized and neutral metallophthalocyanines adapted from Ohno et al. [57], Orti et al. [58], and Gouterman [56]. No account has been taken of changes in the energies of the molecular orbitals as a result of the different ring occupation.

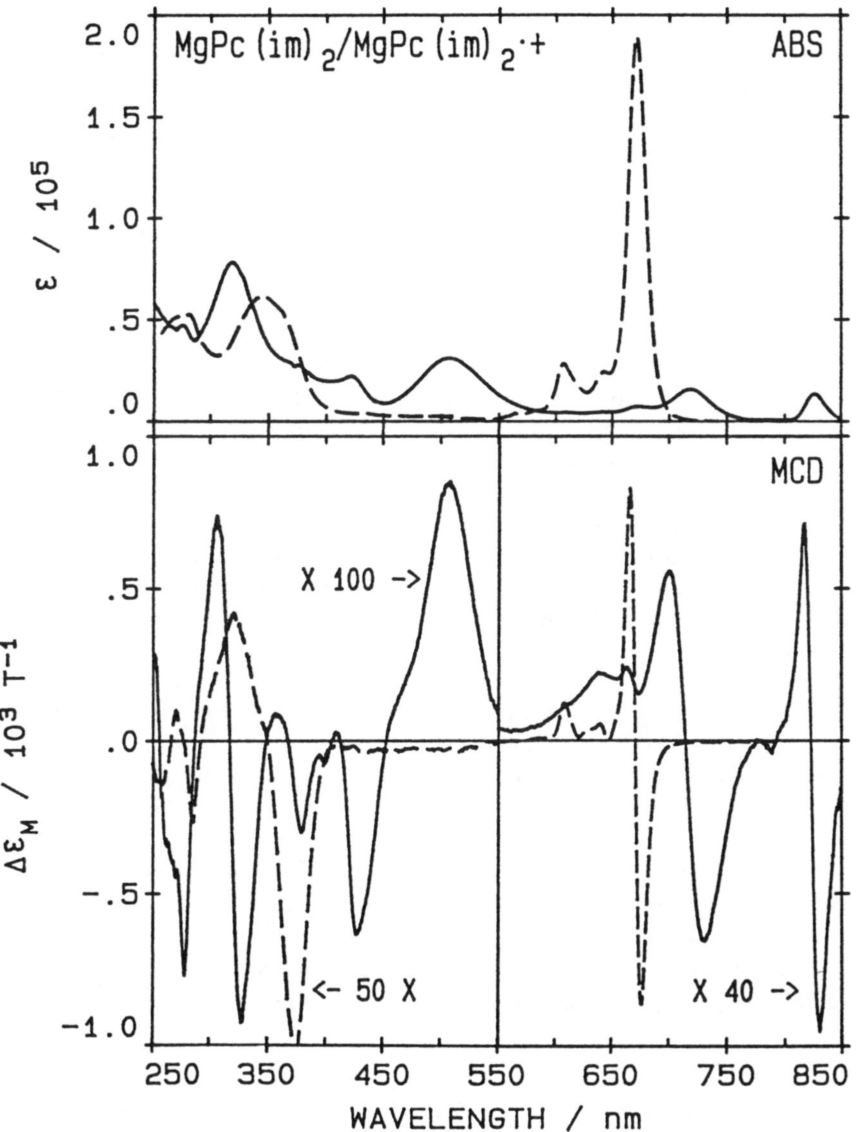

Figure 2. Absorption and MCD spectra of the neutral $(im)_2MgPc$ species (dashed line) and the one-electron oxidized $(im)_2MgPc(-1)$ species (solid line) formed by oxidation with HNO_3, measured at room temperature in methylene chloride. (Reproduced with permission from [61] and [2].)

uv-transparent solvents such as DMF and DCM [2, 7, 66-68], and photochemically in the presence of an appropriate electron acceptor in both room temperature solutions and frozen solutions at 77 K [2, 7]. Thin films can also be oxidized and reduced [22, 69]. With transition metal phthalocyanines, for example, Co(II)Pc, both metal and ring-oxidation can take place, the order depending on the relative reduction half potentials and the solvent and axial ligand. For both porphyrins, for example, CoTPP [42] and Co(II)OEP [36], as well as phthalocyanines, for example, Co(II)Pc, the MCD spectrum clearly identifies when oxidation takes place from the ring compared with metal.

The *Q* band in *monomeric* π cation radical species of MgPc(-1) and ZnPc(-1) has been assigned to the band near 825 nm based on the MCD *A*-term feature observed under the absorption band Fig. 2 (solid lines) [2]) in this region. At room temperature, in many solvents, π cation complexes exist in equilibrium between monomers and dimers, with two sets of bands overlapping throughout the uv-visible wavelength region.

The redox chemistry of the rare earth diphthalocyanines, such as $Lu(Pc)_2$, involves a quite beautiful series of complexes with oxidation states for the rings ranging from $[LuPc(-1)Pc(-1)]^{+1}$ to $[LuPc(-3)Pc(-3)]^{-3}$; see as examples the spectroelectrochemical studies of L'Her et al. [71] and Riou and Clarisse [69].

Reduction using sodium mirrors in tetrahydrofuran solutions has been used by several groups to obtain successive negatively charged MPc species, for example, [54, 72]. Electrochemical reduction using a mercury pool electrode with solutions of octacyano-substituted phthalocyanines in DMF [73] and photochemical reduction of MgPc [15, 16, 74] have also been described.

The absorption and MCD spectra of ring-reduced species do not resemble those of neutral phthalocyanines and have not been assigned to date.

A representative set of spectral band maxima from absorption and MCD spectra of ring-oxidized and ring-reduced complexes are listed in Table 2.

iii. Background to the Use of Magnetic Circular Dichroism Spectroscopy

It has long been shown that MCD spectra provide the additional assignment criteria necessary to assign the spectra of phthalocyanine and porphyrins [64]. In our laboratory, we use the analysis of MCD spectra, first to determine the number of bands actually present, then to estimate band center energies, and finally to identify polarizations of transitions, which allows some characterization of specific bands to particular state to state transitions to be carried out [2, 61, 62]. In this way we used the observation of *A* terms at 810 and 725 nm to assign the *Q* band of monomeric and dimeric radical cation species, respectively (see Fig. 2 for the absorption and MCD data (solid line) for MgPc(-1) [2]). A fairly detailed review of the use of MCD spectra in the study of neutral phthalocyanines appeared in Volume 1 of this series [1]. Examples of the use of MCD spectra in studies of neutral [75, 76] and oxidized porphyrins [36], as well as heme proteins [75, 77], generally illustrate the value of the technique in providing detailed ligand coordination and oxidation state information.

The MCD spectral intensity arises from a combination of factors. In the presence of a magnetic field, transitions between a ground and excited state will differentially absorb left or right circularly polarized light if either of the states are orbitally-degenerate or the field can induce mixing between close-lying excited states. An MCD spectrum will then be measured in the same spectral region as the absorption spectrum because the same transitions result in the absorption spectrum. First, and most important in spectroscopic studies of the phthalocyanines, it is assumed that the band center and band width of MCD bands are the same as the corresponding band in the absorption spectrum [78]. Second, we must consider molecules with a non-orbitally-degenerate ground state separately in analysis from molecules with orbitally degenerate ground states. Recognizing that spin-orbit coupling can complicate this division, we can generally state that with a diamagnetic metal, the ground state of the dianion is nondegenerate. Under these conditions, the MCD spectrum will arise from transitions to degenerate excited states, with a temperature-independent spectral feature called a Faraday "*A*" term, and to nondegenerate states, which are coupled to other states under the influence of the magnetic field intensity, to give a temperature-independent Faraday "*B*" term. The spectral intensity of these bands is essentially independent of temperature. If the ground state is orbitally degenerate, then the MCD spectrum will be dominated by Faraday "*C*" terms, bands that will increase in intensity as the temperature decreases. The shapes of the *A*, *B*, and *C* terms are shown in Fig. 3. A full theoretical description of the origins of MCD spectral features has been given by Piepho and Schatz [78].

For MCD spectral studies of phthalocyanine cations and anions, the first step is to measure room temperature spectra. With the ground-state degeneracy known (see Tables 3-5) band deconvolution analysis [2, 61, 62, 79] can be used to determine the number and type of MCD bands present. Associating spectral features in the MCD spectrum with bands in the absorption spectrum is a necessary part of the spectral analysis. Apparent *A* terms frequently turn out to be a pair of overlapping, oppositely signed *B* terms. Accurately measured, digital absorption and MCD spectral data are

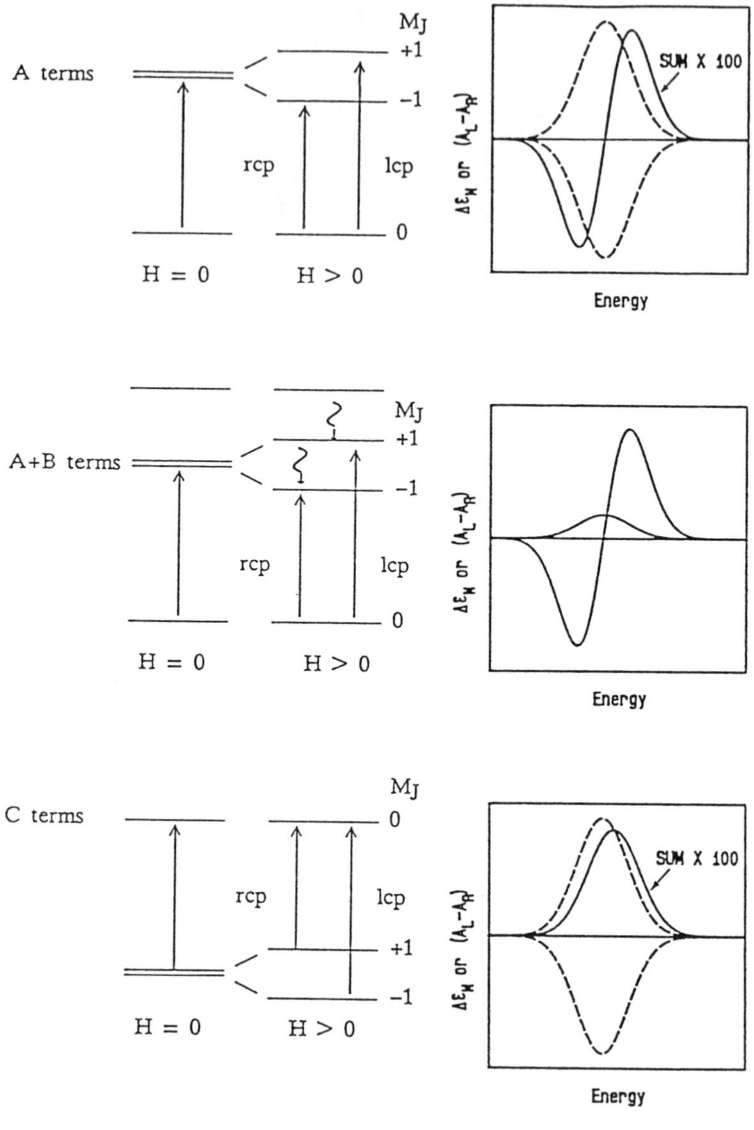

Figure 3. Origin of signal intensity in the MCD spectrum, the Faraday *A*, *B* and *C* terms. The Faraday *A* term is only observed as a completely symmetric derivative signal when *B* term intensity is low. Because *B* term intensity arises from magnetic-field-induced coupling between close-lying states, symmetric *A* terms are observed for transitions to degenerate excited states that are well separated from other states [78].

necessary in order to analyze an MCD spectrum. Examples of the use of extensive deconvolution analysis of both neutral and oxidized phthalocyanines have been published by our group, use of moments analysis on isolated bands is also valuable in extracting theoretically useful parameters, but the bands must be well isolated if the analysis is to be accurate [80]. Finally, Thomson's group at the University of East Anglia has pioneered the use of magnetization curves in the analysis of spectral data from molecules with degenerate ground states (see as an example [81]).

MCD spectra at room temperature are obtained within high magnetic field intensity, usually, but not restricted to, a field of 4-7 tesla from a superconducting magnet. The magnet is located within a circular dichroism spectrometer which records the differential absorption of left or right circularly polarized light as a function of wavelength. Low-temperature MCD spectra must be measured from strain-free glasses, crystals or doped polymers. A temperature range of at least 50 K is required in order to extract ground state degeneracy data, and it is preferable to be able to cool to below 4 K so that the onset of field-induced saturation can be measured [81].

iv. Reviews of Phthalocyanine Spectral Data

Our discussion of the spectral data of phthalocyanine complexes based on the Pc(-2) dianion [1] summarized some of the more comprehensive reviews of the spectral properties of the phthalocyanines and describes a range of absorption and MCD spectral features for complexes of the dianionic π ring, MPc(-2). To our knowledge, the only wide-ranging reviews of radical cation and anion spectral data for the phthalocyanines have been the survey of spectral properties of radicals discussed by Minor, Gouterman, and Lever [82] and the data contained in an extensive table published in Russian by Luk'yanets [83]. Several papers describe the spectral data of ring-oxidized cationic [7, 13, 65, 84-86] or ring-reduced anionic [54, 55, 68] phthalocyanine species in detail. As mentioned above, the papers by Dodd and Hush [54] and Clack and Yandle [55] compile considerable spectral data for both radical anions and cations.

Molecular-orbital calculations have been reported specifically for porphyrin π cation radical species by Edwards and Zerner [87], but only recently have calculations been described for radical cations of the phthalocyanines, Orti et al. [58] have reported results from a valence effective Hamiltonian calculation for LiPc(-1).

B. ABSORPTION AND MCD SPECTRA OF CATION RADICAL COMPLEXES

Oxidation or reduction of metalloporphyrin and metallophthalocyanines can take place at either the metal or the ring [2, 33, 34, 36, 54, 66, 88-92]. In this review we will only discuss optical data for redox reactions of the ring. Extensive spectral analyses of optical data of the radical cations have been reported by our group for MgPc(-1), RuPc(-1) and ZnPc(-1), by Lever's group for ring-substituted CoXPc complexes and by Homborg et al. for a range of phthalocyanine radical cations, including MgPc(-1), HPc(-1), CoPc, and FePc(-1) [2, 7, 62, 65, 84, 85, 92-94].

Ring-oxidation yields initially the cation radical species, $[MPc(-1)]^{\cdot +}$. This species is paramagnetic unless coupling between the radical on the ring and a paramagnetic metal, or dimerization, which occurs even at very low concentrations, yields the diamagnetic complex. Fig. 2 shows typical absorption and MCD spectra of the product formed following one-electron oxidation of $(im)_2MgPc(-2)$ to $[MgPc(-1)]^{+\cdot}$ [2]. The significant features are: (1) two or more weak bands to the red of the *Q* band, (2) a broad band centered on 510 nm that fills the "window" region of neutral MPc species, and series of overlapping bands below 380 nm. Absorption, MCD, and EPR spectral data for MgPc(-1) and ZnPc(-1) [2, 7, 65] indicate that phthalocyanine cation radical species partially dimerize in solutions at room temperature, with complete dimerization to a diamagnetic species occurring at low temperatures [2] or in polar solvents, such as ethanol [65].

Dimerization is also observed for porphyrin π cation radical species, for example, ZnOEP(-1) [95]. Homborg [65] describes the differences in the monomer/dimer equilibrium between nonpolar (monomer) and polar (dimer) solvents, from the spectral properties of HPc(-1), MgPc(-1), and CuPc(-1). Manivannan et al. [67] report that with bulky substituents dimerization is hindered, and EPR signals are recorded for radical cations of polynuclear ZnPc species even at 77 K.

Considerable spectral data have been reported for neutral phthalocyanines as thin films on quartz disks [1, 96]. While under these conditions temperature dependence of the spectral data can be studied readily, the observed absorption envelope clearly comprises overlap of extremely broad bands [1]. Our analysis of the absorption and MCD spectra of a number of neutral MPc complexes [96] has suggested that extensive Davydov or exciton coupling both broadens and splits the degenerate excited states that are responsible for the main phthalocyanine absorption spectrum [1,96]. Myers et al. [90] have demonstrated that oxidation of thin films of

FePc and CoPc leads to the π cation radical species.

C. MAIN GROUP COMPLEXES

i. HPc

Oxidation of HPc in DCM or CH_3NO_2 results in monomeric and dimeric radical cations, respectively (Fig. 4) [65], which exhibit absorption spectra that closely resemble those of MgPc(-1) (Fig. 2). Homborg has shown that the monomer-dimer equilibrium is solvent dependent. The monomer formed in DCM (upper trace) is characterized by a strong band at 888 nm and a weak band at 775 nm. While the dimer forms in nitromethane (lower trace) is characterized by a strong absorption at 1020 and 760 nm.

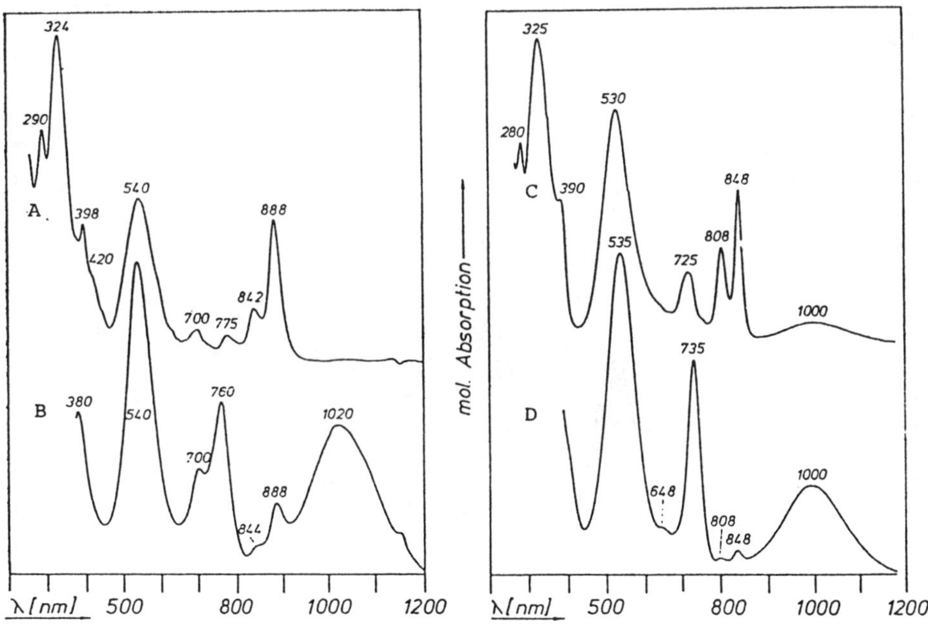

Figure 4. Absorption spectra of (a) HPc(-1) in CH_2Cl_2, (b) HPc(-1) in CH_3NO_2, (c) $CuNO_3Pc$(-1) in CH_2Cl_2, and (d) $CuNO_3Pc$(-1) in CH_3NO_2. (Reproduced with permission from [65].)

ii. LiPc

Homborg and Kalz [86] report the absorption spectrum of LiPc(-1) in 1-chloronaphthalene (with bands at 423, 481, 612, 682, 816, and 944 nm) and also in a KBr pellet at 10 K (Fig. 5). As the band at 682 nm is not coincident with the the Q band of Li_2-Pc(-2), at 667 nm in DMA [97], it is probable that the 682 nm band is a cation radical dimer band (although somewhat blue shifted from the wavelength observed for MgPc(-1) where the 712 nm band is from the monomeric radical [2]). The 481 nm band is clearly the nondegenerate π-π band found for other radical cation species. Homborg and Teske [98] also have reported the spectra of complexes of LiPc(-1) in Nujol mulls. These spectral data show clearly that both monomeric and dimeric complexes also form in the solid state, with the monomer characterized by a strong absorption at 806 nm and the dimer characterized by a strong band at 690 nm.

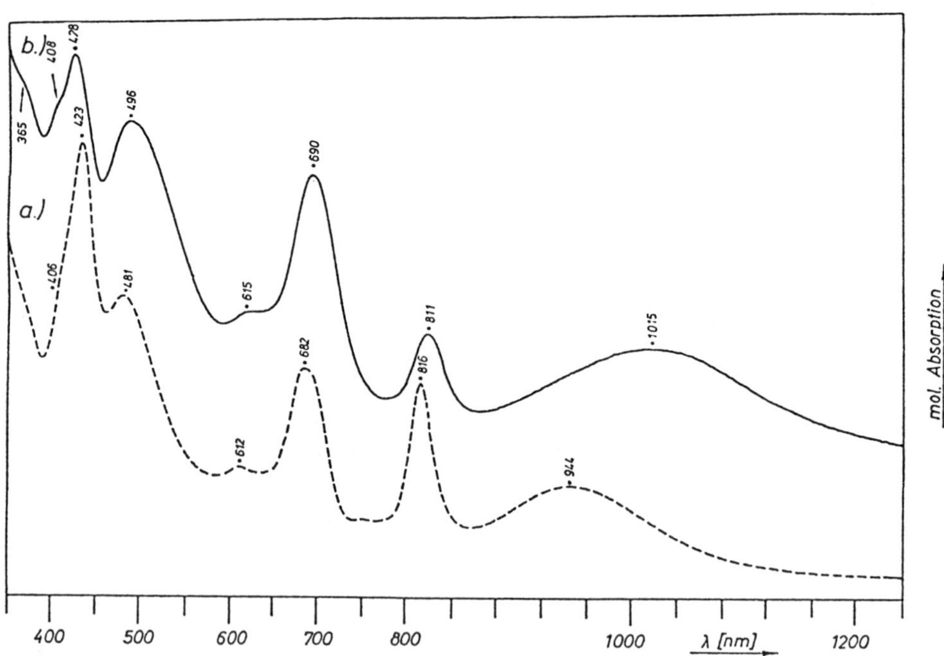

Figure 5. Absorption spectra of (a) LiPc(-1) in chloronaphthalene and (b) LiPc(-1) in a KBr pellet at 10 K. (Reproduced with permission from [86].)

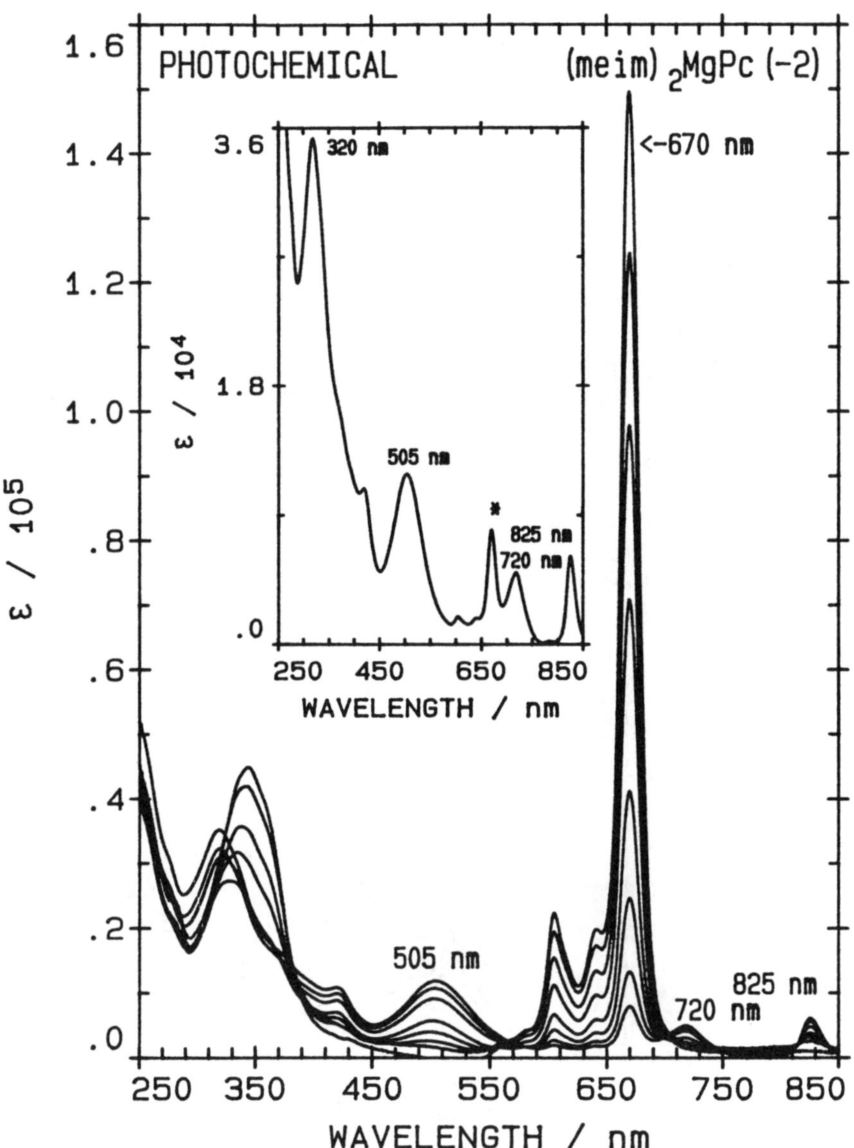

Figure 6. Absorption spectra recorded during the photochemical oxidation of $(meim)_2MgPc(-2)$ to the cation radical species, $(meim)_2MgPc(-1)$. The inset shows the final spectrum (note the small amount of remaining neutral compound marked by *). (Reproduced with permission from [2].)

iii. MgPc

Ough et al. [2] describe the photochemical, electrochemical and chemical formation of MgPc(-1). Absorption and MCD spectra at room temperature and 200 K, show that at room temperature in DCM, both monomers and dimers are present. At low temperatures, the spectra are dominated by the dimer, whereas at high temperatures in dichloroethane the monomer predominates [2]. Because the sequence of spectra reported in this paper generally apply to other radical cation complexes, we reproduce a series of figures here to illustrate the analysis. Fig. 6 shows the series of absorption spectra recorded during photolysis of $(meimid)_2MgPc(-2)$ in DCM, using CBr_4 as the electron acceptor as in our previous studies of the photo-oxidation of phthalocyanines and porphyrins [6, 7, 42]. New bands appear isosbestically at 320, 505, 720, and 825 nm. Although there is

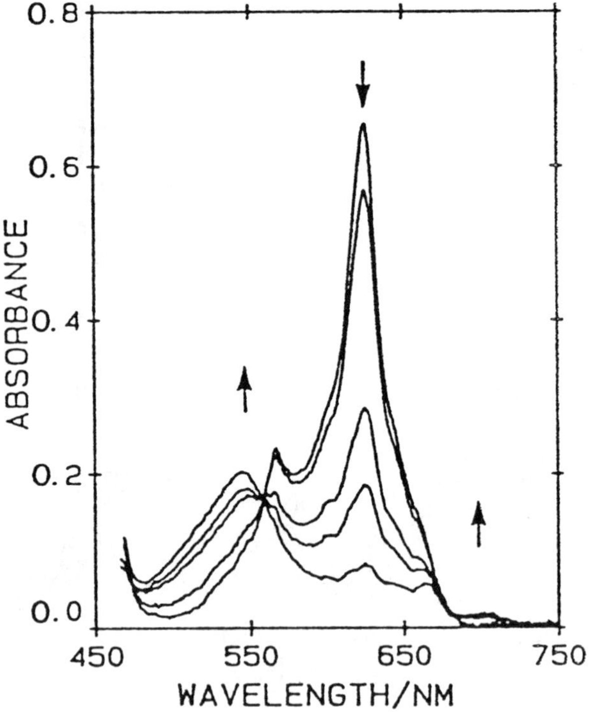

Figure 7. Absorption spectra recorded during the photochemical oxidation of $(py)_2RuPc(-2)$ in methyltetrahydrofuran to the cation radical species, $(py)_2RuPc(-1)$ at 77 K. The electron acceptor was dichlorodicyanoquinone. (Reproduced with permission from [6].)

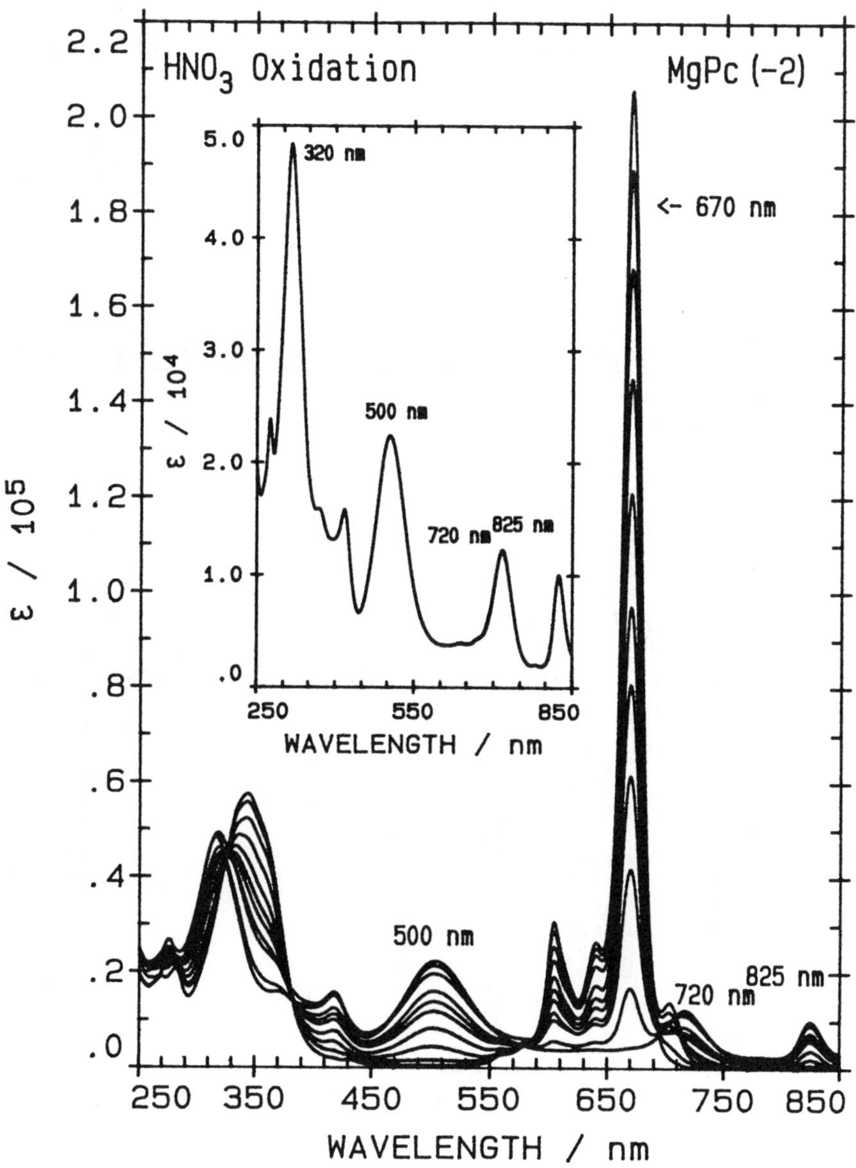

Figure 8. Absorption spectra recorded during the oxidation of $(meim)_2MgPc(-2)$ to the cation radical species, $(meim)_2MgPc(-1)$ with HNO_3. The inset shows the final spectrum. (Reproduced with permission from [2].)

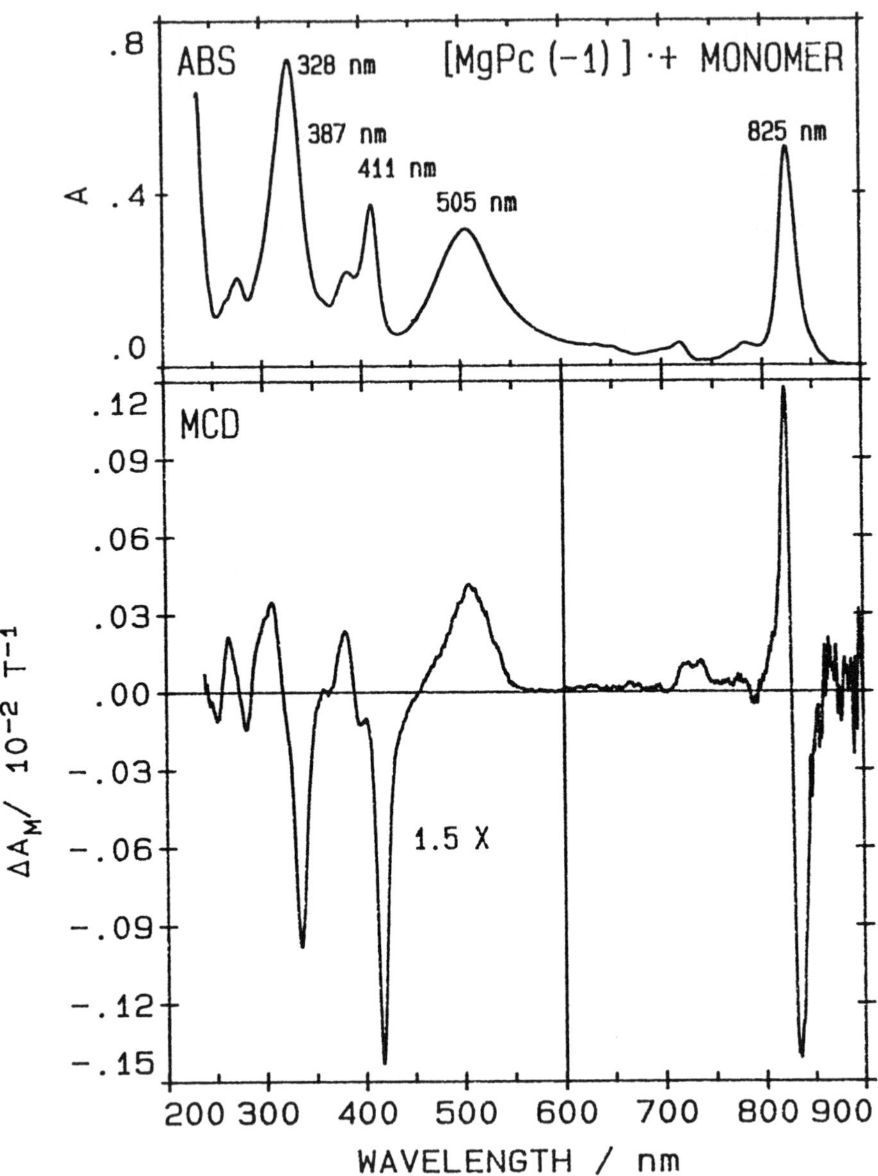

Figure 9. The absorption and MCD spectra of the monomeric, one-electron oxidized $(im)_2MgPc(-1)$ species. (Reproduced with permission from [2].)

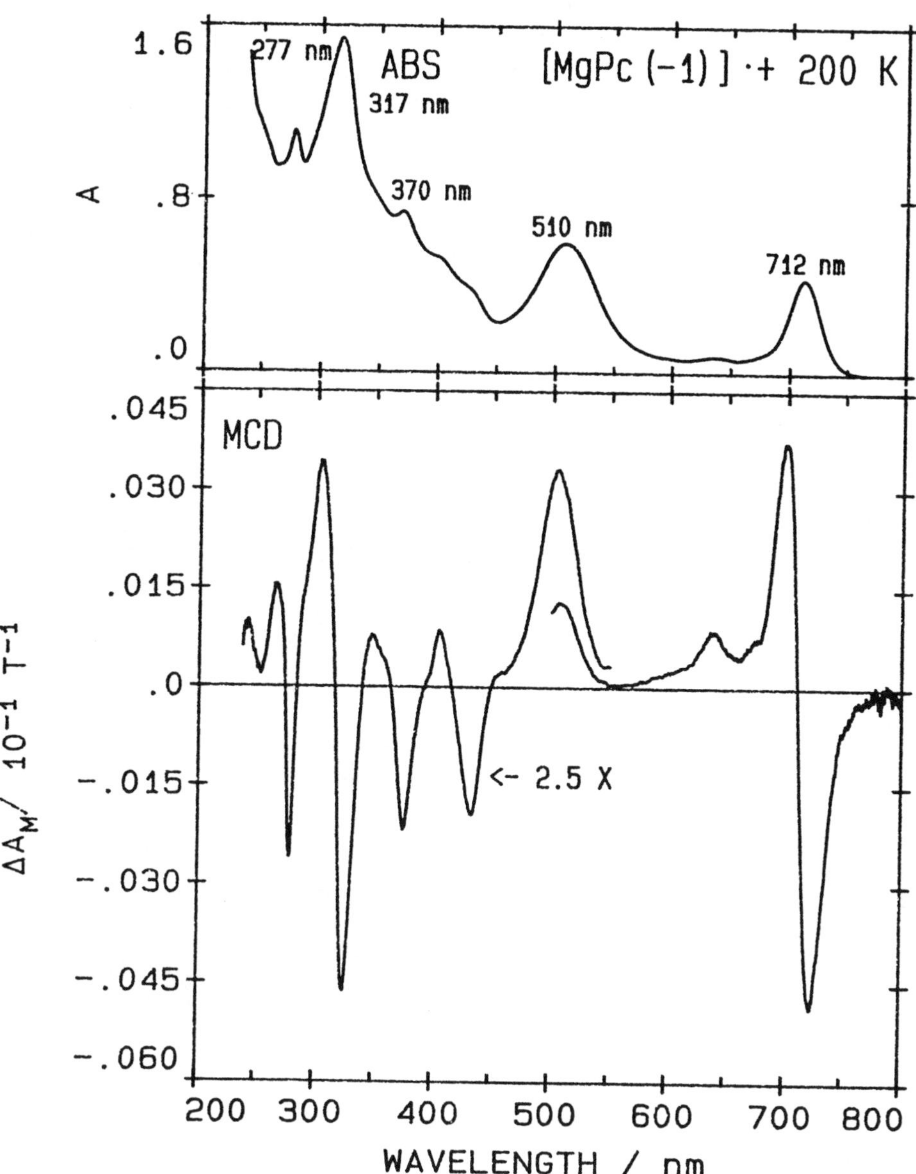

Figure 10. The absorption and MCD spectra of the dimeric, one-electron oxidized $(im)_2MgPc(-1)$ species recorded at 200 K. (Reproduced with permission from [2].)

little evidence of side reactions, it is difficult to carry the photooxidation to completion at room temperature, although at 77 K, we have been able to oxidize RuPc(-2) more fully, Fig. 7 [6]. Chemical oxidation using HNO_3 successfully oxidizes MgPc(-2) quantitatively to the radical cation, Fig. 8. The isosbestic points are sharper, and there is no indication of a band at 670 nm from the neutral species, although the radical exhibits similar band center energies. Fig. 2 compares the absorption and MCD spectra, measured at room temperature, of the neutral and oxidized $(im)_2$MgPc species. This figure emphasizes the major changes in the spectrum that take place in the visible region. The two *A* terms, at 825 and 720 nm, and the *B* term at 505 nm, characterize room temperature, solution spectral data of radical cation phthalocyanines.

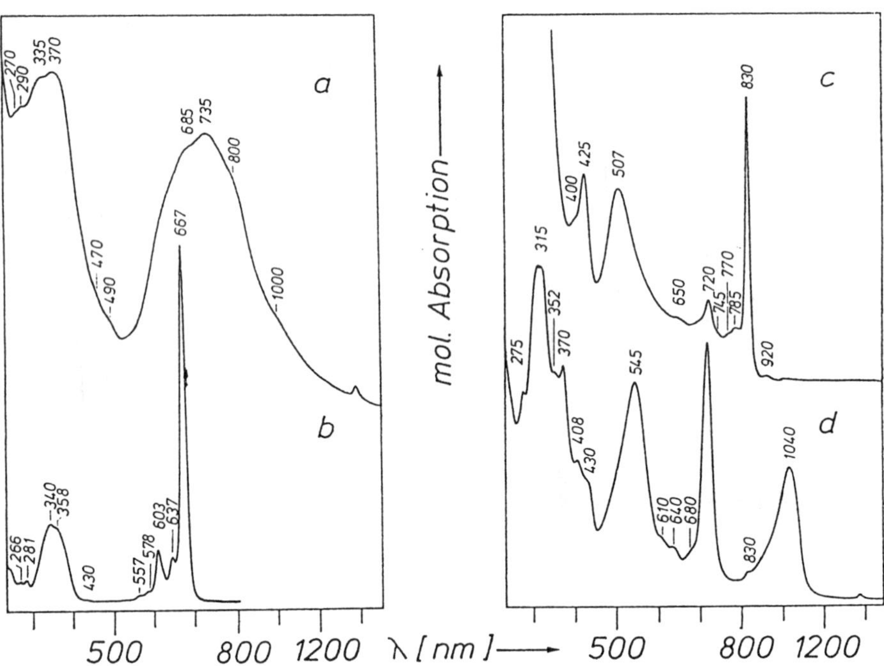

Figure 11. Absorption spectra of (a) MgPc(-2).HCl in THF, (b) MgPc(-2).$2H_2O$ in THF, (c) MgClPc(-1) in chloronaphthalene, and (d) MgClPc(-1) in CH_3CH_2OH. (Reproduced with permission from [65].)

However, the room temperature solution spectra comprise a mixture of monomers and dimers. Fig. 9 shows the absorption and MCD spectra of pure *monomeric* MgPc(-1). The *A* term at 825 nm is assigned as the *Q* band. Fig. 10 shows the spectrum of the *dimeric* species measured at 200 K. The *Q* band and its associated vibronic structure is now observed at 712 nm. Interestingly, the *B* band at 510 nm is largely unaffected by dimerization. Homborg [65] describes the spectra of monomeric and dimeric MgPc(-1) in 1-chloronaphthalene and ethanol solutions. The band at 830 nm is assigned as the monomer *Q* band, and bands at 720 and 1040 nm are assigned to the effects of dimerization (Fig. 11). Homborg [65] has proposed a scheme to account for the effects of dimerization on the magnetic and spectral properties (Fig. 12). While monomeric π cation radical species exhibit EPR signals, the dimer does not.

Band deconvolution of the absorption and MCD spectra of the monomeric and dimeric species provides a quantitative estimate of the energies of the band centers. Fig. 13 shows the results of these calculations for the monomeric MgPc(-1) carried out by Ough et al. [2]. Significantly, the fitting requires 6 *A* terms, which can be associated with 6 degenerate excited states. Fits of the spectral data for the dimeric species yield a similar series

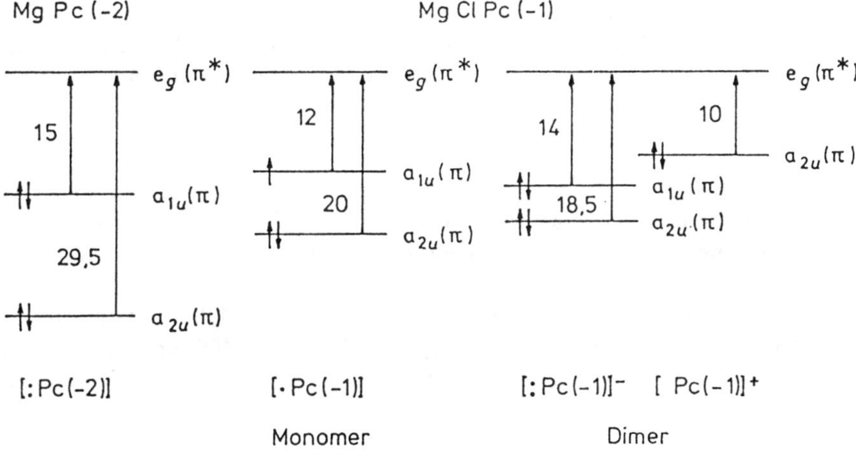

Figure 12. Schematic interpretation of the origins of the lowest energy absorption transitions in neutral, monomeric MgPc(-1) and dimeric $[MgPc(-1)]_2$ as proposed by Homborg et al. (Reproduced with permission from [65].)

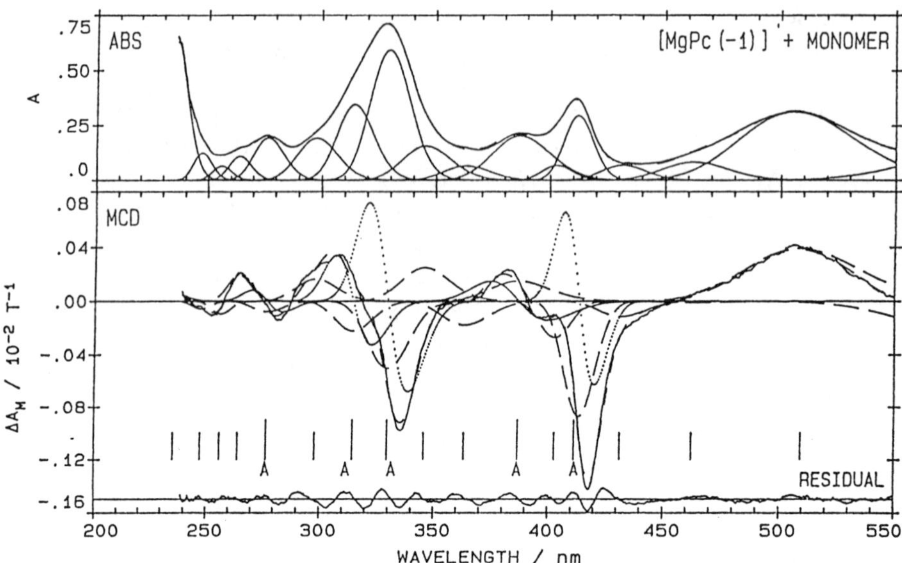

Figure 13. Deconvolution analysis of the absorption and MCD spectra of the monomeric, one-electron oxidized $(im)_2MgPc(-1)$ species in the 200-550 nm region. The five transitions to degenerate excited states are marked with an "A", and are indicated by the *A*-term signal drawn as a dotted line. The band near 500 nm is completely accounted for by a single *B* term, indicating that there is no degeneracy in this excited state. (Reproduced with permission from [2].)

of degenerate states [2]. The energies of the degenerate states, plus the 505 nm nondegenerate state, are shown in Fig. 14. This figure illustrates how there is a general trend to lower energies for the *Q* and *B* bands compared with the neutral MgPc(-2).

Ohno et al. [4] report formation of AlPc(-1) in DMA, with bands (s, strong; w, weak) at 420 (s), 540 (s), 730 (w), and 845 nm (s), following photoinduced, oxidative electron transfer to 2,5-dichloro-1,4-benzoquinone.

D. TRANSITION METAL COMPLEXES

Myers et al. [90] have reported spectral data for oxidized thin films of CrPc, MnPc, FePc, CoPc, and ZnPc. Except for the appearance of a broad band near 550 nm, these radical cation spectra do not closely resemble data from complexes in solution.

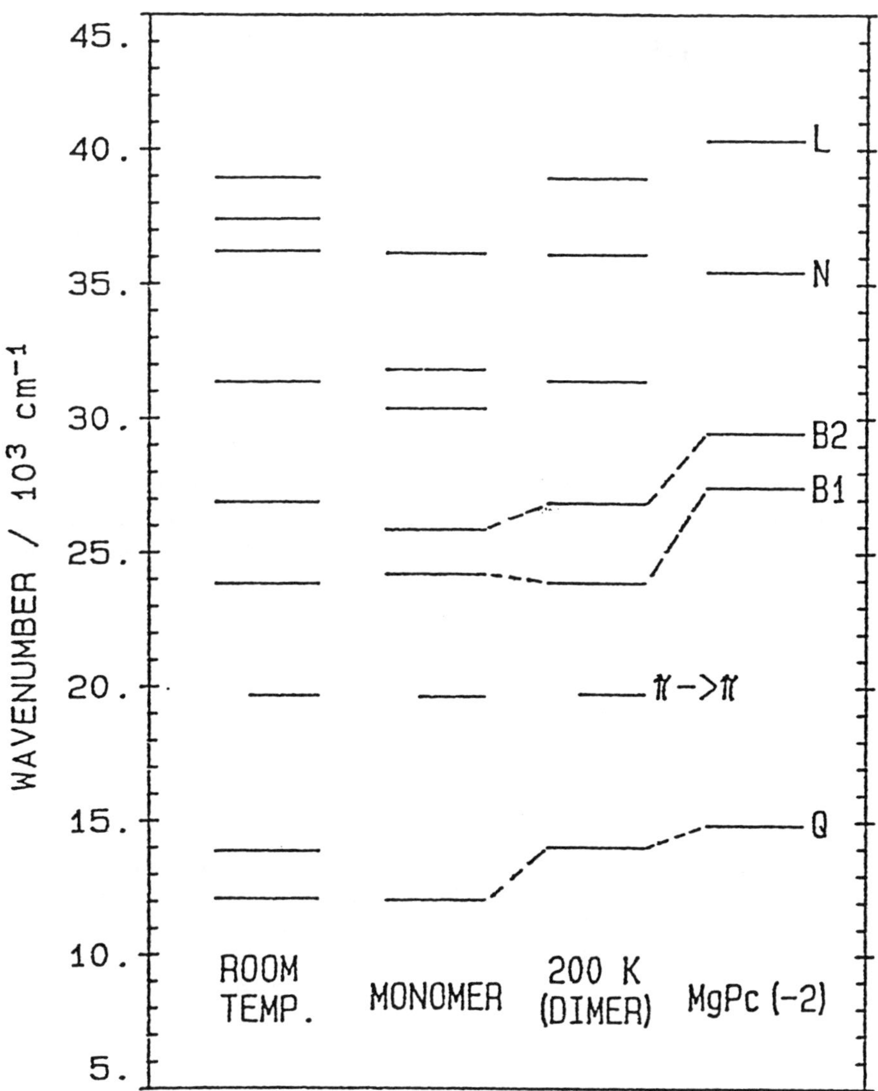

Figure 14. An energy-level diagram constructed from the deconvolution of the absorption and MCD spectra of the neutral MgPc(-2), and the monomeric and dimeric forms of the cation radical MgPc(-1) species. The energies shown are taken from the transition energies calculated by the deconvolution program, rather than from the band maxima in the spectra. (Reproduced with permission from [2].)

i. FePc

Kalz et al. [84] report spectral data for a range of $(L)_2Fe(III)Pc(-2)$ complexes. When $(CN)_2Fe(III)Pc(-2)$ reacts with CF_3COOH in DCM, the π cation radical $(CF_3COO)(CN)Fe(III)Pc(-1)$ forms, with band maxima at 320, 403, 520, and 795 nm. The 520 nm band is typical of π cation radicals, and the spectral data are displayed as Fig. 40(E) in Volume I [1]. Fe(II)Pc(-2) reacts electrochemically to form first Fe(III)Pc(-2), then Fe(III)Pc(-1) [99] (Fig. 15). Bakshi et al. [100] describe Mossbauer data, but not optical data, for Fe(IV)Pc(-1)-C-(Fe(IV)Pc(-1), a species that can be considered as a model for the heme group in compound I of horseradish peroxidase [36, 49, 52].

ii. CoPc

Absorption spectra for $(X)_2Co(III)Pc(-1)$, X = Cl^- or Br^-, have been reported by Homborg and Kalz [85]. The absence of a band near 825 nm, and the strong band near 720 nm, suggests that this species is almost completely dimerized in solution. The difference between the spectra of these two complexes was explained in terms of X $\rightarrow$ Co charge transfer.

iii. CoTNPc

Nevin et al. [93] and Bernstein and Lever [94] have reported spectral data for the oxidation of the peripherally-substituted Co(II) tetraneopentoxy-phthalocyanine, Co(II)TNPc(-2), electrochemically, by Cl_2 gas and by $SOCl_2$, to form Co(III)TNPc(-1). The strong 710 nm region bands and lack of a band at 820 nm in the spectrum of this species (when oxidized by Cl_2, band maxima are located at 366, 404, 540, 680, and 744 nm) suggests that the radical is highly aggregated.

iv. CuPc

Homborg [65] has reported spectral data for the monomeric and dimeric forms of CuPc(-1) (Fig. 4). Unlike the MgPc(-1) described above, the monomer is characterized by bands at 808 and 848 nm; however, the dimer exhibits the more usual narrow 735 nm and broad 1000 nm set.

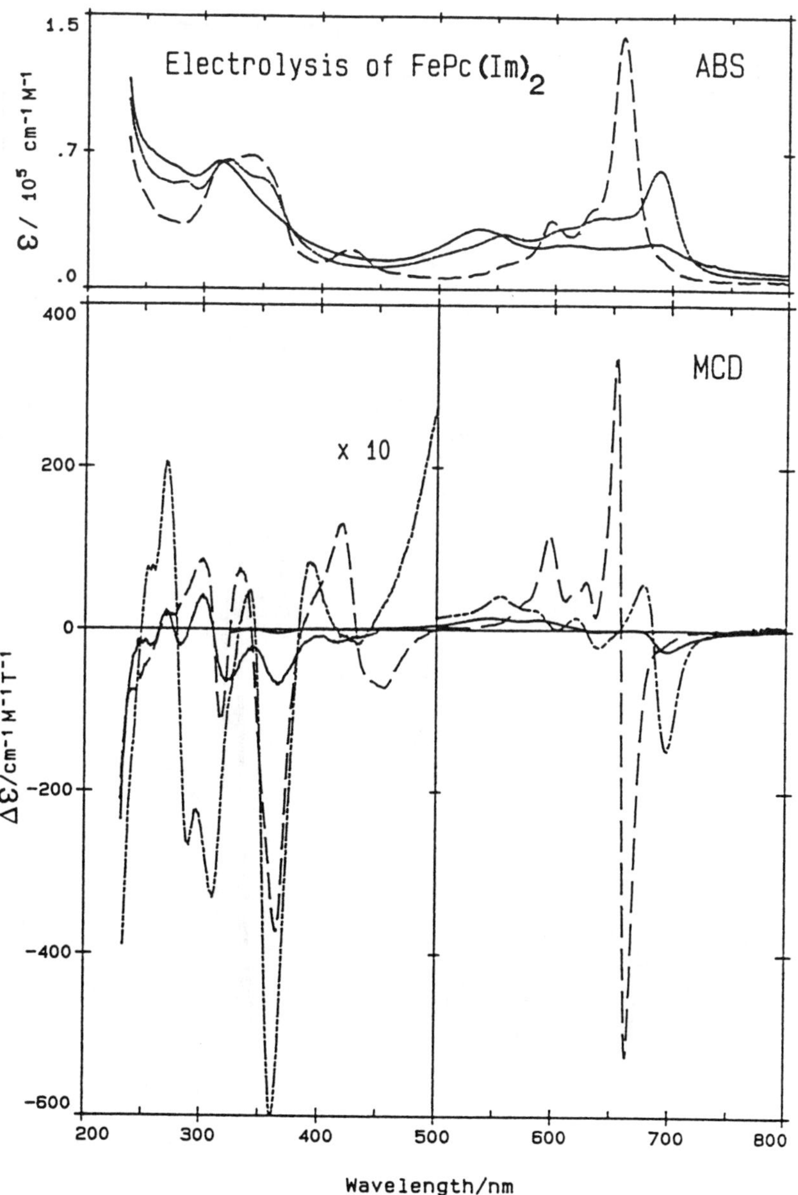

Figure 15. Absorption and MCD spectra recorded during the electrochemical oxidation of Fe(II)Pc (- - -) in CH_2Cl_2. The neutral complex (- - -) is oxidized first to the Fe(III)Pc(-2) species (-- - --), which is then further oxidized to the π-cation radical species (——). (Reproduced with permission from [99].)

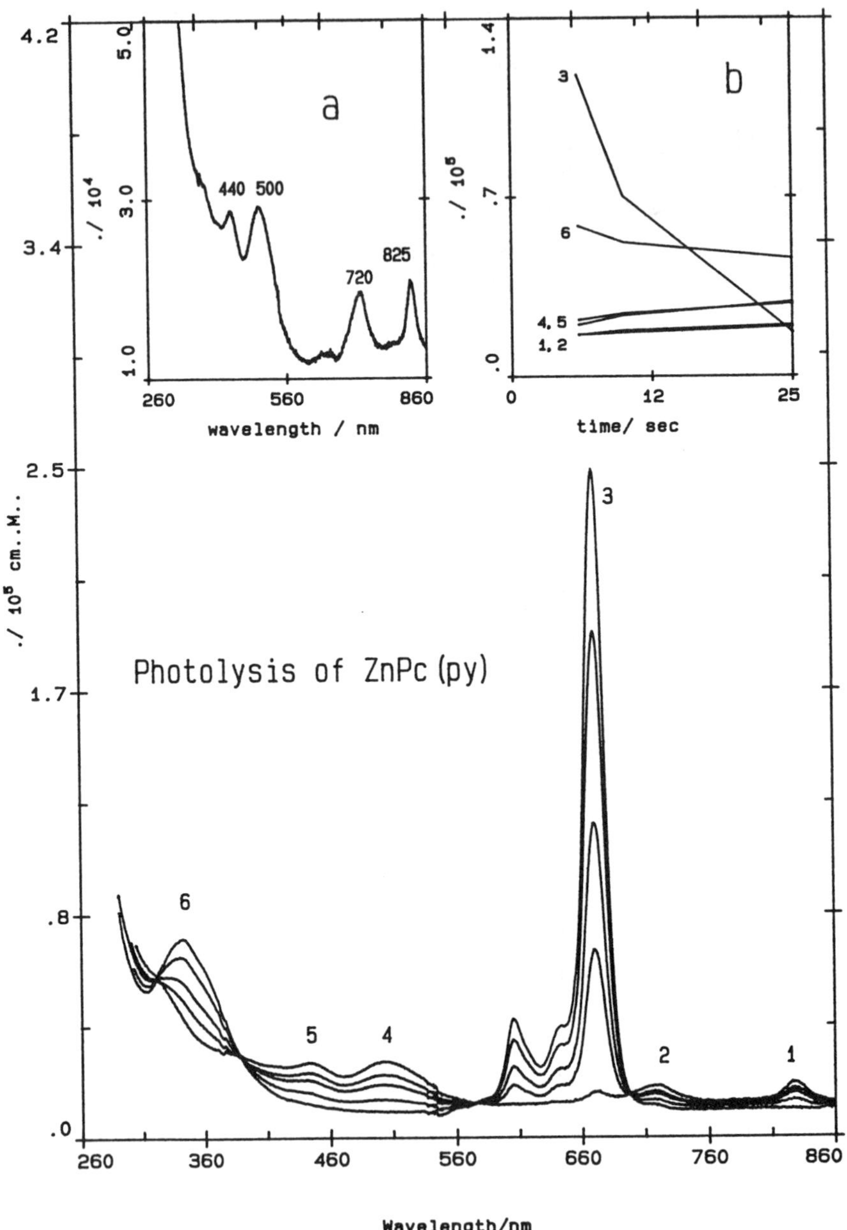

Figure 16. Absorption spectra recorded during the photochemical oxidation of (py)ZnPc(-2) in CH_2Cl_2 to the π-cation radical species. (Reproduced with permission from [7].)

v. ZnPc

ZnPc(-1) can be prepared electrochemically in DCM at a controlled potential of +0.71 V versus SCE [7, 62] and photochemically at room temperature in DCM, using visible region light, with CBr_4 as an electron acceptor. The quantum yield for photooxidation was less than 1%. Band maxima are observed at 440, 500, 720, and 825 nm following photolysis (Fig. 16). Deconvolution analysis of the visible and uv region absorption and MCD spectra provides more accurate band center information. Transitions to degenerate states are located at: 278, 317, 370, 424, 714, and 822 nm [62]. From our more recent studies, we now know that the band at 822 nm arises from the monomeric ZnPc(-1), while the band at 714 nm arises from the dimeric $[ZnPc(-1)]_2$ [2]. Manivannan et al. [67] have described the oxidation of zinc tetraneopentoxyphthalocyanines (ZnTNPc) as monomeric and tetrameric units. Like the unsubstituted ZnPc, oxidation results in the appearance of a strong band near 510 nm, and the spectra for the tetrameric $[ZnTNPc]_4$ shown in Fig. 17 have band maxima at 336, 388, 514, 660, 730, 860, and 1080 nm. From the analysis of spectrum of MgPc(-1) [2], we can interpret the 730 nm band as the *Q* from dimeric species, and the 860 nm band as the *Q* from monomeric species. It is interesting to consider the origin of this dimeric band in the tetrameric complex.

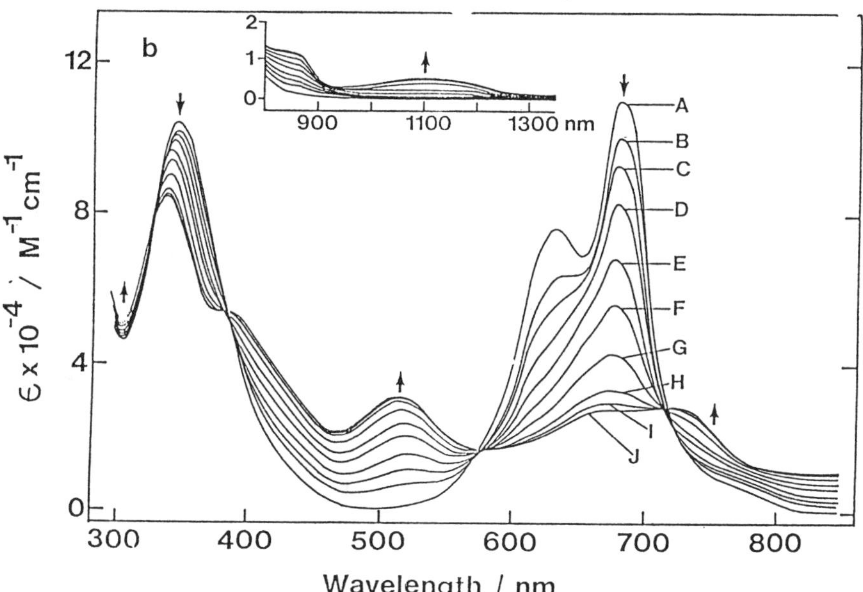

Figure 17. Absorption spectra of the tetramer $[ZnTNPc(-1)]_4$. (Reproduced with permission from [67].)

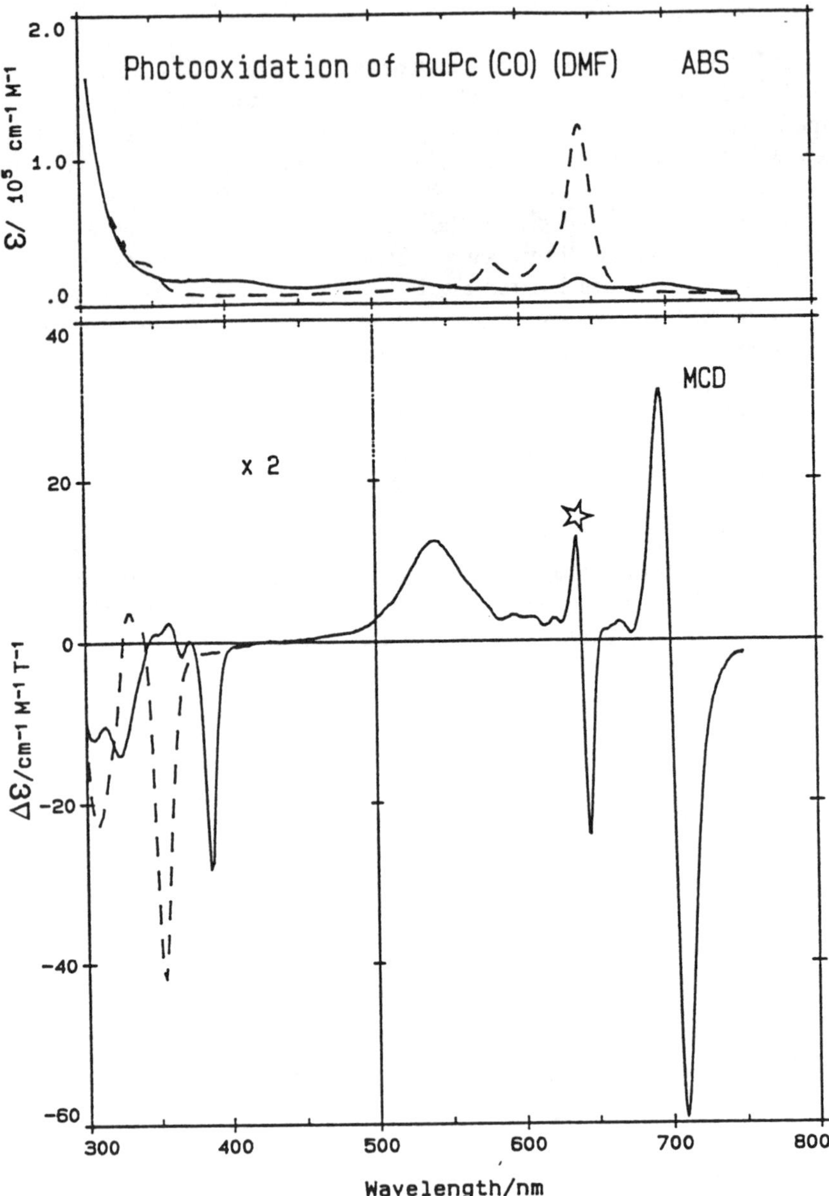

Figure 18. Absorption and MCD spectra recorded during the photochemical oxidation of (DMF)(CO)Ru(II)Pc in CH_2Cl_2 containing 0.01 mol L^{-1} CBr_4 as the electron acceptor. The spectrum of the neutral complex (dashed line) complex is replaced by the spectrum of the oxidized, RuPc(-1), solid line, which is characterized by the presence of the 510 and 710 nm bands. (Reproduced with permission from [6].)

Absorption spectra measured after the electrochemical oxidation of thin films of α-phase ZnPc on In_2O_3-coated glass slides follow the same trends as oxidation in solution, with new bands near 520 and 720 nm [101].

vi. RuPc

Electrochemical oxidation at +0.92 V of (CO)(py)Ru(II)Pc(-2) in DCM yields the π cation radical species [102], which is characterized by bands at 280, 381, 529, and 703 nm. EPR signals at g=2.0006 measured at 77 K suggested that the metal was not oxidized. Nyokong et al. [6] have reported absorption and MCD spectra for RuPc(-1) formed following the photoinduced oxidative electron transfer to CBr_4 in solution at room temperature (Fig. 18) and to 2,3-dichloro-5,6-dicyanobenzoquinone in a frozen glass at 79 K (Fig. 7). At room temperature, in DCM, DMF, and DMSO solutions, the significant absorption at 710 nm indicates a high concentration of dimer is present. At 79 K, on the other hand, this band is absent, which we suggest indicates that when the radical cation is formed within the frozen glass, dimerization demonstrated for MgPc(-1) at low temperatures [2] is inhibited and only the monomeric radical is formed.

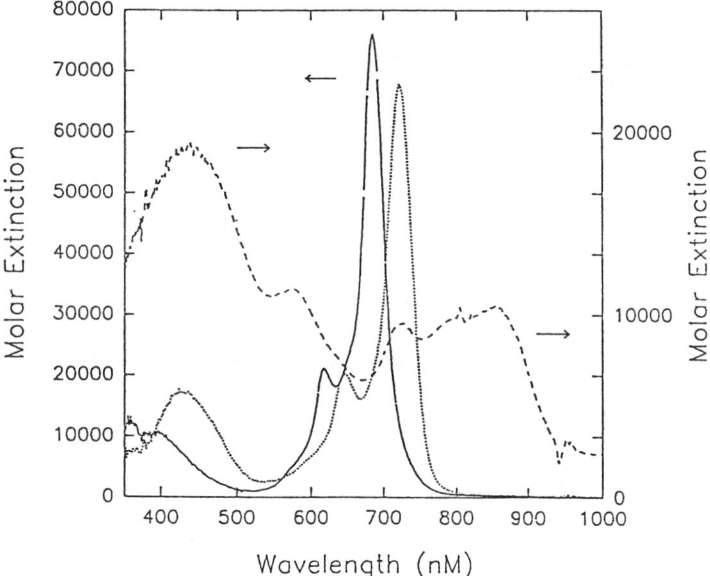

Figure 19. Absorption spectra of Ag(II)TNPc (solid line) and the electrochemically generated $[Ag(III)TNPc(-2)]^+$ (dotted line) and $[Ag(III)Pc(-1)]^{2+}$ in dichlorobenzene solutions. (Reproduced with permission from [66].)

vii. RhPc

Ferraudi et al. [10] report the formation and spectral properties of Rh(III)Pc(-1). The absorption spectrum is characterized by bands at 530 and 710 nm in CH_3CN.

viii. AgTNPc

Fu et al. [66] have reported spectroelectrochemical data for silver tetraneopentoxyphthalocyanine, Ag(II)TNPc(-2), a readily soluble phthalocyanine species. The two-electron oxidized complex is assigned as a Ag(III)TNPc(-1) π cation radical, based on the absorption spectrum (Fig. 19), which exhibits broad bands with maxima at 439, 574, 649, 721, and 854 nm. These broad bands, and the presence of the 721 nm band, suggests that the radical undergoes significant dimerization. The 854 nm band is considerably broader than observed for MgPc(-1) (Fig. 2 earlier), which might also indicate that charge-transfer bands may lie in this region.

E. LANTHANIDE AND ACTINIDE COMPLEXES

Metallation with lanthanide and actinide elements often results in the formation of diphthalocyanine complexes, $M(Pc)_2$, for example, the x-ray structure of $Lu(Pc)_2$ and $Nd(Pc)_2$ [103, 104]. The neutral complex, for example, $Lu(Pc)_2$ [71, 105], involves a sandwich between a dianion and a radical cation monoanion. EPR spectra, and a band near 500 nm in the green solutions of $Lu(Pc)_2$, support this electronic assignment. These diphthalocyanines are readily oxidized and reduced, with striking color changes occurring, see as examples the spectral data reported in these papers [69, 69, 71]. As the spectral data for complexes with different rare earths are quite similar, with no significant spectral shifts, especially for thin films, we will choose the data for only one $M(Pc)_2$ complex as a representative example when several metals have been studied. The redox properties of a wide range of rare earth bisphthalocyanine sandwich complexes have been studied both in solution and as thin films [69, 106, 107]. See also Chapter 5 in this volume by Nicholson.

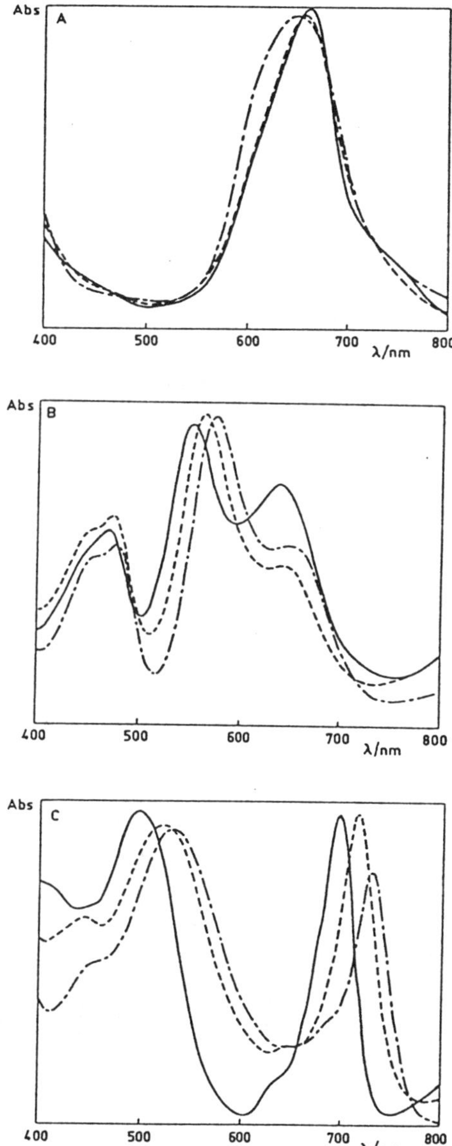

Figure 20. Absorption spectra of rare earth diphthalocyanine thin films formed on transparent electrodes. (A) After one-electron reduction to form the blue, $[RE(Pc(-2)_2]^-$, (B) after further reduction to form the purple anionic form, and (C) oxidation to the orange $[REPc(-1)_2]^+$. Data for three complexes are shown: $Lu(Pc)_2$ (solid line), $Gd(Pc)_2$ (dashed line), and $Nd(Pc)_2$ (dotted-dashed line). (Reproduced with permission from [69].)

i. $Ln(Pc)_2$

Absorption spectra that are representative of the neutral, oxidized and reduced forms have been reported by M'Sadak et al. [108] for thin films of $Dy(Pc)_2$ and $Nd(Pc)_2$; by Markovitsi et al. [107] for complexes of Tm, Lu, Yb, Dy, and Lu; by Castaneda et al. [109] for solutions and Besbes et al. [24] for thin films of alkyl substituted $Lu(Pc)_2$; by Chang and Marchon [110] for solutions of $Lu(Pc)_2$; by Kasuga et al. [111] for solutions of $Y(Pc)_2$ and $Y_2(Pc)_3$; and finally, by Corker et al. [112] for $Lu(Pc)_2$, who also report EPR data for the different redox species.

We describe the following spectral data in more detail as representative samples of the many data that have been reported. Kasuga et al. [113] report

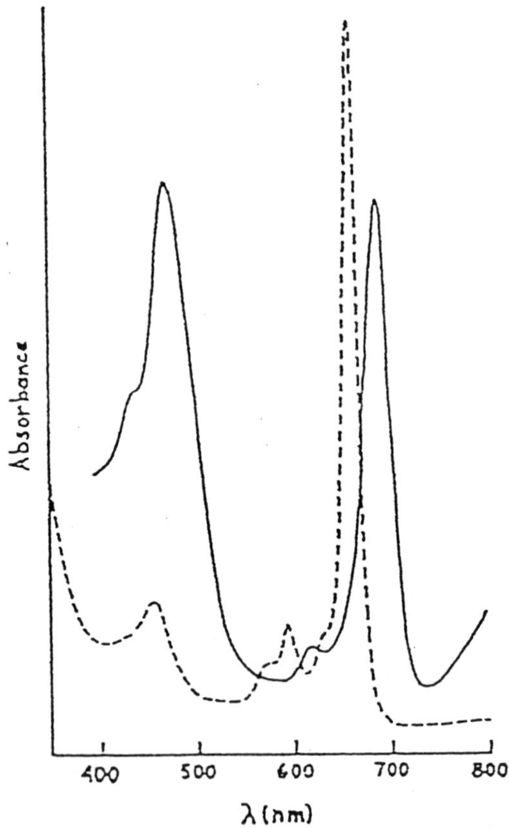

Figure 21. Absorption spectra for Lu[Pc(-1)Pc(-2)] (dashed line) and [Lu(Pc(-1)Pc(-1)]($SbCl_6$) in DCM. (Reproduced with permission from [110].)

spectra of the [Pc(-1)NdPc(-2)] π cation radical; although new bands at 470 and 676 nm grow isosbestically when p-benzoquinone is added, these band centers are different when compared with the band centers of monomeric or dimeric monophthalocyanines; see as an example Fig. 2 for $[MgPc(-1)]^{+\cdot}$, which may result from the significant overlap of the charge over the two rings in the diphthalocyanine. Riou and Clarisse [69] show spectral data for neutral, oxidized and reduced thin films of $Lu(Pc)_2$, $Gd(Pc)_2$, and $Nd(Pc)_2$ (Fig. 20). Reduction of these complexes forms complexes of dianionic π rings, while the one electron oxidized complexes involve two Pc(-1) π cation radical rings. The spectra in Fig. 20(C) show strong absorption at 500 and near 700 nm, both bands that are typical of the π cation radical species. The 700 nm band has been used with the solution complexes to suggest the dimers are present. In this case, the 800-1100 nm region spectrum is not available, so we cannot extend the assignment to thin films. Absorption spectra of thin films of $Lu(Pc)_2$ recorded during the electrochemical oxidation and reduction through the range -1000 to +1000 mV clearly show

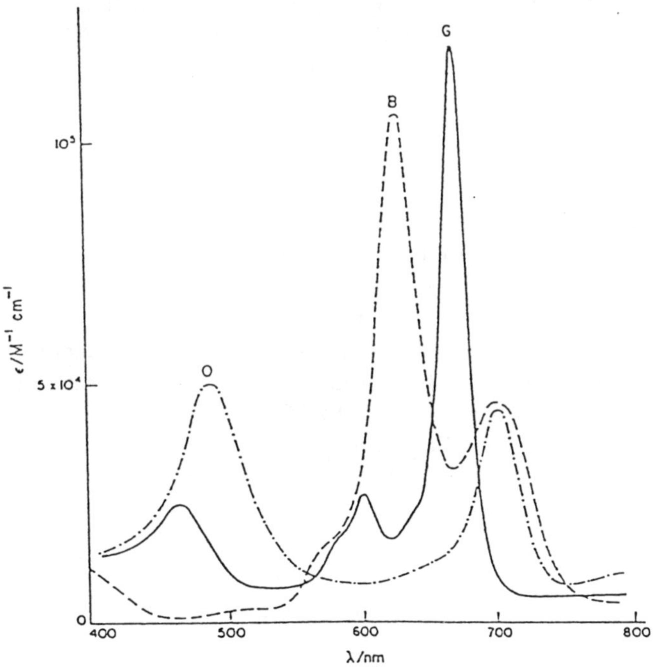

Figure 22. Absorption spectra of the peripherally substituted $Lu[(C_{12})_8Pc]_2$ in DCM. The three spectra represent the green initial form (G), the orange oxidized form (O) and the blue reduced form (B). (Reproduced with permission from [109].)

the formation of the π radical cation and π radical anion complexes [25].

Kasuga et al. [111] describe the spectra of a triple-decker, $Y_2(Pc)_3$ species, which rather closely resembles the spectra of blue, rare earth$(Pc)_2$ species. Recently, from our laboratory. we have studied the spectroelectrochemistry of Ce(IV)$(OEP)_2$ using absorption and MCD spectra to identify changes in transition energies and polarizations following oxidation of the rings [122].

To summarize the spectral and electrochemical data reported by the authors cited above, the characteristic green form of $RE(Pc)_2$, the neutral complex, exhibits a *Q* band near 670 nm and a less intense band at 460 nm (a g = 2.002 EPR signal is observed). Near-IR bands at 904 and 1382 nm are also observed [107]. Oxidation to the di-π-cation radical, MPc(-1)Pc(-1), results in a color change to orange, with bands at 700 (Q) and 500 nm (assigned by us as probably π-π) (Fig. 21) [110]. Reduction to the blue form is accompanied by bands near 620 and 700 nm. Further reduction to the violet form is assigned as rare earth-Pc(-3)Pc(-2), which results in a typical radical anion spectrum being observed (maxima are at 510 and 680 nm) and a g=2.003 EPR signal returns. Fig. 22, taken from Castaneda et al. [109], shows the neutral green form (G), the orange oxidized form (O), and the blue reduced form (B), for alkyl-substituted $Lu(Pc)_2$.

ii. AcPc

Unlike the rare earth phthalocyanines, the actinide complexes often involve coordination by additional ligands, like acetylacetonate (acac), to fill the coordination requirements of the metal. Guilard et al. [114] report that the absorption spectra for the π cation radicals species ThPc(-1)$(acac)_2$ and UPc(-1)$(acac)_2$, measured in benzonitrile, are very similar except for the 729 nm band, with band maxima (for U) near 350, 537, 624, 691, 729 and 852 nm. The significant absorbance at both 729 and 852 nm, suggests that the radical cation species of UPc partially dimerize under these conditions, whereas the ThPc(-1) does not appear to dimerize as much because there is no absorbance near 720 nm.

F. THEORETICAL INTERPRETATION OF THE OPTICAL DATA

Orti et al. [58], Henriksson-Enflo [115] and Ohno et al. [57] have most recently calculated the energies of the lowest singlet excited states for metallophthalocyanines. Transitions to these states contribute the majority of the absorbance measured for neutral phthalocyanines. Like Gouterman's model [56], two of these recent calculations [57, 58] predict that the two lowest transitions are $2a_{1u} \rightarrow 6e_g$ (the *Q* band) and $4a_{2u} \rightarrow 6e_g$ (the *B* band). Ring-oxidation leads to the $^2A_{1u}$ ground state. Unlike the case with the porphyrins [36], there is no ambiguity about the identity of the top filled molecular orbital for phthalocyanine radical cations.

Fig. 23(A) shows the ground and possible doublet excited states for the radical monocation. Because the $\Delta m_S=0$ rule will effectively forbid transitions other than doublet to doublet, we do not need to consider the quartet excited states. Possible configurations for excited states are listed in Table 3.

The calculations for LiPc(-1) and LuPc(-2)Pc(-1) [58] both show that degeneracy in the molecular orbitals is retained. For LiPc(-1), the top filled orbital is still $2a_{1u}$, however, the $2b_{2u}$ orbital is predicted to lie just above the

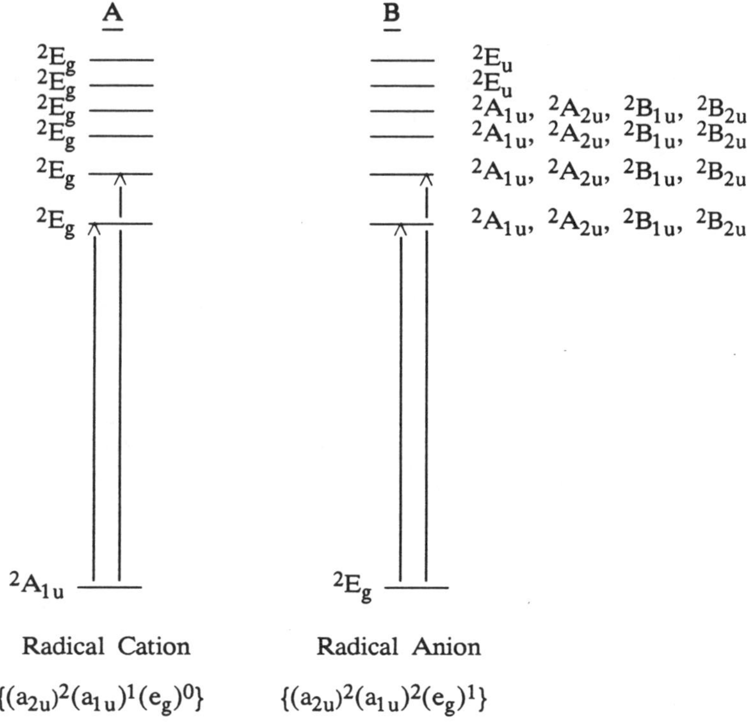

Figure 23. Accessible excited states for (A) cation radical and (B) anion radical phthalocyanine species. See Tables 2-4 for details of the origins of these states.

$4a_{2u}$ orbital. Transitions predicted to lie in the 300 - 800 nm region for LiPc(-1) are shown in Fig. 24. Because the experimental data for $[LiPc(-1)]^{+\cdot}$, Fig. 5 [86], closely resemble the spectral data for MgPc(-1) and ZnPc(-1), we suggest that this predicted absorption spectrum for LiPc(-1) [58] is applicable to any MPc(-1) species for which charge transfer does not add new bands. We can use the MCD spectral data for MgPc(-1) [2] and ZnPc(-1) [62] (Figs. 9 and 16) to determine polarizations of the transitions. The degenerate transitions found by deconvolution of the spectral data for monomeric MgPc(-1) [2] are at: 277, 315, 330, 387, 413 and 826 nm (Q). A prominent nondegenerate band is observed near 500 nm for each of the cation radical species that we have studied: (MgPc(-1) [2], ZnPc(-1) [62], RuPc(-1) [6], and FePc(-1) [13]). Clearly, the *Q* band, identified at 826 nm for MgPc(-1), is at 811 nm in LiPc(-1) [86]. Of greater difficulty is reconciling the intense 500 nm *z* polarized band. None of the bands predicted by the calculation are *z* polarized, yet this band dominates almost all absorption spectra of radical cations.

The dimerization involved in $Lu(Pc)_2$ greatly increases the number of allowed transitions. The spectra of rare earth diphthalocyanines extend from 2,000 nm to the uv region. Transitions are predicted at 319 (B), 443, 602 (Q), 626 (Q), 721 (Q), 734 (Q), and 1494 nm, where all the transitions will be x,y polarized except for the 1490 nm band which will be *z* polarized [58]. The complexity in the number of bands predicted to lie in the *Q*-band region, with four component transitions, arises from the effects of dimerization, and the presence of the Pc(-2) and Pc(-1) rings. While many absorption spectra have been reported for rare earth diphthalocyanines, we have not found any reports of MCD spectral data. MCD data are essential in order to confirm the predictions made with these calculations.

G. ABSORPTION AND MCD SPECTRA OF ANION RADICAL COMPLEXES

Phthalocyanine ring-based anions are much more oxygen sensitive than the corresponding cations [55]; the formation of anionic porphyrins and phthalocyanines is also subject to considerably more extensive side reactions [54,55]. This is particularly true for metals for which multiple oxidations states are accessible, for example, the electrochemical reduction of μ-oxo-bis(iron(III) tetraphenylporphyrin) [34]. Dodd and Hush [54] report monoanion and dianion absorption data for a series of ring-reduced porphyrins and phthalocyanines. Cambell et al. [116], Rollman and Iwamoto [68], and Clack and Yandle [55], report first and second reduction potentials for a range of phthalocyanines. Clack and Yandle [55] describe the colors of each negative ion as follows: monoanions, based on the Pc(-3) ring, tend to be blue in color; dianions, Pc(-4), are purple; trianions, Pc(-5), are blue; and tetraanions, Pc(-6), are blue-green in color. Guilard et al. report reduction potentials and absorption spectra for 1:1 actinide phthalocyanine complexes [117]. The only MCD data previously published, to our knowledge, are by Linder et al. [72], who describe the absorption and MCD spectra for MgPc and the mono-, di-, tri- and tetra-anionic species. Results from a Pariser-Parr-Pople self-consistent-field (PPP-SCF) molecular-orbital calculation were also reported and compared with the monoanion data [72]. Mack et al. [28, 123] have reproduced the MCD spectral data for $[MgPc(-3)]^{\cdot -}$, greatly improving the signal-to-noise ratio and also using spectral envelope deconvolution calculations to determine the type of MCD bands that are present. In general, far more spectral data are available for oxidized rings than for reduced rings.

A typical absorption spectrum for a one-electron, ring-reduced phthalocyanine exhibits prominent bands at 360, 423, 563, and 642 nm (for $[MgPc(-3)]^{\cdot -}$, Fig. 25 [28]). The presence of EPR signals [72, 73], and the similarity between room- and low-temperature absorption spectral data, suggest that phthalocyanine monoanions do not dimerize in solution as readily as the radical cations do. Only weak EPR signals are found for the dinegative ion, the MPc(-4) species [72, 73].

H. MAIN GROUP ANIONIC COMPLEXES

i. MgPc

Linder et al. [72] reported absorption and MCD spectra for the first four negative ions of MgPc. Their absorption data have been confirmed by Clack and Yandle [55]. Mack et al. [28] and Mack and Stillman [123] have measured absorption and MCD spectra for the first two negative ions of MgPc. Fig. 26 shows a series of absorption spectra recorded during the electrochemical reduction of MgPc(-2) in DMF, first to the monoanion (A), then to the dianion (B). The spectra in both of these steps are related isosbestically, which suggests that for these complexes changes in ligation take place rapidly and at all stages throughout the reduction. The monoanion, Pc(-3), is characterized by strong new bands at 312, 355, 423, 561, and 642 nm, as well as much weaker bands near 780 and 860 nm. The dianion, Pc(-4), is characterized by bands at 442, 508, 621, 723, and 785 nm.

Figs. 25 and 27 show absorption and MCD spectra for MgPc(-3) and

MgPc(-4), respectively. The MCD spectra were recorded in a thin layer electrochemical cell placed in the room temperature bore of an Oxford Instruments 5.5 T superconducting magnet. The MCD spectra of [MgPc(-3)]$\cdot^-$ (Fig. 25) feature a series of three derivative-like bands that become more resolved from 930 to 590 to 390 nm. These result from transitions into a zero-field split, nondegenerate orbitals. The observation of EPR signals at low temperatures suggests that [MgPc(-3)]$\cdot^-$ does not dimerize at room temperatures. The MCD spectrum shown in Fig. 25 of the monomeric, monoanion is dominated by MCD *B* terms, arisng from the Jahn-Teller split ground state [28,123].

[MgPc(-3)]$\cdot^-$ has also been formed photochemically when illuminated with visible region light from MgPc(-2), dissolved in methylamine [74], dissolved in piperidine at 253 K [16], and dissolved in a 1:1 mixture of tetrahydrofuran and diethyl ether [15] at 77 K or in DMF at 223 K [15]. In each case, the formation of the [MgPc(-3)]$\cdot^-$ was monitored by the appearance of the distinctive absorption spectrum, which was essentially the same at room temperature [74], 253 K [16], and 223 K [15], where the band centers were observed at 425, 570, 640, and 960 nm. A strong EPR signal near g=2.0025 was detected at 77 K following reduction at 293 K [16]. The presence of the EPR signal at low temperatures and the lack of temperature dependence in the absorption spectrum also suggests that [MgPc(-3)]$\cdot^-$ is monomeric in these solvents.

The MCD spectrum of the dianion (Fig. 27) is of the opposite sign to that published by Linder et al. [72]. Because of contamination by traces of the monoanion when these data are recorded, the strong, negatively signed band at 640 nm remains visible, so that we know that the spectra shown here are internally consistent. (The sign in the MCD experiment depends on the phase of the CD spectrometer and on the direction of the magnetic field parallel with the light beam, and the sign must be checked against a known standard. We use the negative sign of the 510 nm band of $CoSO_4$ in aqueous solutions to ensure consistency.)

ii. AlPc

Clack and Yandle [55] describe spectral data for chemically reduced mono-, di-, tri-, and tetra anions of ClAlPc; these spectral data resemble the spectra of the anions of ZnPc. Ohno et al. [4] report formation of AlPc(-3) in DMA, with bands at 410, 580 (s), 625 nm (s), and a weak, broad band to the red of 750 nm, following photoinduced, reductive electron transfer from a range of electron donors, including EDTA.

I. TRANSITION METAL COMPLEXES

Clack and Yandle [55] report spectral data for the first four negative ions of MnPc, FePc, CoPc, NiPc, and ZnPc, while Rollman and Iwamoto [68] describe spectral data for one- and two-electron reduced complexes of Ni, Cu, and H_2-tetrasulfonated phthalocyanine (MTsPc).

i. MnPc

Like Co(II)Pc, Mn(II)Pc can be reduced at both metal and ring sites. Clack and Yandle [55] report spectral band maxima for four reduced ions. Because the data for each of the reduced species resemble those for Mg(II)Pc(-4), they suggest that if the Mn(II) is reduced to Mn(I) in the first reduction step, it must convert back to Mn(II) for the dinegative ion to give the Mn(II)Pc(-4) species.

ii. FePc

Lever and Wilshire [118], Kadish et al. [119], and Clack and Yandle [55] all describe electrochemical reduction reactions of Fe(II)Pc(-2). Lever and Wilshire and Kadish et al. suggest that Fe(I)Pc(-2) forms first and that this is the pink solution of Clack and Yandle

iii. CoPc

Co(II)Pc is readily reduced to Co(I)Pc(-2), a species that exhibits a prominent charge transfer band near 480 nm and a red shifted *Q* band near 700 nm [70, 120, 121]. However, the MCD signature for the *Q* band persists, which supports the characterization of a Co(I) species, rather than a Pc(-3) species [28, 70]. Clack and Yandle [55] describe five reduced states for CoPc in tetrahydrofuran, including the Co(I)Pc(-2). Stillman and Thomson have reported absorption and MCD spectra for Co(II)Pc, Co(I)Pc, and Co(III)Pc [70].

	Transition	Energy	Oscillator strength	Polarization
PcH_2	$4a_u \rightarrow 6b^*_{2g}$	1.81 (685)	5.1	y
	$4a_u \rightarrow 6b^*_{3g}$	1.98 (625)	4.7	x
	$4a_u \rightarrow 7b^*_{2g}$	3.90 (318)	0.4	y
	$4a_u \rightarrow 7b^*_{3g}$	3.91 (317)	0.3	x
	$7b_{1u} \rightarrow 6b^*_{2g}$	4.12 (301)	1.6	x
	$6b_{1u} \rightarrow 6b^*_{2g}$	4.15 (299)	0.3	x
	$7b_{1u} \rightarrow 6b^*_{3g}$	4.29 (289)	1.6	y
	$6b_{1u} \rightarrow 6b^*_{3g}$	4.32 (287)	0.3	y
	$3a_u \rightarrow 6b^*_{2g}$	4.38 (283)	1.2	y
PcLi	$2a_{1u} \rightarrow 6e^*_g$	1.62 (764)	5.6	x,y
	$5e_g \rightarrow 2a_{1u}$	2.30 (538)	1.2	x,y
	$4e_g \rightarrow 2a_{1u}$	2.42 (512)	1.6	x,y
	$3e_g \rightarrow 2a_{1u}$	3.01 (411)	1.5	x,y
	$2b_{2u} \rightarrow 6e^*_g$	3.91 (317)	0.3	x,y
	$4a_{2u} \rightarrow 6e^*_g$	3.92 (316)	1.5	x,y
	$1a_{1u} \rightarrow 6e^*_g$	3.98 (311)	0.9	x,y
	$3b_{1u} \rightarrow 6e^*_g$	4.05 (306)	1.2	x,y

Figure 24. VEH electronic transitions calculated for PcH_2 and PcLi. Transitions more energetic than 4.5 eV or with oscillator strengths lower than 0.1 are not included. Energies are in eV (nm). (Reproduced with permission from [58].)

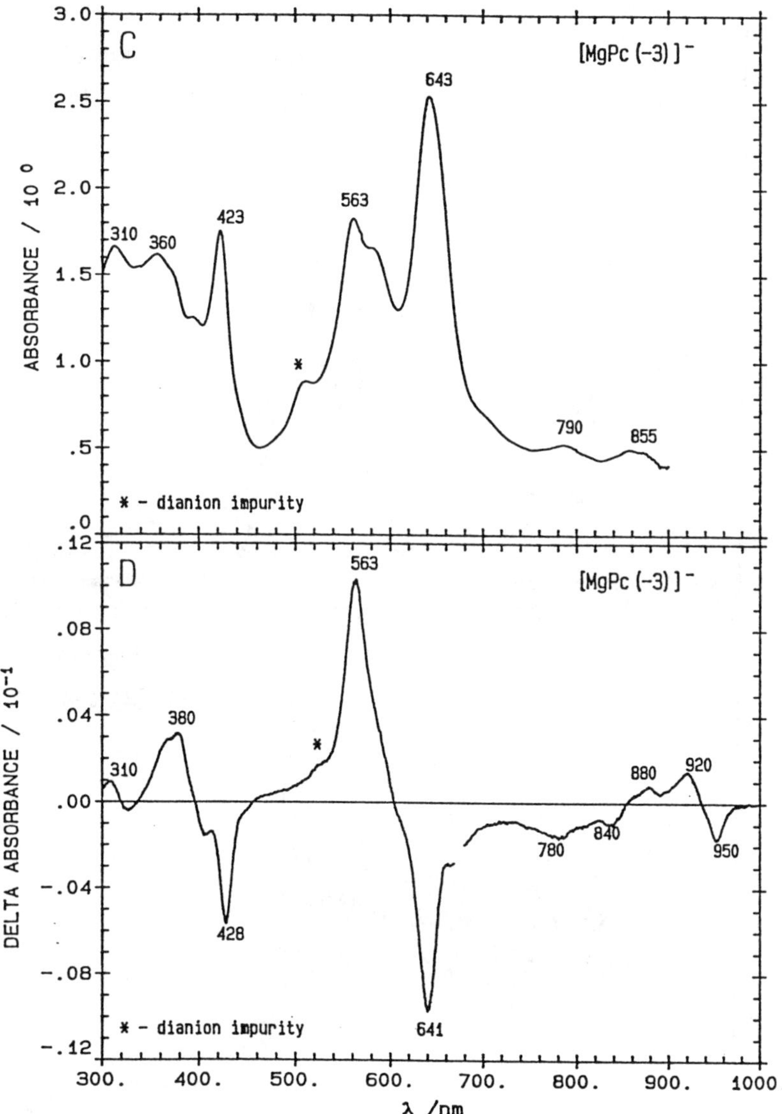

Figure 25. Absorption and MCD spectra of the MgPc(-3) monoanion formed electrochemically in dimethylformamide. Both spectra were recorded from electrochemical cells located inside the spectrometer. (Reproduced with permission from [28].)

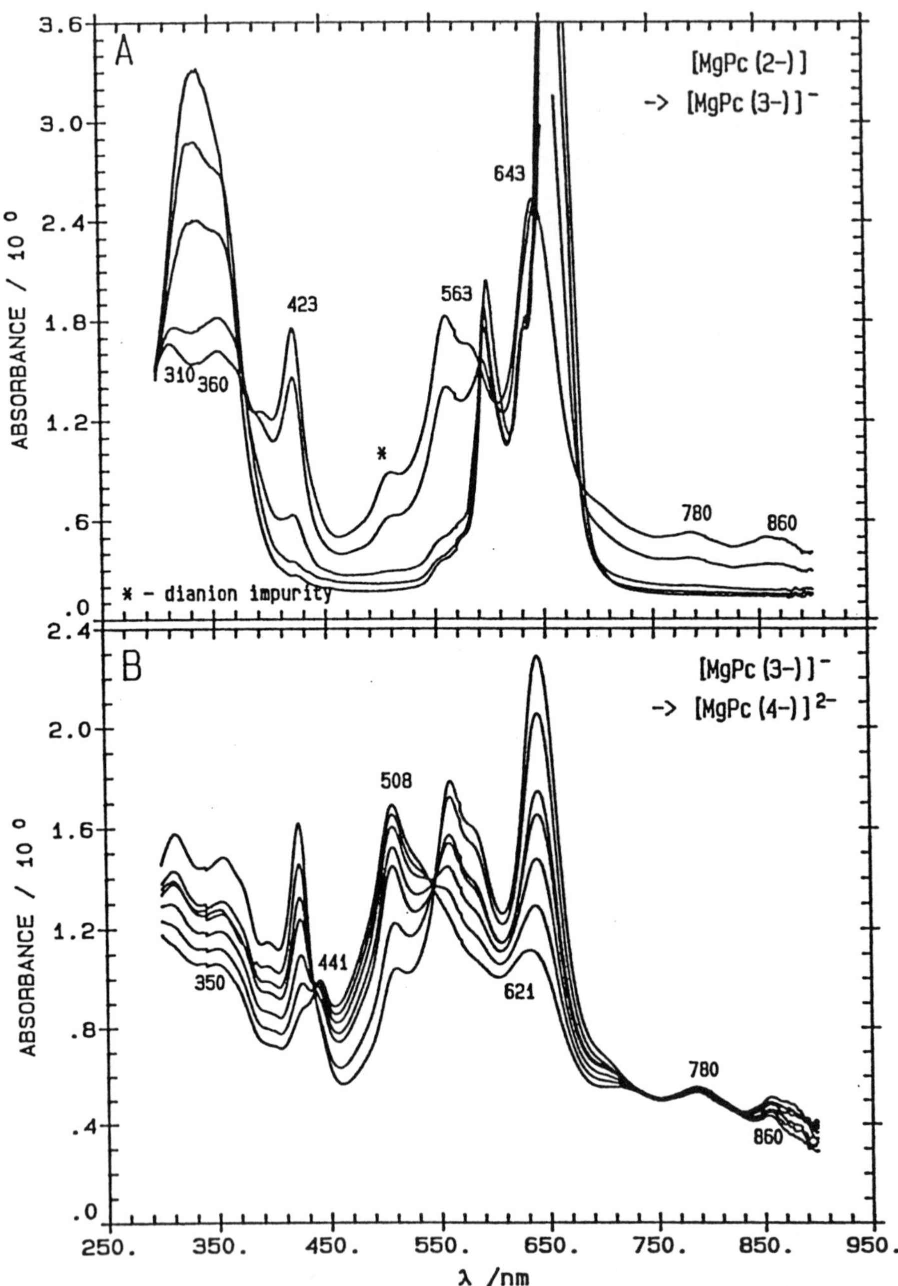

Figure 26. Absorption spectra recorded during the electrochemical reduction of MgPc(-2) in dimethylformamide, first to the monoanion (upper), then to the dianion (lower). (Reproduced with permission from [28].)

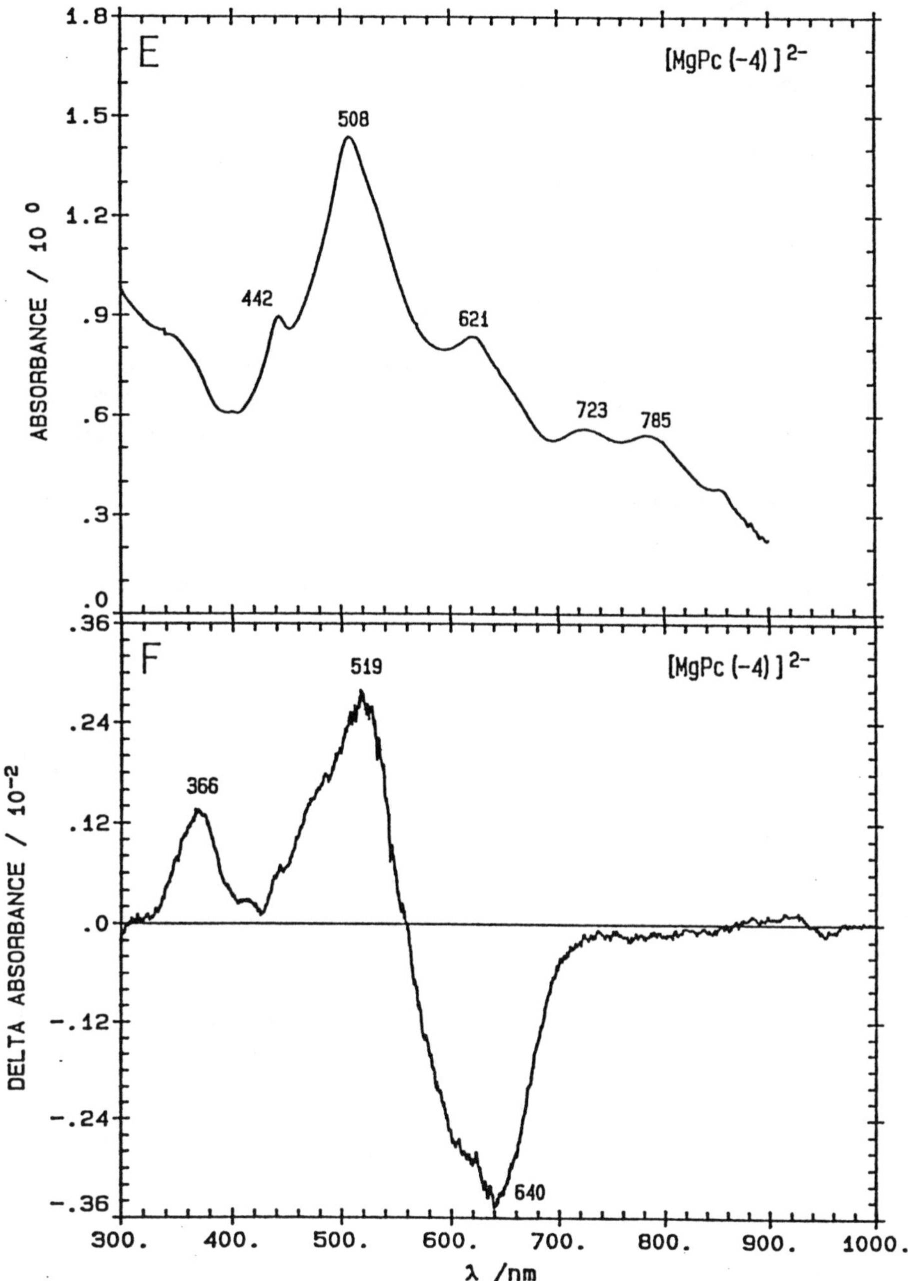

Figure 27. Absorption and MCD spectra of the MgPc(-4) dianion in dimethylformamide. Both spectra were recorded from electrochemical cells located inside the spectrometer. (Reproduced with permission from [28].)

iv. CoXPc

Nevin et al. [27,93] have reported spectral data for the reduction of the mononuclear and tetranuclear complexes based on the peripherally substituted Co(II) tetraneopentoxyphthalocyanine, Co(II)TNPc(-2). Electrochemical methods were used to form Co(I)TNPc(-2) and the π anion radical Co(I)TNPc(-3). While the band centers are rather similar for these two species, the relative absorbances for the 708:471 nm bands are quite different. Co(I)Pc(-2) species are characterized by an isolated, rather weak, but still prominent, *Q* band near 705 nm in addition to the 480 nm band [27, 70, 93]. The π anion species is characterized by a very much weaker 703 nm band than the 480 nm band, and a near-IR band at 955 nm. Fig. 28 [93] shows absorption spectra for the mononuclear and dinuclear Co(I)TNPc(-3) species based on the ring substituted with four neopentoxy groups. Similarly, reduction of aggregated tetrasulfonated CoPc gives first the Co(I)Pc(-2) species, then the Co(I)Pc(-3) species.

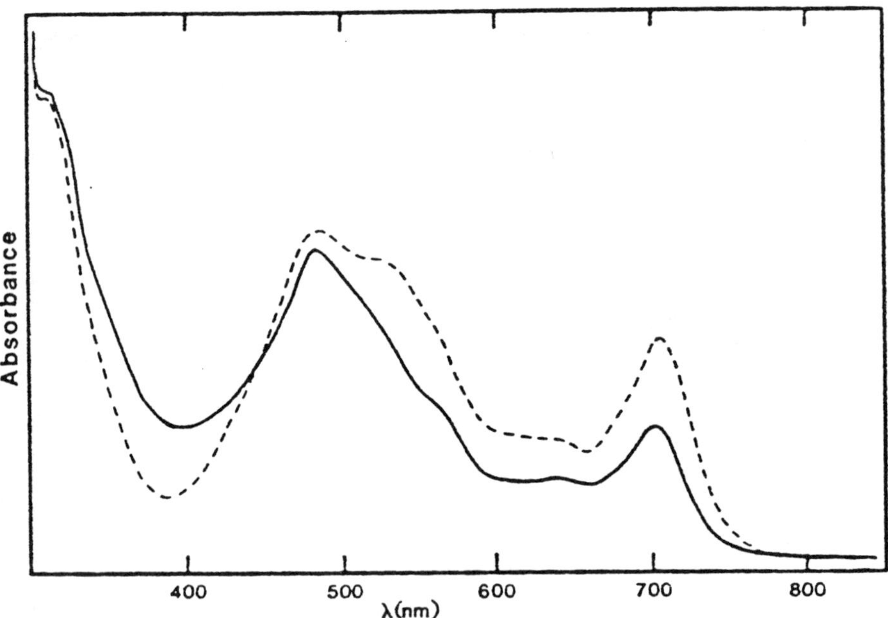

Figure 28. Absorption spectra of electrochemically generated $[Co(I)TNPc(-3)]^{2-}$ (solid line) and $[Co(I)TrNPc(-3)]_2^{4-}$ (dashed line) in DCB. (Reproduced with permission from [93].)

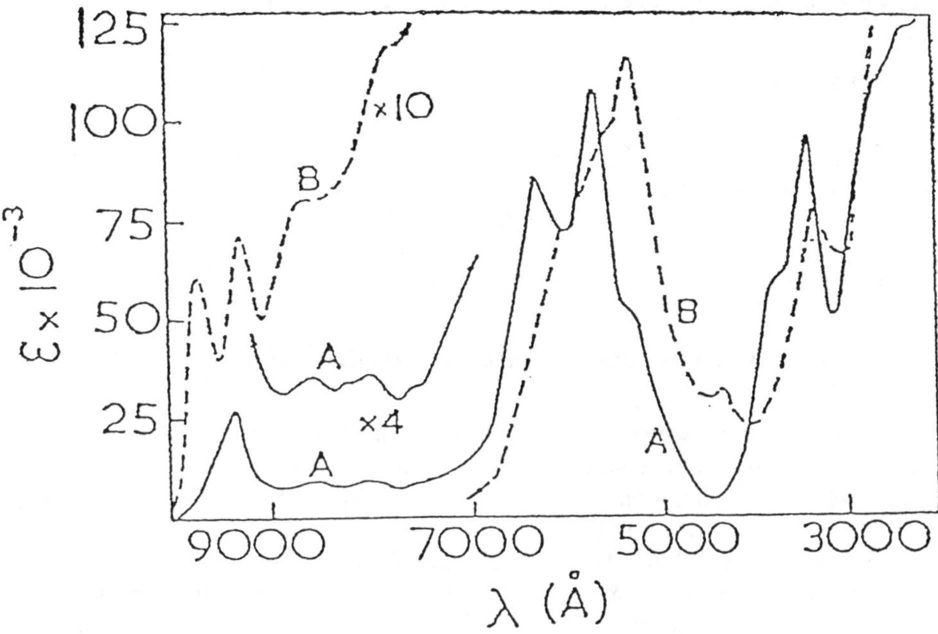

Figure 29. Absorption spectra of the monoanion (A) and dianion (B) of CuPc formed chemically from sodium metal in methyltetrahydrofuran. (Reproduced with permission from [54].)

v. CuPc

Dodd and Hush [54] report spectral data for the mono- (A) and dinegative (B) ions of CuPc in 2-methyltetrahydrofuran (Fig. 29). The mononegative ion, CuPc(-3), is characterized by weak bands near 950 nm, then the usual pair of strong bands near 600 nm, and finally a series of bands starting near 380 nm and extending below 250 nm. The dianion exhibits a similar series of bands, but as with MgPc(-4), is distinguished from the monoanion by the absence of the 640 nm band. Similar spectra are observed for the tetrasulfonated species, CuTsPc(-3) and CuTsPc(-4) [68].

vi. NiPc

The spectra for NiTsPc(-3) and NiTsPc(-4) [68] in DMSO closely resemble those for other metals; the monoanion exhibits bands near 330, 375, 575, 640, and 930 nm. The spectrum of the dianion is simpler with a uv region band at 360 nm and a visible region band near 540 nm.

vii. PdPc

Photoreduction of PdTsPc to the monoanion takes place following irradiation at 600 nm in the presence of cysteine and the micelle CTAC (hexadecyltrimethylammonium chloride) at pH 6.7. New bands appear at 380, 564, and to the red of 700 nm, in accordance with the spectra of a reduced π ring complex.

viii. Complexes of the Lanthanides

As with the cation radicals spectra [1], there are very many more spectra of various redox states of $Lu(Pc)_2$ than other lanthanides [69, 71]. Electrochemical reduction of the green Lu(Pc(-1)Pc(-2)) leads to the turquoise blue $[Lu(Pc(-2))_2]^{1-}$, the dark blue $[Lu(Pc(-2)(Pc(-3))]^{2-}$ and the violet $[Lu(Pc(-3)Pc(-3))]^{3-}$ [71]. Riou and Clarisse [69] have studied the spectroelectrochemical properties of thin films of lanthanide diphthalocyanines. Like the spectra of the neutral thin films [1], the bands of $[Lu(Pc(-3)Pc(-3))]^{2-}$ while of the same energy as for other Pc(-3) complexes, are very much broader. Guilard et al. [117] report that the absorption spectra for the π anion and dianion radical species of $ThPc(acac)_2$ and $UPc(acac)_2$ formed electrochemically in benzonitrile, with band maxima (for U) near 352, 591, 620, 652, and 677 nm for the monoanion, and near 349, 549, and 646 nm for the dianion.

J. THEORETICAL INTERPRETATION OF THE OPTICAL SPECTRA OF RING-REDUCED COMPLEXES

The major change on the absorption spectrum following one-electron reduction is the appearance of the 423, 563, and 643 bands, together with two weaker bands in the 800 nm region, (Fig. 25) [28, 123]. Similar spectral changes have been reported for the anion radical of bacteriochlorophyll c and bacteriopheophorbide c, d, and e by Fajer et al. [32], with new bands for the radical anion appearing near 640 and 810 nm. The Soret region exhibits a general reduction in absorbance rather than growth of new bands. Fajer et al. also report transition energies from a PPP-SCF-MO calculation based on the chlorin structure; these transitions quite closely matched the observed anion radical spectral data. Linder et al. [72] carried out a calculation based on the PPP-SCF molecular-orbital approach to assign the absorption and MCD spectra of the MgPc(-3) monoanion and MgPc(-6) tetra-anion species. We would caution that without band deconvolution it is very difficult to determine which bands are B/C terms and which are A terms for spectra as complicated as those of the negative ions [123]. Analysis of the absorption and MCD spectra of monoanion radicals of MgPc and ZnPc show that a similar nondegenerate ground state exists in both species [28,123].

ACKNOWLEDGMENTS

It is a pleasure to be able to acknowledge the many members of my group who have contributed to the work on the phthalocyanines and porphyrins described here. I particularly wish to thank Tebello Nyokong, Bill Browett, Zbyszek Gasyna, Stanislaw Radzki, Scott Kirkby, John Heggtveit, Edward Ough, and John Mack, for their extensive studies on the spectroscopic properties of the π cation and anions of phthalocyanines and porphyrins. I am very grateful to Edward Ough and John Mack for helping in the preparation of figures prior to publication and to Tebello Nyokong for help in preparing the table of band maxima. Continuing support by the Centre for Chemical Physics and the Photochemistry Unit at the University of Western Ontario is gratefully acknowledged. Financial support from NSERC of Canada, the Academic Development Fund at the University of Western Ontario, the Center for Chemical Physics at the University of Western Ontario, and Imperial Oil is very much appreciated. This is publication number 454 of the Photochemistry Unit.

REFERENCES

1. M. J. Stillman and T. Nyokong, in *Phthalocyanines: Properties and Applications*, Ch. 3., C. C. Leznoff and A. B. P. Lever, Eds., VCH, New York, 1989, p. 133.
2. E. Ough, Z. Gasyna, and M. J. Stillman, *Inorg. Chem.*, 30 (1991) 2301.
3. J. R. Darwent, I. McCubbin, and G. Porter, *J. Chem. Soc.*, 78 (1982) 903.
4. T. Ohno, S. Kato, A. Yamada, and T. Tanno, *J. Phys. Chem.*, 87 (1983) 773.
5. D. K. Geiger, G. Ferraudi, K. Madden, J. Granifo, and D. P. Rillema, *J. Phys. Chem.*, 89 (1985) 3890
6. T. Nyokong, Z. Gasyna, and M. J. Stillman, *Inorg. Chim. Acta*, 112

(1986) 11.

7. T. Nyokong, Z. Gasyna, and M. J. Stillman, *Inorg. Chem.*, 26 (1987) 548.
8. G. Ferraudi, *Inorg. Chem.*, 18 (1979) 1005.
9. H. Ohtani, T. Kobayashi, T. Ohno, S. Kato, T. Tanno, and A. Yamada, *J. Phys. Chem.*, 88 (1984) 4431.
10. G. Ferraudi, S. Oishi, and S. Muraldiharan, *J. Phys. Chem.*, 88 (1984) 5261.
11. A. B. P. Lever, S. Licoccia, B. S. Ramaswamy, S. A. Kandil, and D. V. Stynes, *Inorg. Chim. Acta*, 51 (1981) 169.
12. J. R. Darwent, P. Douglas, A. Harriman, G. Porter, and M-C. Richoux, *Coord. Chem. Rev.*, 44 (1982) 83.
13. T. Nyokong, Z. Gasyna, and M. J. Stillman *ACS, Symp. Series*, 321 (1986) 309.
14. H. Ohtani, T. Kobayashi, T. Tanno, A. Yamada, D. Wohrle, and T. Ohno, *Photochem. Photobiol.*, 44 (1986) 125.
15. A. P. Bobrovskii, V. G. Maslov, A. N. Sidoov, and V. E. Kholmogoov, *Dokl. Akad. Nauk. USSR*, 195 (1970) 384.
16. A. P. Bobrovski and V. E. Kholmogorov, *J. Phys. Chem. USSR*, 47 (1973) 983.
17. J. D. Spikes, *Photochem. Photobio.*, 43 (1986) 691.
18. E. Ciliberto, K. A. Doris, W. J. Pietro, G. M. Reisner, D. E. Ellis, I. Fragala, F. H. Herbstein, and M. A. Ratner, *J. Am. Chem. Soc.*, 106 (1984) 7748.
19. C. J. Schramm, R. P. Scaringe, D. R. Stojakovic, B. M. Hoffman, J. A. Ibers, and T. J. Marks, *J. Am. Chem. Soc.*, 102 (1980) 6702.
20. P. D. Hale, W. J. Pietro, M. A. Ratner, D. E. Ellis, and T. J. Marks, *J. Am. Chem. Soc.*, 109 (1987) 5943.
21. D. W. DeWulf, J. K. Leland, B. L. Wheeler, A. J. Bard, D. A. Batzel, D. R. Dininny, and M. E. Kenney, *Inorg. Chem.*, 26 (1987) 266.
22. P. Turek, P. Petit, J-J. Andre, J. Simon, R. Even, B. Boudjema, G. Guillaud, and M. Maitrot, *J. Am. Chem. Soc.*, 109 (1987) 5119.
23. F. Castaneda and V. Plichon, *J. Electroanalytical Chem.*, 236 (1987) 163.
24. S. Besbes, V. Plichon, J. Simon, and J. Vaxiviere, *J. Electroanal. Chem.*, 237 (1987) 61.
25. G. C. S. Collins and D. J. Schiffron, *J. Electroanal. Chem.*, 139 (1982) 335.
26. A. L. Balch, L. Latos-Grazynski, and M. W. Renner, *J. Am. Chem. Soc.*, 107 (1985) 2983.
27. W. A. Nevin, W. Liu, S. Greenberg, M. R. Hempstead, S. M. Marcuccio, M. M. Melnik, C. C. Leznoff, and A. B. P. Lever, *Inorg. Chem.*, 26 (1987) 891.

28. J. Mack, S. Kirkby, E. Ough, and M. J. Stillman, *Inorg. Chem.*, 31 (1992) 1717.
29. Y. Orii, H. Shimada, T. Nozawa, and H. Mashahiro, *Biochem. Biophys. Res. Commun.*, 76 (1977) 983.
30. D. R. English, D. N. Hendrickson, and K. S. Suslick, *Inorg. Chem.*, 24 (1985) 121.
31. Y. O. Su, R. S. Czernuszewicz, L. A. Miller, and T. G. Spiro, *J. Am. Chem. Soc.*, 110 (1988) 4150.
32. J. Fajer, I. Fujita, A. Forman, L. K. Hanson, G. W. Craig, D. A. Goff, L. A. Kehres, and K. M. Smith, *J. Am. Chem. Soc.*, 105 (1983) 3837.
33. X. H. Mu and K. M. Kadish, *Inorg. Chem.*, 28 (1989) 3743.
34. K. M. Kadish, G. Larson, D. Lexa, and M. Momenteau, *J. Am. Chem. Soc.*, 97 (1975) 282.
35. D. Ostovic and T. C. Bruice, *J. Am. Chem. Soc.*, 111 (1989) 6511.
36. Z. Gasyna and M. J. Stillman, *Inorg. Chem.*, 29 (1990) 5101.
37. D. Dolphin, Z. Muljiani, K. Rousseau, D. C. Borg, J. Fajer, and R. H. Felton, *Ann. New York Acad. Sci.*, 206 (1973) 177.
38. G. Barth, R. E. Linder, E. Bunnenberg, and C. Djerassi, *Ann. New York Acad. Sci.*, 206 (1973) 223.
39. J. Fajer, D. C. Borg, A. Forman, R. H. Felton, L. Vegh, and D. Dolphin, *Ann. New York Acad. Sci.*, 206 (1973) 349.
40. K. M. Kadish and D. G. Davis, *Ann. New York Acad. Sci.*, 206 (1973) 495.
41. Z. Gasyna, W. R. Browett, and M. J. Stillman, *Inorg. Chem.*, 24 (1985) 2440.
42. Z. Gasyna, W. R. Browett, and M. J. Stillman, *Inorg. Chem.*, 23 (1984) 382.
43. Z. Gasyna, W. R. Browett, and M. J. Stillman *ACS Symp. Ser.*, 321 (1986) 298.
44. Z. Gasyna, W. R. Browett, and M. J. Stillman, *Inorg. Chem.*, 27 (1988) 4619.
45. W. R. Browett and M. J. Stillman, *Inorg. Chim. Acta*, 49 (1981) 69.
46. J. H. Fuhrhop, P. Wasser, D. Riesner, and D. Mauzerall, *J. Am. Chem. Soc.*, 94 (1972) 7996.
47. K. A. Macor, R. S. Czernuszewicz, and T. G. Spiro, *Inorg. Chem.*, 29 (1990) 1996.
48. R. H. Felton, G. S. Owen, D. Dolphin, A. Forman, D. C. Borg, and J. Fajer, *Ann. New York Acad. Sci.*, 206 (1973) 504.
49. W. R. Browett and M. J. Stillman, *Biochim. Biophys. Acta*, 660 (1981) 1.
50. V. Palaniappan and J. Terner, *J. Biol. Chem.*, 264 (1989) 16046.
51. Z. Gasyna, W. R. Browett, and M. J. Stillman, *Biochemistry*, 27 (1988) 2503.

52. W. R. Browett, Z. Gasyna, and M. J. Stillman, *J. Am. Chem. Soc.*, 110 (1988) 3633.
53. L. R. Furenlid, M. W. Renner, K. M. Smith, and J. Fajer, *J. Am. Chem. Soc.*, 112 (1990) 1634.
54. J. W. Dodd and N. S. Hush, *J. Chem. Soc.*, (1964) 4607.
55. D. W. Clack and J. R. Yandle, *Inorg. Chem.*, 11 (1972) 1738.
56. M. Gouterman, in *The Porphyrins. Physical Chemistry, Part 4A*, D. Dolphin, Ed., Academic Press, New York, 1978.
57. O. Ohno, N. Ishikawa, H. Matsuzawa, Y. Kaizu, and H. Kobayashi, *J. Phys. Chem.*, 93 (1989) 1713.
58. E. Orti, J. L. Bredas, and C. Clarisse, *J. Chem. Phys.*, 92 (1990) 1228.
59. A. Henriksson and M. Sundbom, *Theoret. Chim. Acta*, 27 (1972) 213.
60. A. Henriksson, B. Roos, and M. Sundbom, *Theoret. Chim. Acta*, 27 (1972) 303.
61. E. Ough, T. Nyokong, K. A. M. Creber, and M. J. Stillman, *Inorg. Chem.*, 27 (1988) 2725.
62. T. Nyokong, Z. Gasyna, and M. J. Stillman, *Inorg. Chem.*, 26 (1987) 1087.
63. J. W. Owens and C. J. O'Connor, *Coord. Chem. Rev.*, 84 (1988) 1.
64. P. J. Stephens, W. Suetaka, and P. N. Schatz, *J. Chem. Phys.*, 44 (1966) 4592.
65. H. Homborg, Z. anal. chem., 507 (1983) 35.
66. G. Fu, Y. Fu, K. Jayaraj, and A. B. P. Lever, *Inorg. Chem.*, 29 (1990) 4090.
67. V. Manvannan, W. A. Nevin, C. C. Leznoff, and A. B. P. Lever *J. Coord. Chem., 19 (1988) 139.*
68. L. D. Rollmann and R. T. Iwamoto, *J. Am. Chem. Soc.*, 90 (1968) 1455.
69. M. T. Riou and C. Clarisse, *J. Electroanal. Chem.*, 249 (1988) 181.
70. M. J. Stillman and A. J. Thomson, *J. Chem. Soc. Farad. Trans.* II, 70 (1974) 790.
71. M. L'Her, Y. Cozien, and J. Conrtot-Coupez, *C. R. Acad. Sci. Paris*, 302 (1986) 9.
72. R. E. Linder, J. R. Rowlands, and N. S. Hush, *Mol. Phys.*, 21 (1971) 417.
73. A. Louati, M. E. l. Meray, J. J. Andre, J. Simon, K. M. Kadish, M. Gross, and A. Giraudeau, *Inorg. Chem.*, 24 (1985) 1175.
74. M. de Backer, P. Jacquot, F. X. Sauvage, B. van Vlierberge, and G. Lepoutre, *J. chimie Phys.*, 84 (1987) 429.
75. J. H. Dawson and D. M. Dooley, in *Iron Porphyrins, Part 3, Physical Bioinorganic Chemistry Series.*, A. B. P. Lever and H. B. Gray, Eds., VCH Pub. Inc., New York, 1989.
76. H. Kobayashi, *Adv. Biophys.*, 8 (1975) 191.

77. M. Hatano and T. Nozawa, *Adv. Biophys,* 11 (1978) 95.

78. S. B. Piepho and P. N. Schatz, *Group Theory in Spectroscopy - With Applications to MCD*, J. Wiley, New York, 1983.

79. W. R. Browett and M. J. Stillman, *Comp. Chem.*, 11 (1987) 241.

80. K. Schmitt, P. W. M. Jacobs, and M. J. Stillman, *J. Phys. C,* 16 (1983) 603.

81. D. G. Eglinton, P. M. A. Gadsby, G. Sievers, J. Peterson, and A. J. Thomson, *Biochim. Biophys. Acta,* 742 (1983) 648.

82. P. C. Minor, M. Gouterman, and A. B. P. Lever, *Inorg. Chem.*, 24 (1985) 1894.

83. E. A. Luk'yanets, *Catalog of Electronic Spectra of Phthalocyanines, and Related Compounds*, Moscow Scientific Manufacturing Union, Division of the Scientific Research Institute for Technoeconomic Research, 1989.

84. W. Kalz, H. Homborg, H. Kuppers, B. J. Kennedy, and K. S. Murray, *Z. Naturforsch.*, 39B (1984) 1478.

85. H. Homborg and W. Kalz, *Z. Naturforsch.*, 39B (1984) 1490.

86. H. Homborg and W. Kalz, *Z. Naturforsch.*, 33B (1978) 1067.

87. W. D. Edwards and M. C. Zerner, *Can. J. Chem.*, 63 (1985) 1763.

88. L. O. Spreer, A. Leone, A. C. Maliyackel, J. W. Otvos, and M. Calvin, *Inorg. Chem.*, 27 (1988) 2401.

89. D. R. Prasad and G. Ferraudi, *Inorg. Chem.*, 21 (1982) 2967.

90. J. F. Myers, G. W. R. Canham, and A. B. P. Lever, *Inorg. Chem.*, 14 (1975) 461.

91. R. Taube, *Pure Applied Chem.*, 38 (1974) 427.

92. A. B. P. Lever, M. R. Hempstead, C. C. Leznoff, W. Liu, M. Melnik, W. A. Nevin, and P. Seymour, *Pure Applied Chem.*, 58 (1986) 1467.

93. W. A. Nevin, M. R. Hempstead, W. Liu, C. C. Leznoff, and A. B. P. Lever, *Inorg. Chem.*, 26 (1987) 570.

94. P. A. Bernstein and A. B. P. Lever, *Inorg. Chem.*, 29 (1990) 608.

95. H. Song, C. A. Reed, and W. R. Scheidt, *J. Am. Chem. Soc.*, 111 (1989) 6867.

96. B. R. Hollebone and M. J. Stillman, *J. Chem. Soc. Farad. Trans.* II, 74 (1978) 2107.

97. M. J. Stillman and A. J. Thomson, *J. Chem. Soc. Faraday Trans.* II, 70 (1974) 805.

98. H. Homborg and C. L. Teske, *Z. anal. chem.*, 527 (1985) 45.

99. T. Nyokong, *Spectroscopic and Electrochemical Studies of Phthalocyanines.* PhD awarded by the University of Western Ontario, London, Ontario, Canada, 1986.

100. E. N. Bakshi, C. D. Delfs, K. S. Murray, B. Peters, and H. Homborg, *Inorg. Chem.*, 27 (1988) 4318.

101. J. M. Green and L. R. Faulkner, *J. Am. Chem. Soc.*, 105 (1983) 2950.

102. D. Dolphin, B. R. James, A. L. Murray, and J. R. Thornback, *Can. J. Chem.*, 58 (1980) 1125.
103. H. Sugimoto, T. Higashi, A. Maeda, M. Mori, H. Masuda, and T. Taga, *J. Chem. Soc., Chem. Commun.*, (1983) 1234.
104. K. Kasuga, M. Tsutsui, R. C. Petterson, K. Tatsumi, N. Van Opdenbosch, G. Pepe, and E. F. Meyer, *J. Am. Chem. Soc.*, 102 (1980) 4835.
105. D. Walton, B. Ely, G. Elliott, and J. C. Marchon, *J. Electrochem. Soc.*, 128 (1981) 2479.
106. P. N. Moskalev, G. N. Shapkin, and N. I. Alimova, *Russian J. Inorg. Chem.*, 27 (1982) 794.
107. Markovitsi, T. Tran-Thi, R. T. Even, and J. Simon, *Chem. Phys. Lett.*, 137 (1987) 107.
108. M. M'sadak, J. Roncali, and F. Garnier, *J. Electroanalytical Chem.*, 189 (1985) 99.
109. F. Castaneda, C. Iiechocki, V. Plichon, J. Simon, and J. Vaxiviere, *J. Electroanalytical Chem.*, 31 (1986) 131.
110. A. T. Chang and J-C. Marchon, *Inorg. Chim. Acta*, 53 (1981) L241.
111. K. Kasuga, M. Ando, H. Morimoto, and M. Isa, *Chem. Lett.*, (1986) 1095.
112. G. A. Corker, B. Grant, and N. J. Clecak, *J. Electroanalytical Chem.*, 126 (1979) 1339.
113. K. Kasuga, M. Ando, and H. Morimoto, *Inorg. Chim. Acta*, 112 (1986) 99.
114. R. Guilard, A. Dormond, M. Belkalem, J. E. Anderson, Y. H. Liu, and K. M. Kadish, *Inorg. Chem.*, 26 (1987) 1410.
115. A. Henricksson-Enflo, *Int. J. Quantum Chem.*, 37 (1990) 547.
116. R. H. Campbell, G. A. Heath, G. T. Hefter, and R. C. S. McQueen, *J. Chem. Soc., Chem. Commun.*, (1983) 1123.
117. R. Guilard, A. Dormond, M. Belkalem, J. E. Anderson, Y. H. Liu, and K. M. Kadish, *Inorg. Chem.*, 26 (1987) 1410.
118. A. B. P. Lever and J. P. Wilshire, *Inorg. Chem.*, 17 (1978) 1145.
119. K. M. Kadish, L. A. Bottomley, and J. S. Cheng, *J. Am. Chem. Soc.*, 100 (1978) 2731.
120. P. Day, H. A. O. Hill, and M. G. Price, *J. Chem. Soc. (A)* (1968) 90.
121. W. A. Nevin, W. Liu, M. Melnik, and A. B. P. Lever, *J. Electroanalytical Chem.*, 213 (1986) 217.
122. S. Radzki, J. Mack, and M. J. Stillman, *New J. Chem.*, (1992) 0000.
123. J. Mack, and M. J. Stillman, *Submitted to Inorg. Chem.*, (1992).

TABLE 1 Abbreviations used in Table 2

ABBREVIATIONS	
SOL	solvent used
axial ligand	the phthalocyanine may be 5 or 6 coordinate
Ring Ox State	oxidation state of the phthalocyanine ring
Band Maxima	taken from tables or data
Comments	"FIG": a figure is shown; "Table" tabular data also reported

Abbreviation for peripherally substituted rings:

MTNPc: metallotetraneopentoxyphthalocyanine; from Lever's laboratory, ref. [27,93]

MTrPc: the single MTNPc part of a polynuclear molecule; from Lever's laboratory, ref. [27,93]

MTsPc: metallotetrasulfonatedphthalocyanine

SOLVENTS USED	
ACN	acetonitrile
ClN	chloronaphthalene
DCB	dichlorobenzene
DCM	dichloromethane
DMF	dimethylformamide
DMSO	dimethyl sulfoxide
MTM	CH_3NO_2, nitromethane
PhCN	phenylnitrile
TCM	$CHCl_3COOH$, trifluoroacetic acid

PHTHALOCYANINE RING OXIDATION STATE CODES	
-2 =	Pc(-2) the dianion in most neutral metallophthalocyanines; written as M(II)Pc(-2)
-1 =	Pc(-1) the "radical cation"; $[M(II)Pc(-1)]^{+\cdot}$
0 =	Pc(0) dication; $[M(II)Pc(0)]^{+2}$
-3 =	Pc(-3) the "radical anion"; $[[M(II)Pc(-3)]^{-\cdot}$
-4 =	Pc(-4) $[M(II)Pc(-4)]^{-2}$
-5 =	Pc(-5) $[M(II)Pc(-5)]^{-3}$

TABLE 2 Band Centers from the Absorption Spectra of Phthalocyanine Cation and Anion Complexes

Central Element	SOL	Axial Ligand	Ring Ox State	Band Maxima	REF	Comments
H2Pc						
Radical cation						
H	DCM		-1	888 842 775 700 540 420 398 324 290	[65]	Fig & Table
				Note: chemical oxidation		
H	MTN		-1	1020 888 844 760 700 540 380	[65]	Fig & Table
				Note: chemical oxidation		
Ag						
Radical cation						
Ag3+	DCM		-1	854 721 649 574 439	[66]	Fig & Table
				Note: substituted ring, Ag(III)TNPc(-1); electrochemical oxidation		
Al						
Radical anion						
ClAl	DMF		-3	975 900 820 724 618 575 329	[55]	Fig & Table
				Note: electrochemical reduction; blue solution		
MPc(-4), MPc(-5), MPc(-6)						
ClAl	DMF		-4	900 800 518 327	[55]	Fig & Table
				Note: electrochemical reduction; purple solution		
ClAl	DMF		-5	1042 750 587 332	[55]	Fig & Table
				Note: chemical reduction; blue solution		
ClAl	DMF		-6	810 604 380 345	[55]	Fig & Table

				Note: chemical reduction; blue-green solution'		
Co						
Dication						
Co^{3+}	DCB		0	620 430 330	[93]	
				Note: Electrochemical oxidation; Substituted ring: CoTNPc		
Cation Radical						
Co^{3+}	DCM		-1	730 510 450 410	[13]	photochem
				Note: Photochemical oxidation		
Co^{3+}	TCM	Cl^-	-1	722 516 437 410 335 257	[85]	Fig & Table
				Note: Chemical oxidation; [(Cl)2Co(III)Pc(-1)]		
Co^{3+}	TCM	Br^-	-1	727 511 476 335 316 275	[85]	Fig & Table
				Note: Chemical oxidation; [Cl)2Co(III)Pc(-1)]		
Co^{3+}	DMF		-1	795 680 600 405	[93]	Fig & Table
				Note: Electrochemical oxidation; Substituted ring: CoTNPc		
Co^{3+}	film	Cl^-	-1	748 556 443 350 280	[90]	Fig & Table
				Note: Chemical oxidation		
Co^{3+}	film	Br^-	-1	1395 856 747 558 467 350 276	[90]	Fig & Table
			-1	580 520 400 380 320	[93]	Fig & Table
				Note: Electrochemical oxidation; Substituted ring: CoTNPc		
Co^{2+}	DCB		-1	686 620 590 495 405 360 320	[93]	Fig & Table
				Note: Electrochemical oxidation; Substituted ring: CoTNPc		
Co^{3+}	DCB		-1	780 554 370 330	[27]	Fig & Table
				Note: Electrochemical oxidation		
				Note: Tetramer of ring substituted CoPc: $[Co(III)TrNPc(-1)]_4$		
Co^{2+}	DCB		-1	975 690 590 496 365 320	[27]	Fig & Table

				Note: Electrochemical oxidation		
				Note: Tetramer of ring substituted CoPc: $[Co(III)TrNPc(-1)]_4$		
Co^{3+}	DCB	L	-1	755 679 540 399	[94]	Fig & Table
				Note: Monomeric, tetraneopentoxyphthalocyanine, Co(III)TNPc		
				Note: Chemical oxidation with SOCl2; L,L = Cl^-		
Co^{3+}	DCB	L	-1	742 580 520 400 380	[94]	Fig & Table
				Notes: Monomeric, tetraneopentoxyphthalocyanine, Co(III)TNPc; Electrochemical oxidation; L,L = ClO_4^-		
Radical Anion						
Co^{1+}	DCB		-3	955 703 640 560 480 345 315	[93]	Fig & Table
				Note: Electrochemical reduction; Substituted ring: CoTNPc		
Co^{1+}	H_2O		?	680 610 540 448 310	[121]	Co(I)TsPc
Co^{1+}	DMF		-3	920 681 625 476 319	[55]	Fig & Table
				Note: Electrochemical reduction; red solution		
Co^{1+}	H_2O		-3	810 700 570 490 460 365 310	[121]	Fig & Table
				Note: Electrochemical reduction		
				Co(I)TsPc(-3); at pH 2; see paper for pH dependence		
Co^{1+}	DCB		-3	672 520 476 315	[27]	Fig & Table
				Note: Electrochemical reduction		
				Tetramer of ring substituted CoPc: [Co(I)TrNPc(-3)]		
MPc(-4), MPc(-5), MPc(-6)						
Co^{1+}	DMF		-4	1179 905 510 326	[55]	Fig & Table
				Note: Electrochemical reduction; violet solution		
Co^{1+}	DMF		-5	1000 870 572 422 317	[55]	Fig & Table
				Note: Chemical reduction; blue solution		

Co^{1+}	DMF		-6	885 690 540 357	[55]	Fig & Table
				Note: Chemical reduction; green solution		
Radical cation						
Cr^{3+}	film Cl^-		-1	797 753 576 445 362	[90]	Table
				Note: Chemical oxidation		
Cu						
Radical cation						
Cu^{2+}	DCM	NO_3	-1	1000 848 808 725 530 390 325 280	[65]	Fig & Table
				Note: Chemical oxidation; $[(NO_3)Cu(II)Pc(-1)]$		
Cu^{2+}	MTN	NO_3	-1	1000 848 808 735 648 535	[65]	Fig & Table
				Note: Chemical oxidation; $[(NO_3)Cu(II)Pc(-1)]$		
Cu^{2+}	H_2O		-1	720 620 650 625	[122]	
				Note: Tetrasulfonated		
Fe						
Radical cation						
Fe^{3+}	DCM	py	-1	720sh 690 625 520	[13,99]	Fig & Table
				Note: Electrochemical oxidation; photochemical oxidation		
Fe^{3+}	(a) (b)		-1	885 795 775 735 695 669 520 403 320	[84]	Fig & Table
				Note: Chemical oxidation; $[(CN)(CF_3COO)$ Fe(III)Pc(-1)]		
				(a) solvent: DCM/TFA mixture;		
				(b) axial ligands CN^-/CF_3COO-		
Fe^{3+}	film Cl^-		-1	818 765 700 597 436 370 275	[90]	Fig & Table
		Note: Chemical oxidation				
Radical anion						
Fe^{2+}	DMF		-3	800 665 596 515 326	[55]	Table

Note: Electrochemical reduction; pink solution

MPc(-4), MPc(-5), MPc(-6)

Fe^{2+} DMF	-4	740 625 506 395 340		[55]	Table
		Note: Electrochemical reduction; purple solution			
Fe^{2+} DMF	-5	530 373 338		[55]	Table
		Note: Chemical reduction; violet solution			
Fe^{2+} DMF	-6	768 601 504 366		[55]	Table
		Note: Chemical reduction; green-blue solution			

Li

Radical cation

Li^+ nujol mull	-1	1020 690 615 500 435 365 310		[98]	Fig & Tab
		Note: Chemical oxidation			
		LiPc.H20; spectral data also shown for m-LiPc, t-LiPc, LiPcX, X=Cl, Br^- & I			
Li^+ CLN	-1	944 815 682 612 481 423 406		[86]	Fig & Table
		Note: Electrochemical and chemical oxidation			
Li^+ KBR	-1	1015 811 690 615 496 428 408		[86]	Fig & Table
		Note: Electrochemical and chemical oxidation			

Lu

Radical cation

Lu^{3+} DCM	-2/-1	1382 904 660 460 316	$[Lu(Pc)_2]$	[107]	Fig & Table
Lu^{3+} DCM	-1/-1	700 620 350	$[Lu(Pc)_2]$	[107]	Fig & Table
Lu^{3+} DCM	-2/-2	855 700 450 350	$[Lu(Pc)_2]$	[107]	Fig & Table
		Note: Spectral data shown for near IR region for Lu, and also Yb, Tm and Dy [107]			
Lu^{3+} DCM	-2/-1	668 (120,000)	$[Lu(Pc)_2]$	[109]	Fig & Table

Metal	Solvent	Ligand	Charge	Bands	Ref.	Data
				Note: $[(C_8)_8Pc]_2Lu$; green solution		
Lu^{3+}	DCM		-1/-1	701 (44,500) 488 (49,500)	[109]	Fig & Table
				Note: Electrochemical oxidation; $[(C_8)_8Pc]_2Lu$; orange solution		
Lu^{3+}	DCM		-2/-2	627 (106,000)	[109]	Fig & Table
				Note: Electrochemical reduction; $[(C_8)_8Pc]_2Lu$; blue solution		
				Note: Data reported also for the $[(C_{12})_8Pc]_2Lu$ complex in [109]		
Lu^{3+}	DCM		-1/-1	690 610 450 430	[110]	Fig
				Note: Electrochemical oxidation; $[Lu(Pc)_2]$		
Lu^{3+}	DCM		-2/-3	680 510 410 350	[110]	Fig
				Note: Electrochemical reduction; $[Lu(Pc)_2]$		
Lu^{3+}	DCM		-2/-2	700 620 410 350	[110]	Fig
				Note: Electrochemical reduction; $[Lu(Pc)_2]$		
Mg						
Radical cation						
Mg^{2+}	ClN	Cl^-	-1	920 830 785 770 720 650 507 425 400	[65]	Fig & Table
Mg^{2+}	ENL	Cl^-	-1	1040 830 720 680 640 610 545 430 408 370 352 315 275	[65]	Fig & Table
Mg^{2+}	DCM	H_20	-1	826 714 506 423 320 278	[2]	MCD Fig
				deconvolution calculations with absorption and MCD		
				all transitions were degenerate, except for the		
				508 nm band, which is very broad and nondegenerate		
				824Q-monomer 716Q-dimer 508(p-p) 422B 371N 319L 275C		
Mg^{2+}	DCM	imid	-1	827 720 507 422 319 276	[2]	MCD Fig
				deconvolution calculations with absorption and MCD		
				all transitions were degenerate, except for the		

Metal	Solvent	Ligand	Charge	Data	Ref.	Type
				508 nm band, which is very broad and nondegenerate		
				825Q-monomer 717Q-dimer 507(p-p) 422B 372N 319L 276C		
Mg^{2+}	DCM	L	-1	similar values to those listed above have been reported for L=MeIm, py, Mepy, CN^-	[2]	Fig & Table
				Note: Chemical, electrochemical and photochemical oxidation absorption, MCD and band deconvoluted spectra shown		
Mg^{2+}	DCM	H20	-1	825 505 411 387 328 275	[2]	MCD Fig
				Note: absorption and MCD spectra for the pure monomeric cation radical at 300 K		
Mg^{2+}	DCM	H20	-1	712 510 370 317 277	[2]	MCD Fig
				Note: absorption and MCD spectra for the pure dimeric cation radical at 200 K		
Radical Anion						
Mg^{2+}	DMF		-3	960 860 798 638 562 420 340	[55]	Table
				Note: Electrochemical redcution; blue solution		
Mg^{2+}	DMF		-3	960 645 570 425	[15]	Fig & Table
				Note: Photochemical reduction at -50C with hydrazine, also with piperidine [16]		
MPc(-4), MPc(-5), MPc(-6)						
Mg^{2+}	DMF		-4	900 800 615 520 335	[55]	Table
				Note: Electrochemical reduction; purple solution		
Mg^{2+}	DMF		-5	1115 825 590 339	[55]	Table
				Note: Chemical redcution; blue solution		
Mg^{2+}	DMF		-6	840 620 412 305	[55]	Table
				Note: Chemical reduction; blue-green solution		
Mg^{2+}	THF			Absorption and MCD data have been reported for the negative ions of MgPc	[72]	MCD Figs.

Metal	Solvent	Charge	Bands (nm)	Ref.	Comment
Mn					
Radical anion					
Mn?	DMF	-?	835 740 630 580 512 353 329	[55]	Table
			Note: Electrochemical reduction to $Mn^{1+}Pc(-2)$ or $Mn(^{2+}Pc(-3)$ (?); purple solution		
MPc(-4), MPc(-5)					
Mn^{2+}	DMF	-4	708 522 331	[55]	Table
			Note: Electrochemical reduction; purple solution		
Mn^{2+}	DMF	-5	1100 566 340	[55]	Table
			Note: Chemical reduction; blue solution		
Mn^{2+}	DMF	-6	832 620 370	[55]	Table
			Note: Chemical reduction; blue-green		
Ni					
Radical anion					
Ni^{2+}	DMF	-3	915 845 810 630 567 415 366	[55]	blue 1- Fig
Ni^{2+}	DMSO	-3	920 640 580 380 340	[68]	Fig; NiTsPc
			Note: Electrochemical reduction		
MPc(-4), MPc(-5), MPc(-6)					
Ni^{2+}	DMF	-4	910 880 850 536 333	[55]	Fig & Table
			Note: Electrochemical reduction; purple solution		
Ni^{2+}	DMSO	-4	650sh 540s 360s	[68]	Fig
			Note: Electrochemical reduction; NiTsPc		
Ni^{2+}	DMF	-5	1085 890 590 333	[55]	Fig & Table
			Note: Chemical reduction; blue solution		
Ni^{2+}	DMF	-6	880 672 407 365 311	[55]	Fig & Table
			Note: Chemical reduction; green solution		

Radical cation						
Rh^{3+}	ACN	L	-1	710 530	[10]	Fig
				Note: Photochemical oxidation; (CH3OH)(Cl)RhPc		
Ru						
Radical cation						
Ru^{2+}	DCM	Py	-1	710 534 368B 341N 302L	[99]	Fig
Ru^{2+}	MTHF	Py	-1	670w 540s	[6]	Fig
				Note: Low temperature photochemical oxidation reaction with CBr_4 as electron acceptor		
Ru^{2+}	DCM	MePy	-1	710 534 362 306	[6,99]	Fig
				Note: Photochemical oxidation with CBr_4 as electronacceptor		
Ru^{2+}	DCM	DMA	-1	710 530 368 302	[99]	
Ru^{2+}	ACN	CN^-	-1	700 520	[99]	
Ru^{2+}	DCM	BPy	-1	700 530 360 302	[99]	
Ru^{2+}	DCM	Bpy	-1	690 530 300	[102]	Fig & Table
				Note: Electrochemical oxidation		
Ru^{2+}	DCM	DMF/CO	-1	710 510 360	[6,99]	MCD Fig & Table
				Note: Photochemical oxidation with CBr_4 as electron acceptor		
Ru^{2+}	DCM	DMF	-1	695 (25,000) 529	[102]	Table
				Note: Electrochemical oxidation		
Ru^{2+}	DCM	pip	-1	720 540 366 304	[99]	
Ru^{2+}	DCM	Im-	-1	720 530 364 304	[99]	
Ru^{2+}	DCM	DMSO	-1	700 530 400 292	[6,99]	MCD Fig
				Note: Photochemical oxidation with CBr_4 as electron		
Ru^{2+}	DCM	DMSO	-1	688 (37,000) 519 379	[102]	Table

			Note: Electrochemical oxidation		
Ru^{2+}	DCM	DMF/CO	-1 704 534 430 362B 297N 275L	[99]	
			Note: Electrochemical oxidation		
Ru^{2+}	DCM	DMF/CO	-1 704 (43,000) 510 (37,000)	[102]	Table
			Note: Electrochemical oxidation		
Ru^{2+}	DCM	DMSO/CO	-1 695 (41,000) 516 383	[102]	Table
			Note: Electrochemical oxidation		
Ru^{2+}	DCM	MePy/CO	-1 702 503 420 385 274	[99]	
			Note: Electrochemical oxidation		
Ru^{2+}	DCM	Py	-1 700 530 300	[102]	Table
			Note: electrochemical oxidation		
Ru^{2+}	DCM	Py/CO	-1 703 (35,000) 529 381 280	[102]	Fig & Table
			Note: Electrochemical oxidation		
Ru^{2+}	DCM	MePy	-1 690 535 300	[102]	Table
			Note: Electrochemical oxidation		
Tb					
Radical cation					
Tb^{3+}	BNZ	DPM	-1 680 620 500 330 ;these bands also for:	[123]	monomers; Fig
			Note: Eu, Er, Dy, Gd, Ho, Lu, Nd, Sm, Tm, Y, Yb: generally M^{3+}Pc(-1)(dpm)3		
Th					
Radical cation					
Th^{6+}	PhCN	acac	-1 845 (7,000) 684 (63,000) 617 516 351	[114]	Fig
			Note: Electrochemical oxidation		
Radical anion					
Th^{6+}	PhCN	acac	-3 679 (41,000) 651 617 587 352	[114]	Fig

				Note: Electrochemical reduction		
MPc(-4)						
Th^{6+}	PhCN	acac	-4	649 (22,000) 544 351	[114]	Fig
				Note: Electrochemical reduction		
U						
Radical cation						
U^{6+}	PhCN	acac	-1	852 (5,000) 729 691 (61,000) 624 537 350	[114]	Fig
				Note: Electrochemical oxidation		
Radical anion						
U^{6+}	PhCN	acac	-3	677 (37,000) 652 620 591 352	[114]	Fig
				Note: Electrochemical reduction		
MPc(-4)						
U^{6+}	PhCN	acac	-4	646 (23,000) 549 349	[114]	Fig
				Note: Electrochemical reduction		
Y						
Y^{3+}	ClBz		-1	660 460 [Y2Pc3]	[111]	Fig
Zn						
Radical cation						
Zn^{2+}	DCM	Im	-1	822 714 424 370 317 278	[62,99]	MCD Fig
				Note: Electrochemical oxidation		
Zn^{2+}	DCM	Im	-1	results of deconvolution calculations on MCD and absorption spectra 822Q-monomer 714Q-dimer 691 664 633 595 511 475 424 388 370 353 334 317 295 287 278 268 255 see the paper for A/B terms	[62]	MCD deconv

Zn^{2+} DCM Im -1 825Q-monomer 720Q-dimer 500 440 321 273 [7]

Zn^{2+} DCM py -1 825 720 500 440 [7,13,99] Fig & Table

Note: Photochemical reaction with CBr_4 as electron acceptor; absorption and MCD spectra

Zn^{2+} film -1 830 720 550 335 [122]

Zn^{2+} film Cl^- -1 1095 743 660 550 445 379 323 [90] Table

Note: Chemical oxidation

This compound was diamagnetic, which suggests dimerization occurs

Zn^{2+} DCB -1 1080 860 724 680 514 400 340 [67] Fig & Table

Note: Electrochemical oxidation; monomeric, tetraneopentoxyphthalocyanine, ZnTNPc

Zn^{2+} DCB -1 1080 860 730 660 514 388 336 [67] Fig & Table

Note: Electrochemical oxidation; tetrameric, trineopentoxyphthalocyanine, $[ZnTNPc]_4$

Radical anion

Zn^{2+} DMF -3 948 850 785 636 562 412 323 [55] Fig & Table

Note: Electrochemical reduction; blue solution

MPc(-4), MPc(-5)

Zn^{2+} DMF -4 772 522 443 335 [55] Fig Table

Note: Electrochemical oxidation; purple solution

Zn^{2+} DMF -5 1100 570 340 [55] Fig & Table

Note: Chemical reduction; blue solution

Zn^{2+} DMF -6 821 620 385 304 [55] Fig & Table

Note: Chemical reduction; blue-green solution

Table 3. Ground and Accessible Excited States for the Phthalocyanine Monocation and Dication.

Ground-state configuration/symmetry	Excited-state configuration[a]	State symmetry
Monocation MPc(−1)		
$\{(e_g)^4(e_g)^4(e_g)^4(a_{2u})^2(a_{1u})^1(e_g)^0\}^b = {}^2A_{1u}$		
the Q band of MPc(−2) spectra: $\pi \rightarrow \pi^*$	$\{(a_{2u})^2(a_{1u})^0(e_g)^1\}$	^2E_g
the B band of MPc(−2) spectra: $\pi \rightarrow \pi^*$	$\{(a_{2u})^1(a_{1u})^1(e_g)^1\}$	^2E_g
new transitions: $\pi \rightarrow \pi^b$	$\{(e_g)^4(e_g)^4(e_g)^3(a_{2u})^2(a_{1u})^2(e_g)^0\}$	^2E_g
new transitions: $\pi \rightarrow \pi$	$\{(e_g)^4(e_g)^3(e_g)^4(a_{2u})^2(a_{1u})^2(e_g)^0\}$	^2E_g
new transitions: $\pi \rightarrow \pi$	$\{(e_g)^3(e_g)^4(e_g)^4(a_{2u})^2(a_{1u})^2(e_g)^0\}$	^2E_g
Dication MPc(0)		
$\{(e_g)^4(e_g)^4(e_g)^4\ (a_{2u})^2(a_{1u})^0(e_g)^0\}^b = {}^1A_{1g}$		
there is no equivalent of the Q band of MPc(−2) spectra		
the B band of MPc(−2) spectra: $\pi \rightarrow \pi^*$	$\{(a_{2u})^1(a_{1u})^0(e_g)^1\}$	^1E_u
new transitions: $\pi \rightarrow \pi^b$	$\{(e_g)^4(e_g)^4(e_g)^3(a_{2u})^2(a_{1u})^1(e_g)^0\}$	^1E_u
new transitions: $\pi \rightarrow \pi$	$\{(e_g)^4(e_g)^3(e_g)^4(a_{2u})^2(a_{1u})^1(e_g)^0\}$	^1E_u
new transitions: $\pi \rightarrow \pi$	$\{(e_g)^3(e_g)^4(e_g)^4(a_{2u})^2(a_{1u})^1(e_g)^0\}$	^1E_u

[a] Accessible by allowed optical transitions; this assumes that $\Delta m_S=0$ and that only transitions that obey the g<−>u parity rule give significant intensity.

[b] Note that because the a_{1u} orbital is partially filled or empty for cationic species, transitions into this orbital are from e_g orbitals that do not usually appear in transitions in the uv−visible region of the absorption spectrum.

Table 4. Accessible Excited States for the Monoanion.

Ground-state configuration/symmetry	Transition	Excited-state configuration/symmetry
$\{(a_{2u})^2(a_{1u})^2(e_g)^1\}$ ^2E_g *($^2B_{2g}$, from MCD)*	$\pi \rightarrow \pi^*$	$\{(a_{2u})^2(a_{1u})^1(e_g)^2\}$
the Q band in MPc(−2) spectra		$^2A_{1u}$, $^2A_{2u}$, $^2B_{1u}$, $^2B_{2u}$
	$\pi \rightarrow \pi^*$	$\{(a_{2u})^1(a_{1u})^2(e_g)^2\}$
the B band in MPc(−2) spectra		$^2A_{1u}$, $^2A_{2u}$, $^2B_{1u}$, $^2B_{2u}$
	$\pi \rightarrow \pi^*$	$\{(b_{1u})^2(b_{2u})^1(a_{2u})^2(a_{1u})^2(e_g)^2\}$
		$^2A_{1u}$, $^2A_{2u}$, $^2B_{1u}$, $^2B_{2u}$
	$\pi \rightarrow \pi^*$	$\{(b_{1u})^1(b_{2u})^2(a_{2u})^2(a_{1u})^2(e_g)^2\}$ $^2A_{1u}$, $^2A_{2u}$, $^2B_{1u}$, $^2B_{2u}$
	$\pi^* \rightarrow \pi^*$	$\{(a_{2u})^2(a_{1u})^2(e_g)^1(b_{1u})^1(b_{2u})\}$
new transitions not observed for MPc(−2)		^2E_u
	$\pi^* \rightarrow \pi^*$	$\{(a_{2u})^2(a_{1u})^2(e_g)^1(b_{1u})(b_{2u})^1\}$
new transitions not observed for MPc(−2)		^2E_u

Table 5. Accessible Excited States for Phthalocyanine Anions.

Ground-state configuration			Accessible excited states[a]	
Dianion MPc(−4)				
$\{(a_{2u})^2(a_{1u})^2(e_g)^2\}$	$^1B_{1g}$	$\pi \rightarrow \pi^*$	^1E_u	Q
		$\pi \rightarrow \pi^*$	^1E_u	B
		$\pi^* \rightarrow \pi^*$	^1E_u	
The Jahn–Teller effect is responsible for the singlet spin state [82]				
Trianion, MPc(−5)				
$\{(a_{2u})^2(a_{1u})^2(e_g)^3\}$	^2E_g	$\pi \rightarrow \pi^*$	$^2A_{1u}$	Q
		$\pi \rightarrow \pi^*$	$^2A_{2u}$	B
		$\pi^* \rightarrow \pi^*$	$^2A_{1u}$, $^2A_{2u}$, $^2B_{1u}$, $^2B_{2u}$	
Tetraanion, MPc(−6)				
$\{(a_{2u})^2(a_{1u})^2(e_g)^4\}$	$^1A_{1g}$	$\pi \rightarrow \pi^*$	none	Q
		$\pi \rightarrow \pi^*$	none	B
		$\pi^* \rightarrow \pi^*$	^1E_u	

[a] Accessible by allowed optical transitions; this assumes that $\Delta m_s = 0$ and that only transitions that obey the g<->u parity rule give significant intensity.

Index